Neuromethods

Series Editor
Wolfgang Walz
University of Saskatchewan
Saskatoon, SK, Canada

For further volumes:
http://www.springer.com/series/7657

Neuromethods publishes cutting-edge methods and protocols in all areas of neuroscience as well as translational neurological and mental research. Each volume in the series offers tested laboratory protocols, step-by-step methods for reproducible lab experiments and addresses methodological controversies and pitfalls in order to aid neuroscientists in experimentation. Neuromethods focuses on traditional and emerging topics with wide-ranging implications to brain function, such as electrophysiology, neuroimaging, behavioral analysis, genomics, neurodegeneration, translational research and clinical trials. Neuromethods provides investigators and trainees with highly useful compendiums of key strategies and approaches for successful research in animal and human brain function including translational "bench to bedside" approaches to mental and neurological diseases.

More information about this series at http://www.springer.com/series/7657

Developmental, Physiological, and Functional Neurobiology of the Inner Ear

Edited by

Andrew K. Groves

Department of Neuroscience, Baylor College of Medicine, Houston, TX, USA

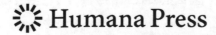

Editor
Andrew K. Groves
Department of Neuroscience
Baylor College of Medicine
Houston, TX, USA

ISSN 0893-2336   ISSN 1940-6045   (electronic)
Neuromethods
ISBN 978-1-0716-2024-3   ISBN 978-1-0716-2022-9   (eBook)
https://doi.org/10.1007/978-1-0716-2022-9

Cover Caption: Confocal image of the basal region of an embryonic day 18 mouse cochlea. The whole mount preparation is stained with antibodies to E-Cadherin (blue), Myosin7a (red) and with fluorescently labeled phalloidin (green). Image provided by Elizabeth Driver, National Institute on Deafness and Other Communication Disorders

This Humana imprint is published by the registered company Springer Science+Business Media, LLC part of Springer Nature.
The registered company address is: 1 New York Plaza, New York, NY 10004, U.S.A.

Preface to the Series

Experimental life sciences have two basic foundations: concepts and tools. The *Neuro-methods* series focuses on the tools and techniques unique to the investigation of the nervous system and excitable cells. It will not, however, shortchange the concept side of things as care has been taken to integrate these tools within the context of the concepts and questions under investigation. In this way, the series is unique in that it not only collects protocols but also includes theoretical background information and critiques which led to the methods and their development. Thus it gives the reader a better understanding of the origin of the techniques and their potential future development. The *Neuromethods* publishing program strikes a balance between recent and exciting developments like those concerning new animal models of disease, imaging, in vivo methods, and more established techniques, including, for example, immunocytochemistry and electrophysiological technologies. New trainees in neurosciences still need a sound footing in these older methods in order to apply a critical approach to their results.

Under the guidance of its founders, Alan Boulton and Glen Baker, the *Neuromethods* series has been a success since its first volume published through Humana Press in 1985. The series continues to flourish through many changes over the years. It is now published under the umbrella of Springer Protocols. While methods involving brain research have changed a lot since the series started, the publishing environment and technology have changed even more radically. Neuromethods has the distinct layout and style of the Springer Protocols program, designed specifically for readability and ease of reference in a laboratory setting.

The careful application of methods is potentially the most important step in the process of scientific inquiry. In the past, new methodologies led the way in developing new disciplines in the biological and medical sciences. For example, Physiology emerged out of Anatomy in the nineteenth century by harnessing new methods based on the newly discovered phenomenon of electricity. Nowadays, the relationships between disciplines and methods are more complex. Methods are now widely shared between disciplines and research areas. New developments in electronic publishing make it possible for scientists that encounter new methods to quickly find sources of information electronically. The design of individual volumes and chapters in this series takes this new access technology into account. Springer Protocols makes it possible to download single protocols separately. In addition, Springer makes its print-on-demand technology available globally. A print copy can therefore be acquired quickly and for a competitive price anywhere in the world.

Saskatoon, SK, Canada   *Wolfgang Walz*

Preface

The inner ear is a sensory organ of exquisite sensitivity, able to detect auditory and balance-related displacements in the sub-nanometer range. It develops from a simple piece of embryonic ectoderm on the side of the head, but as it develops, the inner ear primordium generates all the mechanosensory cells required for hearing and balance, the glial-like supporting cells which nourish and protect them, the neurons that innervate the mechanosensory cells and convey stimuli to the brain, and the non-sensory cells crucial for maintaining ionic homeostasis in the inner ear labyrinth. Loss of mechanosensory hair cells and neurons can lead to sensorineural hearing loss. Moreover, degradation of the 30–40 afferent synapses on each inner hair cell can lead to so-called "hidden hearing loss," in which patients with no patent cell loss nevertheless have hearing problems in noisy environments. Understanding these disorders of the auditory and vestibular periphery and their central pathways, and developing methods to treat them, is therefore of critical importance to the millions of Americans who suffer from hearing and balance disorders.

The inner ear is notable among sensory organs in that it possesses a very small number of receptor cells—numbered in the thousands, compared to the millions of photoreceptors in the eye, or the abundant olfactory receptor neurons. Many techniques used by neuroscientists to analyze the development and function of the brain or other sensory organs do not translate well to the inner ear. As a result, specialized techniques have had to be developed over many years that are specifically suited to this particular sensory organ and its central pathways. This volume represents a forum to compile state-of-the-art methods in inner ear development, analysis of its sensory cells, and characterization and manipulation of the central auditory and vestibular pathways.

Although individual methods used for particular areas of inner ear research have been published in research articles, there are very few books that have attempted to compile protocols across the whole spectrum of auditory and vestibular developmental, physiological, and functional research. Moreover, as auditory and vestibular research becomes more cross-disciplinary, individual laboratories are having to expand the repertoire of their techniques. The chapters in this volume of *Neuromethods* bring together experimental protocols that run the gamut of modern auditory and vestibular research—from the dissection and imaging of the cochlea to manipulation of central auditory pathways and behavioral evaluation of animal models of diseases such as tinnitus. Moreover, insights gained from nonmammalian vertebrates capable of regenerating their hair cells, such as zebrafish and chickens, have proved invaluable to our understanding of hair cell function and regeneration, and some chapters on these important models are also included here. Finally, although the small number of cells in the inner ear historically precluded transcriptomic analyses, recent advances in sequencing technology have now made the sorts of experiments that could only be carried out with millions of cells finally feasible in the ear, and we include two chapters on this important topic.

The contributing author list spans a unique balance of investigators working at the cutting edge of the auditory and vestibular fields who employ different techniques and model systems in their work. I am deeply grateful to all the authors for their willingness to contribute chapters, especially during the COVID-19 pandemic when many of us had to close our labs or restrict our research activities. I am thrilled at the breadth and depth of this book, and I hope that it will serve as a valuable resource to our field for many years to come.

Houston, TX, USA   *Andrew K. Groves*

Contents

Contributors

ROBERTO APONTE-RIVERA • *Section on Sensory Cell Development and Function, National Institutes on Deafness and Other Communication Disorders, National Institutes of Health, Bethesda, MD, USA*

JONATHAN F. ASHMORE • *UCL Ear Institute, UCL, London, UK; Department of Neuroscience, Physiology and Pharmacology, UCL, London, UK*

RANA M. BARGHOUT • *Department of Biology and Neuroscience Program, Amherst College, Amherst, MA, USA*

LESTER TORRES CADENAS • *Section on Neuronal Circuitry, National Institutes of Health, National Institute for Deafness and Other Communication Disorders, Bethesda, MD, USA*

GRAHAM CASEY • *Department of Neurobiology and Anatomical Sciences, University of Mississippi Medical Center, Jackson, MS, USA; Department of Otolaryngology—Head and Neck Surgery, University of Mississippi Medical Center, Jackson, MS, USA*

GUANG-DI CHEN • *Center for Hearing and Deafness, Department of Communicative Disorders and Sciences, University at Buffalo, Buffalo, NY, USA*

NICOLAS DAUDET • *The Ear Institute, University College London, London, UK*

ANTONIO MIGUEL GARCIA DE DIEGO • *UCL Ear Institute, UCL, London, UK; Instituto Teófilo Hernando and Departamento de Farmacología, Facultad de Medicina, Universidad Autónoma de Madrid, Madrid, Spain*

GIOVANNI H. DIAZ • *Department of Otolaryngology–Head and Neck Surgery, Stanford University School of Medicine, Stanford, CA, USA; Institute for Stem Cell Biology and Regenerative Medicine, Stanford University School of Medicine, Stanford, CA, USA; Department of Developmental Biology, Stanford University School of Medicine, Stanford, CA, USA*

ELIZABETH CARROLL DRIVER • *Laboratory of Cochlear Development, National Institute on Deafness and Other Communication Disorders, National Institutes of Health, Bethesda, MD, USA*

RAN ELKON • *Department of Human Molecular Genetics and Biochemistry, Sackler School of Medicine, Tel Aviv University, Tel Aviv, Israel*

GWENAËLLE S. G. GÉLÉOC • *Department of Otolaryngology, Harvard Medical School and Kirby Neurobiology Center, Boston Children's Hospital, Boston, MA, USA*

SUMANA GHOSH • *Department of Neurobiology and Anatomical Sciences, University of Mississippi Medical Center, Jackson, MS, USA; Department of Otolaryngology—Head and Neck Surgery, University of Mississippi Medical Center, Jackson, MS, USA*

JARED J. HARTSOCK • *Department of Otolaryngology, Washington University School of Medicine, St. Louis, MO, USA*

STEFAN HELLER • *Department of Otolaryngology–Head and Neck Surgery, Stanford University School of Medicine, Stanford, CA, USA; Institute for Stem Cell Biology and Regenerative Medicine, Stanford University School of Medicine, Stanford, CA, USA*

RONNA HERTZANO • *Department of Otorhinolaryngology Head and Neck Surgery, University of Maryland School of Medicine, Baltimore, MD, USA; Department of Anatomy and Neurobiology, University of Maryland School of Medicine, Baltimore, MD, USA; Institute for Genome Sciences, University of Maryland School of Medicine, Baltimore, MD, USA*

SAMAN HUSSAIN • *Section on Sensory Cell Development and Function, National Institutes on Deafness and Other Communication Disorders, National Institutes of Health, Bethesda, MD, USA*

ARTUR A. INDZHYKULIAN • *Department of Otolaryngology, Harvard Medical School and Massachusetts Eye and Ear, Boston, MA, USA*

YOICHIRO IWASA • *Department of Otorhinolaryngology, Shinshu University School of Medicine, Matsumoto, Nagano, Japan*

AMANDA JANESICK • *Department of Otolaryngology–Head and Neck Surgery, Stanford University School of Medicine, Stanford, CA, USA; Institute for Stem Cell Biology and Regenerative Medicine, Stanford University School of Medicine, Stanford, CA, USA*

STUART L. JOHNSON • *Department of Biomedical Science, University of Sheffield, Sheffield, UK*

PHILIP X. JORIS • *Laboratory of Auditory Neurophysiology, KU Leuven, Leuven, Belgium*

MATTHEW W. KELLEY • *Laboratory of Cochlear Development, National Institute on Deafness and Other Communication Disorders, National Institutes of Health, Bethesda, MD, USA*

YE-HYUN KIM • *Department of Otolaryngology-Head and Neck Surgery, Johns Hopkins School of Medicine, Baltimore, MD, USA*

KATIE S. KINDT • *Section on Sensory Cell Development and Function, National Institutes on Deafness and Other Communication Disorders, National Institutes of Health, Bethesda, MD, USA*

MILES J. KLIMARA • *Molecular Otolyarngology and Renal Research Laboratories, Department of Otolaryngology, University of Iowa, Iowa City, IA, USA*

JIN-YOUNG KOH • *Molecular Otolaryngology and Renal Research Laboratories, Department of Otolaryngology, University of Iowa, Iowa City, IA, USA*

AMANDA M. LAUER • *Department of Otolaryngology-Head and Neck Surgery, Johns Hopkins School of Medicine, Baltimore, MD, USA*

HSIN-WEI LU • *Laboratory of Auditory Neurophysiology, KU Leuven, Leuven, Belgium*

SENTHILVELAN MANOHAR • *Center for Hearing and Deafness, Department of Communicative Disorders and Sciences, University at Buffalo, Buffalo, NY, USA*

MAGGIE S. MATERN • *Department of Otorhinolaryngology Head and Neck Surgery, University of Maryland School of Medicine, Baltimore, MD, USA; Department of Otolaryngology—Head & Neck Surgery, Stanford University School of Medicine, Stanford, CA, USA*

CONNOR MAUCHE • *Center for Hearing and Deafness, Department of Communicative Disorders and Sciences, University at Buffalo, Buffalo, NY, USA*

BEATRICE MILON • *Department of Otorhinolaryngology Head and Neck Surgery, University of Maryland School of Medicine, Baltimore, MD, USA*

KEVIN K. OHLEMILLER • *Fay and Carl Simons Center for Biology of Hearing and Deafness, Central Institute for the Deaf at Washington University, St. Louis, MO, USA; Department of Otolaryngology, Washington University School of Medicine, St. Louis, MO, USA*

RYOTARO OMICHI • *Department of Otolaryngology–Head and Neck Surgery, Okayama University Graduate School of Medicine, Dentistry and Pharmaceutical Sciences, Kita-Ku, Okayama, Japan*

PAUL T. RANUM • *The Raymond G. Perelman Center for Cellular and Molecular Therapeutics, The Children's Hospital of Philadelphia, Philadelphia, PA, USA; Postdoctoral Training Program in Genomic Medicine, University of Pennsylvania, Philadelphia, PA, USA*

EDWIN W. RUBEL • *Virginia Merrill Bloedel Hearing Research Center, University of Washington, Seattle, WA, USA*

ALEC N. SALT • *Department of Otolaryngology, Washington University School of Medicine, St. Louis, MO, USA*

RICHARD SALVI • *Center for Hearing and Deafness, Department of Communicative Disorders and Sciences, University at Buffalo, Buffalo, NY, USA*

MIRKO SCHEIBINGER • *Department of Otolaryngology–Head and Neck Surgery, Stanford University School of Medicine, Stanford, CA, USA; Institute for Stem Cell Biology and Regenerative Medicine, Stanford University School of Medicine, Stanford, CA, USA*

KATRINA M. SCHRODE • *Department of Otolaryngology-Head and Neck Surgery, Johns Hopkins School of Medicine, Baltimore, MD, USA*

RICHARD J. H. SMITH • *Molecular Otolaryngology and Renal Research Laboratories, Department of Otolaryngology, University of Iowa, Iowa City, IA, USA*

KENDRA L. STANSAK • *Department of Neurobiology and Anatomical Sciences, University of Mississippi Medical Center, Jackson, MS, USA; Department of Otolaryngology—Head and Neck Surgery, University of Mississippi Medical Center, Jackson, MS, USA*

THEA STOLE • *The Ear Institute, University College London, London, UK*

JENNIFER S. STONE • *Department of Otolaryngology/Head and Neck Surgery and the Virginia Merrill Bloedel Hearing Research Center, University of Washington School of Medicine, Seattle, WA, USA*

KIRUPA SUTHAKAR • *Section on Neuronal Circuitry, National Institutes of Health, National Institute for Deafness and Other Communication Disorders, Bethesda, MD, USA*

STEPHEN TERRY • *The Ear Institute, University College London, London, UK*

PUNAM THAPA • *Department of Neurobiology and Anatomical Sciences, University of Mississippi Medical Center, Jackson, MS, USA; Department of Otolaryngology—Head and Neck Surgery, University of Mississippi Medical Center, Jackson, MS, USA*

HANNAH THORNER • *Center for Hearing and Deafness, Department of Communicative Disorders and Sciences, University at Buffalo, Buffalo, NY, USA*

JOSEF G. TRAPANI • *Department of Biology and Neuroscience Program, Amherst College, Amherst, MA, USA*

ERIC VERSCHOOTEN • *Laboratory of Auditory Neurophysiology, KU Leuven, Leuven, Belgium*

DANIEL WALLS • *Molecular Otolaryngology and Renal Research Laboratories, Department of Otolaryngology, University of Iowa, Iowa City, IA, USA*

BRADLEY J. WALTERS • *Department of Neurobiology and Anatomical Sciences, University of Mississippi Medical Center, Jackson, MS, USA; Department of Otolaryngology—Head and Neck Surgery, University of Mississippi Medical Center, Jackson, MS, USA*

MARK E. WARCHOL • *Department of Otolaryngology, Washington University, St Louis, MO, USA*

CATHERINE WEISZ • *Section on Neuronal Circuitry, National Institutes of Health, National Institute for Deafness and Other Communication Disorders, Bethesda, MD, USA*

CODY WEST • *Molecular Otolaryngology and Renal Research Laboratories, Department of Otolaryngology, University of Iowa, Iowa City, IA, USA*

MAGDALENA ŻAK • *The Ear Institute, University College London, London, UK*

Part I

Experimental Manipulation of the Inner Ear

Pou4f3^{DTR} Mice Enable Selective and Timed Ablation of Hair Cells in Postnatal Mice

Jennifer S. Stone, Edwin W. Rubel, and Mark E. Warchol

Abstract

Experimental studies of inner ear development and regeneration, as well as investigations of the influences of sensory input on CNS development, often require a rapid and nearly complete elimination of the hair cells of the inner ear at any postnatal age. Although these cells can be killed by noise trauma or by exposure to ototoxic drugs, both of these interventions are highly variable in their efficacy, resulting in considerable differences in sensory functions among individual animals that receive the same treatment. Furthermore, much current research of the auditory and vestibular systems is conducted using mice, and the ears of mice are relatively resistant to the effects of many ototoxins. In response to these concerns and others, the Rubel and Palmiter labs at the University of Washington developed a transgenic mouse line (called *Pou4f3^{DTR}*) in which the human form the diphtheria toxin receptor (also known as HB-EGF) is expressed under regulation of the *Pou4f3* promoter. Because *Pou4f3* is expressed by all hair cells (and relatively few other cells in the body), this mouse model permits the selective elimination of hair cells via 1–2 systemic injections of diphtheria toxin. This mouse line has been successfully used in studies of auditory CNS development and hair cell regeneration. This chapter provides an overview of this model, as well as detailed protocols for its use.

Key words Hair cells, Cell death, Mouse model, Deafness, Diphtheria toxin, Auditory system, Vestibular system, Regeneration

1 Historical Background

The sensory hair cells of the inner ear detect sound vibrations (in the cochlea) and head position and motion (in the vestibular organs) and convey this information to the brain via synapses upon the eighth cranial nerve. Injury or death of hair cells is relatively common in humans and can lead to permanent hearing loss, disequilibrium, and vertigo. The causes are varied and are likely to consist of a combination of genetic predisposition, along with a history of ototoxic drug exposure, noise exposure, or other forms of injury. Given the high prevalence of these conditions, it is of great interest to understand how hair cells die, how their death (and

Andrew K. Groves (ed.), *Developmental, Physiological, and Functional Neurobiology of the Inner Ear*, Neuromethods, vol. 176, https://doi.org/10.1007/978-1-0716-2022-9_1,

subsequent lack of neural input) affects other regions of the nervous system, and also how hair cells might be replaced or regenerated. Basic research on all of these issues relies on the use of animal models.

Early research on the function and pathology of the inner ear has employed a number of mammalian and non-mammalian models (e.g., [1, 2]). Insights into the operation of the cochlea were largely derived from studies of guinea pigs (which have a large and readily accessible cochlea), chinchillas (which also have an accessible cochlea and a hearing range similar to humans), and cats (which are advantageous for single unit physiology at all levels of the auditory system). Beginning in the 1990s, however, much of the research on the inner ear has involved the use of mice. Adoption of mice as a common animal model is attributable in part to the development of powerful tools for genetic sequencing and manipulation in mice, which has greatly enhanced our knowledge of the genetic and molecular basis of inner ear dysfunction. Mice are also easy to breed and maintain in laboratory environments. Apart from their genetic advantages, however, mice are not an optimal model for the study of hearing. The hearing range of mice (~4–60 kHz) is very different from that of humans (~0.02–20 kHz), which may be indicative of differences in the mechanics of the cochlea. In addition, quantifying the physiological function of the cochlea (e.g., via the recording of cochlear microphonics or the sound-evoked responses of single afferents) in mice is experimentally challenging. Similarly, mice are somewhat ill-suited for the study of vestibular function. Their vestibular sensory organs are small and relatively inaccessible, and their quadrupedal location makes it difficult to detect subtle changes in balance function. Mice also have small laterally positioned eyes, and their vestibulo-ocular reflexes (VORs) are difficult to quantify. Still, mice have emerged as very productive models in the studies of aging, ototoxicity, and acoustic trauma; the genetic homogeneity of inbred strains can minimize the degree of variability that is common for these types of insults.

The initial impetus to develop a transgenic mouse for time-sensitive, targeted ablation of hair cells was to better study the role of experience on shaping the development of structure and function of the brain. These types of studies can be traced back at least to the observations and writings of Aristotle. In the modern era, this includes the contributions of D. O. Hebb, visual system scientists like Austin Riesen, David Hubel and Torsten Wiesel, neuroembryologists like Victor Hamburger and behaviorists like Konrad Lorenz and Gilbert Gottlieb.

Building on the classic studies of auditory system embryology of Ramon y Cajal and Levi-Montalcini, and of visual system deprivation, the laboratory of one author (EWR) studied the role of synaptic activity on development of auditory pathways in birds and mammals for over four decades. One important observation made by this group and others was that while damage to the inner ear prior to the onset of hearing had rapid and dramatic consequences on cochlear nucleus neuron survival, the response was age dependent; the same manipulation a few days later resulted in no or minimal cell death. These studies provided further examples of critical periods wherein normal CNS development relies on synaptic inputs from the periphery (reviewed in [3–7]).

One of the challenges for these studies in the auditory system as well as other pathways has been how to remove or manipulate synaptic activity in areas of the brain in quantifiable ways without destroying other cellular components or creating other pathologies in inner ear or brain regions under investigation. In vivo approaches applied in the auditory field have included raising animals with ear plugs or in environments with abnormal acoustic experiences. While it is obvious that such manipulations change the *pattern* of activity, the unusually high levels of spontaneous activity of auditory neurons were likely not significantly altered under these conditions (e.g., [4, 8]). Another approach to silencing synaptic input to the brain included removal of the cochlea, as first shown by Kiang [9]. However, this manipulation did successfully eliminate excitatory synaptic activity at the level of the ventral cochlear nucleus, and it created pathologies in the inner ear, the nerve, and the cochlear nucleus. Investigators also pharmacologically blocked eighth nerve action potentials (e.g., [10–12]). While this approach was used in mature animals to great advantage, it is cumbersome, requires monitoring, and is difficult to validate in young animals.

Since the most common cause of hearing loss in humans is loss of hair cells, it makes sense to use a method that eliminates hair cells but does not damage other cochlear or CNS cells for altering synaptic input to the brain. Unfortunately, none existed. Therapeutic medicines that killed hair cells also damage other cochlear structures, and we do not know their direct effects on CNS neurons. Some genetic manipulations were becoming available (e.g., [13]) but usually limited to induction only in young animals. The obvious solution was to use genetics to remove or silence hair cells or auditory nerve axons in developing and mature animals.

Discovery of hair cell regeneration in birds [14, 15] brought attention to a new methodological problem. How could one distinguish the difference between a native (original) hair cell, a regenerated hair cell, and a hair cell that had been injured, had changed the expression of marker proteins, and had recovered? While tritiated thymidine or other cell cycle markers were immediately used to distinguish recently divided offspring [16, 17], it was quickly discovered that new hair cells could also arise by direct transdifferentiation (a non-mitotic conversion supporting cells in the sensory epithelia into hair cells [18, 19]) and that this method was the primary way in which hair cells were naturally regenerated in rodents [20–22]. While ototoxins and noise were available to induce hair cell death, these methods had problems (discussed below) due to high rates of animal mortality, high lesion variability, off-target effects, or age restrictions.

Careful consideration revealed that the ideal mouse model for studying the role of synaptic activity on auditory brain development and hair cell regeneration should: (1) be inducible at any age; (2) be highly specific to hair cells (no off-target effects in key cell types in the inner ear or CNS); (3) work rapidly, in a matter of hours or a few days; (4) be reliably quantifiable and consistent across subjects or complete; (5) be non-invasively inducible; and (6) be effective in vitro and in vivo. Fortuitously, Richard Palmiter's group at the University of Washington had developed a method to selectively ablate hypothalamic neurons that control feeding behavior in mice [23, 24], as discussed in more detail in Subheading 3. The application of this method to create $Pou4f3^{DTR}$ mice resulted in a highly useful model for time-controlled and selective ablation of hair cells in postnatal mice.

2   Traditional Methods for Inducing Hair Cell Death in Mice

The exact method used to create a hair cell lesion in mice depends on the goals of the study. If the objective is to identify the cellular mechanisms that underlie hair cell death caused by specific ototoxins (e.g., aminoglycoside antibiotics or the chemotherapy agent cisplatin), then the use of those specific drugs is mandated. On the other hand, studies focused on the consequences of inner ear injury (such as CNS plasticity, effects of sensory deprivation, or hair cell regeneration) will require creation of a hair cell lesion, but the actual method used to kill hair cells may not be critical. Instead, it may be more important that the lesion method is reliable and consistent, so that all experimental animals experience the same kind and extent of hair cell loss, or complete killing of all hair cells in the organ. Furthermore, it is often desirable that other cell types in the ear, such as organ of Corti supporting cells, peripheral and central neurons in auditory and vestibular pathways, or stria vascularis and dark cells, are not adversely affected by the method intended to kill hair cells.

2.1   Cochlear Hair Cells

The most common method for lesioning cochlear hair cells is to expose animals to intense sound. Before employing this technique, several species-specific parameters must be considered. First, the frequency range of hearing sensitivity varies greatly among mammals, so the sound frequency (or frequency band) to be used must be optimized for the specific animal species that is being studied. Also, the cellular effects of noise vary with the intensity and duration of exposure. In most strains of laboratory mice, moderately loud sounds (~90 dB SPL) will induce injury to afferent synapses but will not impact hair cells or supporting cells [25, 26]. In contrast, exposure to higher sound levels ($\geq$106 dB SPL) will kill cochlear hair cells, and intense sound (>116 dB SPL) can disrupt the integrity of the sensory epithelium [27]. Noise-induced hearing loss is common in humans, so identifying the mechanisms by which noise damages the cochlea is of great translational interest. However, the effects of noise on cochlear hair cells can be highly variable, such that the lesion induced by a particular noise regimen can vary from animal to animal. For this reason, unless noise exposure is a critical aspect of the rationale for a particular experiment, noise trauma is probably not the best option for ablating cochlear hair cells.

Another option for inducing cochlear injury involves treatment with ototoxic drugs. Most studies using drugs to kill hair cells in mice have used either aminoglycoside antibiotics or platinum-containing chemotherapeutic drugs that are widely used to treat tumors. Unfortunately, mice are often a suboptimal model for such studies. Killing cochlear hair cells with aminoglycosides requires delivery of multiple injections over 1–3 weeks, with doses that approach systemic toxicity [28, 29]. Aminoglycoside-induced death of cochlear hair cells is greatly enhanced when the antibiotic is administered in combination with a "loop" diuretic (e.g., furosemide, ethacrynic acid, bumetanide; [30]), and this approach has been used to create ototoxic lesions in mice [28, 31, 32]. The precise mechanism of the pharmacological interaction between loop diuretics and aminoglycosides is not fully understood, but it is likely that the diuretic permits increased transport of aminoglycosides across the stria vascularis, leading to higher concentrations of those drugs in the cochlear fluids. From an experimental standpoint, co-administration of loop diuretics and aminoglycosides is a reasonably reliable method for inducing hearing loss in mice. While aminoglycosides (with or without loop diuretics) mainly target outer hair cells and—at higher doses—inner hair cells, these treatments often cause damage or death of other cells in the organ of Corti, in the stria vascularis, and/or amongst spiral ganglion neurons. Also, the extent of damage to hair cells, and probably other cell types, varies dramatically as a function of age.

Cisplatin and other chemotherapeutic drugs have been used to lesion cochlear hair cells since the early 1970s (when those drugs were first developed). However, their use in mice presents similar challenges to those encountered with aminoglycosides. A clinically relevant protocol for cisplatin ototoxicity in mice has recently been developed [33], and it involves giving mice three 4-day courses of systemic injections of cisplatin, with each series separated by a 10-day drug-free "recovery" period. However, creating a cisplatin lesion with this protocol in mice requires 2–3 months and careful attention to detail. Finally, it should be noted that both high doses of cisplatin and intense noise can damage supporting cells [34–37], stria vascularis [38, 39], and/or neurons (e.g., [40–42]) as well as hair cells.

A few transgenic mouse lines have been used to destroy hair cells in mice. Fujioka et al. [43] engineered *Pou4f3-Cre;Mos-iCsp3* mice in which treatment in vivo or in vitro with a drug called AP20187 results in mosaic, partial killing of the auditory hair cell population. This method should also work to achieve partial killing of vestibular hair cells. Additionally, mice generated by crossing *Atoh1-Cre^{ER}* mice to *Rosa26-stop-loxp-DTA* mice show near-complete killing of hair cells in both the cochlea [44] and vestibular organs [45] when tamoxifen is injected in the neonatal period. Unfortunately, this method does not work in adult mice because *Atoh1-Cre^{ER}* expression is lost in hair cells as mice mature.

2.2 Vestibular Hair Cells

The traditional method for killing vestibular hair cells in non-mammals and large rodents (guinea pigs) is to treat animals with a series of subcutaneous injections of aminoglycoside antibiotics. However, this approach is problematic in mice. In adult mice, kanamycin fails to induce any substantial hair cell loss in the utricle [31, 32] and, presumably, other vestibular organs. The high doses of aminoglycosides that are apparently required to kill vestibular hair cells in vivo are lethal to mice, and unlike in the cochlea, supplementation with diuretics such as furosemide does not enhance vestibular hair cell loss [31, 32]. Intralabyrinthine delivery of gentamicin creates near-complete lesions in adult mice [46], but this approach requires surgery.

There have been very few studies examining cisplatin as an inducer of vestibular hair cell damage in mice. Although small changes in a vestibular reflex and motor behaviors have been noted after cisplatin treatment [47], studies have found little or no loss of vestibular hair cells following cisplatin treatments at different doses and schedules [33, 47]. Furthermore, Fernandez et al. [33] detected no change in vestibular stimulus-evoked electrical potentials in the brain stem. There are other disadvantages to using cisplatin to kill hair cells including its lethal effects on supporting cells and neurons (*see* Subheading 2.1). Therefore, similar to aminoglycosides, cisplatin seems to pose technical limitations for studies requiring extensive and precise loss of hair cells.

Another toxin, 3,3′-Iminodipropionitrile (IDPN), has been employed to destroy vestibular hair cells in mice (e.g., [48, 49]). Although IDPN is easily administered to mice, it can cause pathological changes in different regions of the rodent body including the kidney, liver, and brain (e.g., [50–53]). In rats and mice, IDPN induces loss of hair cell-afferent nerve synapses in both vestibular and cochlear organs [54–56] and structural and molecular changes in the hair cell-calyx junction [56]. In addition, at high doses, IDPN can cause supporting cell death [57]. Most studies of IDPN ototoxicity have focused on the vestibular organs and

IDPN's effects on the cochlea are not well-characterized. The off-target effects of IDPN can make it challenging to isolate behavioral or physiological effects of hair cell loss and regeneration, which is important in many studies.

3 Selective Cell Ablation in Mice Expressing the Human Diphtheria Toxin Receptor

To minimize the off-target effects and extend the age range of treatment that accompany the more traditional methods for hair cell ablation, Edwin Rubel and Richard Palmiter engineered the *Pou4f3^DTR* transgenic mouse line [58–60]. In this line, the human diphtheria toxin receptor (hDTR) is expressed specifically on hair cells, thus allowing those cells to be selectively ablated upon administration of diphtheria toxin (DT also called DTx). Diphtheria toxin is a bacterial protein that kills cells by inhibiting protein synthesis [51, 61–63]. The DT molecule has two functional domains: DT-B, which is the receptor-binding and transactivating domain (which facilitates entry into the cell), and DT-A, which is the catalytic, toxic domain. DT binds to a receptor on the cell surface, which has been shown in some cells to be the precursor of heparin-binding epidermal growth factor like growth factor (HB-EGF) [64]. HB-EGF, which is expressed in many cell types, acts as a ligand and regulates a variety of cellular behaviors [65]. However, because HB-EGF can bind to and internalize DT, it is commonly referred to as the "DT receptor" (DTR). Once bound to HB-EGF, DT becomes incorporated into clathrin-coated pits and then endosomes. In the acidic environment of the endosome, the toxin becomes translocated to the cytosol, where it enzymatically alters the structure of eukaryotic elongation factor 2 (EF2), affects the actin cytoskeleton [66], and activates nucleases, resulting in apoptosis. Much of the DT that enters the cell is degraded in the lysosomal pathway. In cases where the extracellular fluid is acidic, DT may be directly transported across the plasma membrane to the cytoplasm.

Diphtheria toxin is very potent; once internalized within a cell, a single molecule of DT appears to be sufficient to induce cell death [67]. However, susceptibility to DT varies across species and cell types and depends on expression of the DTR and affinity of the DTR for DT [62]. Cells of humans and mice differ greatly in their vulnerability to DT, which is a consequence of species-specific differences in the amino acid sequence of HB-EGF that endows the human version with a ~1000× higher affinity for DT than mouse version [23]. Richard Palmiter and other genetic engineers took advantage of these features to generate mice that express the full coding region of the human *DTR* gene (*hDTR*) in specific cell types in mice [23, 24]. *hDTR* expression is driven by gene regulatory elements that are specific to a given cell type. Such mice then receive systemic injection(s) of DT. Since DT interacts very weakly

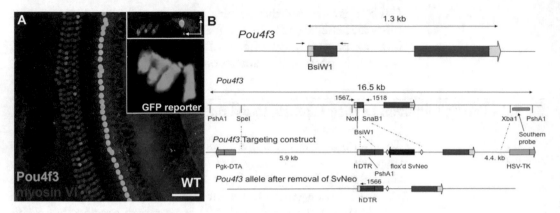

Fig. 1 Strategy for generating *Pou4f3^DTR/+* mice. (**a**) Pou4f3 expression is limited to hair cells in the inner ear, as verified two ways: by labeling whole-mount tissue with the Pou4f3 antibody and using sectioned tissue from the *Pou4f3^GFP* reporter mouse. Pou4f3 immunolabel (main panel, green) is selectively expressed in the nuclei of hair cells reacted for myosin VI (red) from a mature (P56) WT mouse. An orthogonal view from the same tissue is shown in the upper inset. The *Pou4f3^GFP* reporter mouse demonstrates expression in both inner and outer hair cells (lower inset). Scale bar, 50 μm. (**b**) *Pou4f3^DTR/+* mice were genetically engineered to contain the human *DTR* downstream of the *Pou4f3* promoter, creating a mouse model in which sensory hair cells in the inner ear can be selectively ablated after a systemic injection of DT. From [58]

with mouse HB-EGF, this treatment has only minor effects on most cells of these mice. However, cells that express *hDTR* quickly undergo cell death.

To target hair cells for DT-induced ablation, *hDTR* was inserted into exon 1 of the *Pou4f3* gene, generating *Pou4f3^DTR* mice (Fig. 1b; [58, 59, 60, 68]). Pou4f3 is a transcription factor that, at the protein level, is highly expressed in the nucleus of all inner ear hair cells (Fig. 1a), at early and late stages of development and in maturity but is not expressed in other key cell types in the auditory or vestibular periphery such as supporting cells or primary sensory neurons [69–71]. This *hDTR* insertion inactivates the *Pou4f3* coding region. Germline knockout mice that are homozygous null for *Pou4f3* experience hair cell death in the early postnatal period [69, 71]. By contrast, mice heterozygous for *Pou4f3* exhibit apparently normal development of cochlear hair cells, vestibular hair cells, and hearing [72]. Therefore, *Pou4f3^DTR* mice are used as heterozygotes.

A related—but very different—approach to cell killing is to express an inducible form of the gene for the DT-A fragment within a specific cell population (e.g., [73, 74]). This approach has been used to kill supporting cells (e.g., [75]) and hair cells (e.g., [44, 45]) in the mouse inner ear. Another similar approach has been to use mice with Cre-inducible *DTR* expression (e.g., [76]), killing cochlear ganglion neurons.

The first studies to implement the *Pou4f3^{DTR}* mouse line [58, 59, 60, 68] sought to destroy all hair cells in the cochlea and vestibular organs of neonatal and adult mice, in order to assess changes in cellular properties in the cochlear nucleus, hair cell regeneration in the vestibular epithelium, or development of vocalizations in mice.

4 Auditory Hair Cell Ablation Using Pou4f3^DTR Mice

Figure 2 shows phenotypic responses of mature *Pou4f3^{DTR/+}* mice given single IM injections at a dosage of 25 ng/g of diphtheria toxin [58]. A single intramuscular (IM) injection of DT (25 ng/g) is sufficient to kill all cochlear hair cells (Fig. 2a–d). Hair cell loss is evident within 3 days of DT injection. At 5 days after DT treatment at this dosage, no normal-appearing hair cells remain. Assessment of cochlear function by auditory brainstem response (ABR) thresholds indicates that these DT-treated mice fail to show any reliable response to clicks or pure tone stimuli up to 90 dB (SPL) by 5 days after the DT treatment (Fig. 2e). None of these pathologies are seen in wild type mice from the same litters given identical injections of DT nor in *Pou4f3^{DTR/+}* mice given saline injection. Of importance, both qualitative and quantitative analyses indicate that the cochlear pathologies resulting from the DT injection at this dosage (or below) appear entirely specific for hair cells. Other cell types of the cochlea, such as epithelial supporting cells, lateral wall fibrocytes, cells of stria vascularis, and spiral ganglion cells appear unaffected [58, 59, 68, 77]. Interestingly, during the period of hair cell loss, there appears to be a small but significant reduction of the endocochlear potential (EP) that subsequently recovers to normal levels. DT dosage, survival period, and age of injection were also varied by Tong et al. and Kaur et al. Complete loss of inner and outer hair cells is also observed in mature *Pou4f3^{DTR/+}* mice given single injection dosages of 15 ng/g and 5 ng/g, but the timing of hair cell loss is delayed by up to 5 days. For the use of *Pou4f3^{DTR/+}* mice to study the influence of synaptic activity on development of brain structure and function, it is critical to quantify neural activity in the neurons under investigation. Studies cited above lead to the assumption that auditory nerve activity will be severely diminished (*see* also ref. 78), and the expected dramatic reduction of spontaneous activity has been confirmed in young animals in studies conducted in the laboratory of Prof. Rudolf Rubsamen in Leipzig (Fig. 2f). At 6 days after DT injection, spontaneous discharge in AVCN neurons was reduced by >99%. More extensive studies on the changes on ongoing ("spontaneous") activity throughout the auditory pathways in neonatal and mature *Pou4f3^{DTR/+}* mice are needed. Recent developments in methods for in vivo Ca^{2+} imaging should facilitate this.

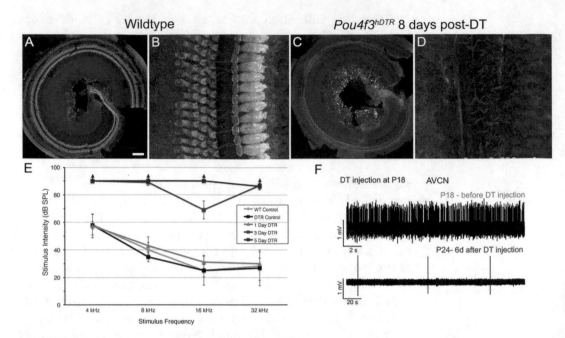

Fig. 2 Summary of auditory phenotype of *Pou4f3*$^{DTR/+}$ mice following DT treatment. (**a–d**) Low and high magnification confocal images of mature cochleas from wild type (WT) mice (**a, b**) and *Pou4f3*$^{DTR/+}$ mice at 8 days following IM injection of 25 ng/g DT (**c, d**). Tissue is reacted for myosin 7a (green), neurofilaments (red), and nuclei (blue). Note complete loss of all hair cells and robust survival of nerve fibers in **c** and **d**. Scale bar in **a** = 100 μm. **b** and **c** from [58]. (**e**) Average (+/− SEM) auditory brainstem response (ABR) thresholds from WT, mice and *Pou4f3*$^{DTR/+}$ mice at 1–5 days after DT treatment. Note total loss of response by 5 days. From [58]. (**f**) Electrophysiological recordings from single neurons in the antero-ventral cochlear nucleus (AVCN) of a young *Pou4f3*$^{DTR/+}$ mouse before and 6 days after 15 ng/g IM injection of DT. Note the difference in time scale on *x*-axis. Action potential frequency is reduced by over 99% by 6 days after DT injection. Data provided courtesy of Prof. Rudolf Rubsamen, Univ. of Leipzig

In summary, the studies cited above indicate that the *Pou4f3*$^{DTR/+}$ mouse appears to be an excellent experimental model for quickly eliminating cochlear hair cells and hearing function, with minimal damage to other cell types. This technique has now been employed in a number of studies, with highly consistent outcomes (e.g., [79, 80]).

As noted above, an important aspect of the *Pou4f3*$^{DTR/+}$ model is the ability to eliminate hair cells at any postnatal age. Several studies have used this property for studies of organ of Corti hair cell regeneration and the effects of hearing loss in neonatal vs mature mice (e.g., [44, 58, 68, 81]). Results of DT injections have been similar to those noted above with a couple of exceptions. First, smaller injections of DT have typically been used (<10 ng/g). Unpublished experience revealed considerable lethality with doses >15 ng/g. With P2–P7 mice, a single dose of 5 mg/g yields complete hair cell loss within 10 days and minimal but detectable rapid changes in organ of Corti supporting cells. On the other hand, the response of spiral ganglion cells was profoundly different than mature mice. When neonatal *Pou4f3*$^{DTR/+}$ mice were injected with DT, they showed profound loss of SGNs as early as 8 days later. This loss progressed such that only 30% of spiral ganglion cell bodies remained 70 days later. Wild type mice injected with DT as neonates did not have any observable SGN cell body loss at any time point [58]. Finally, the neonatal mouse cochlea can be explanted and maintained in organotypici culture. Hair cells in explanted cochleae from *Pou4f3*$^{DTR/+}$ mice can be selectively lesioned by adding DT to the culture medium. Treating such cultures for 3 days with 25 ng/ml DT results in complete loss of inner hair cells, extensive loss of outer hair cells, and survival of supporting cells [58]. Much more detailed and quantitative work needs to be conducted on the use of this model for in vitro studies.

5 Vestibular Hair Cell Ablation Using Pou4f3^{DTR} Mice

Golub et al. [60] sought to destroy all hair cells in the vestibular organs of adult mice (6–9 weeks of age) and then assess subsequent hair cell regeneration over time. This study employed *Pou4f3*$^{DTR/+}$ mice in order to overcome the limitation of partial hair cell ablation that was observed with all prior methods. For studies of hair cell regeneration, it is advantageous to kill all original (native) hair cells, so that any newly produced hair cells can be definitively identified as "replacement" or "regenerated." In contrast, an incomplete hair cell lesion leaves open the possibility for cellular repair or migration of surviving hair cells into the injured area, both of which would confound data interpretation. Therefore, Golub et al. [60] implemented *Pou4f3*^{DTR} mice, with the goal of killing all vestibular hair cells and preserving other cell types. Hair cells were counted in several control mice to assess the specificity of the method. Analysis of utricular hair cell numbers in *Pou4f3*$^{DTR/+}$ mice that did not receive DT injection revealed that vestibular hair cells develop normally in *Pou4f3* heterozygotes [60]. In addition, injection of DT (two intramuscular injections of 25 ng/g, 2 days apart) to *Pou4f3*^{DTR} wild type mice (*Pou4f3*$^{+/+}$) failed to cause hair cell loss. This finding was expected because the low dose of DT should not induce hair cell loss in mice lacking the *hDTR*. However, in *Pou4f3*$^{DTR/+}$ mice, the same DT regimen caused 50% of vestibular hair cells to die by 7 days post-DT and 94% of hair cells to die by 14 days post-DT. A few papers have demonstrated that DT induces apoptosis-like death of vestibular hair cells (e.g., [82, 83]). Terminal deoxynucleotidyl transferase dUTP nick end labeling (TUNEL) at 7 days post-DT revealed wide-spread chromatin degradation characteristic of apoptosis in cells throughout the utricle (Fig. 3a, b). Condensed chromatin could be detected using 4′,6-diamidino-2-phenylindole (DAPI) labeling in small numbers of hair cells as

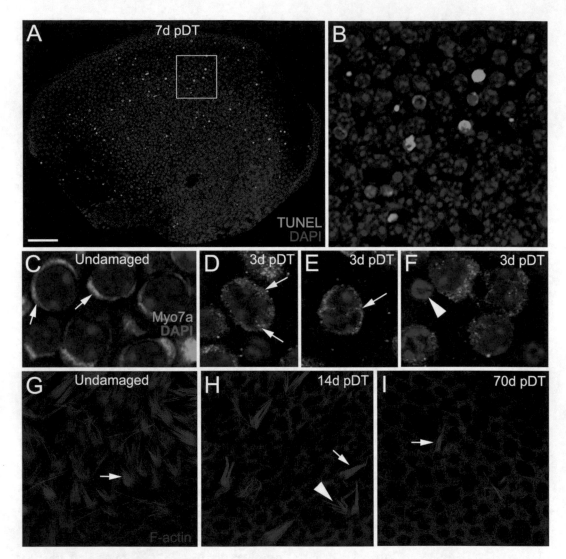

Fig. 3 Hair cell degeneration and death in adult mouse utricles following DT administration. (**a**) TUNEL labeling in a whole-mount utricle from a *Pou4f3^{DTR/+}* mouse at 7 days post-DT. (**b**) Higher magnification of the box shown in **a**. (**c**) Undamaged WT hair cells (arrows) with cytoplasm labeled green (Myosin 7a) and nuclei labeled blue. (**d**)–(**f**) Utricle with similar labeling as **c**, showing hair cells (arrows) from *Pou4f3^{DTR/+}* mice at 3 days post-DT. Arrowhead in F points to nucleus of an apoptotic cell. (**g**) Phalloidin labeling of filamentous (F) actin in stereocilia (arrow) in an undamaged WT utricle. (**h, i**) Stereocilia of surviving hair cells (arrows) at 14 (**h**) and 70 (**i**) days post-DT in *Pou4f3^{DTR/+}* mice. Arrowhead in (**h**) shows a bundle of abnormally splayed stereocilia, indicative of injury. The stereocilia of regenerated hair cells are too small and too lightly labeled in panel **i** at 70 days post-DT to be evident. Scale bar: 100 μm in **a**, 12 μm in **b**, 4 μm in **c–f**, and 20 μm in **g–i**

early as 3 days post-DT (Fig. 3f). Imaging with transmission electron microscopy also provided evidence for pyknotic nuclei at 7 days post-DT (not shown). Additional pieces of evidence of hair cell demise were: (1) deformation of the nucleus; (2) ectopic distribution of myosin 7a (Myo7a) protein, which is normally cytoplasmic (Fig. 3c) but in some cells appeared to penetrate the nucleus (Fig. 3d–f); and (3) degeneration of stereocilia in the apical hair bundle (Fig. 3g–i).

Amniotes have two types of vestibular hair cell—type I and type II (reviewed in [84]). Mouse utricles contain approximately 3800 hair cells, about half of which are type I [85, 86]. Golub et al. [60] showed that ~200 hair cells of both types remained at 14 days post-DT, but over the next several months, cells resembling type I hair cells decreased further in number. In contrast, hair cells with type II characteristics increased in number to approximately 700 by 60 days post-DT, after which their numbers remained the same. These observations are consistent with the hair cell replacement noted in mice or guinea pigs after hair aminoglycoside treatment, indicating the *Pou4f3^DTR* mouse model is a viable tool for regeneration research.

The delay of hair cell destruction, particularly in type I hair cells, indicated that DT uptake, DT trafficking, and/or execution of cell death takes place over several weeks. The design logic for *Pou4f3^DTR/+* mice predicted that DT administration should rapidly kill hair cells in adult mice. Further, it was anticipated that all hair cells in adult mice would die rapidly because they continue to express high levels of Pou4f3 (*see* Fig. 1). The finding that 6% of utricular hair cells remained in *Pou4f3^DTR/+* mice at 14 days post-DT was surprising, and there are several possible interpretations. First, some of these hair cells may have already been regenerated. This interpretation is supported by the observation that many hair cells at this time lacked a well-formed bundle, which is a sign of immaturity. However, this interpretation cannot account for all remaining hair cells, because many of them were type I and are not naturally regenerated in adult mammals (e.g., [21, 22, 46, 82, 87]). A second possibility is that *Pou4f3* promoter activity may vary amongst hair cells, causing some cells to have insufficient human *DTR* expression, which would result in DT resistance. In other mouse lines employing a similar strategy, complete cell death is obtained in a shorter period. For instance, Wu et al. [88] killed 99% of neurons expressing agouti-related protein within 6 days. Third, DT may be processed differently by similar cell types. For instance, DT enters the cell via endosomes and may remain in that compartment longer in some cells, which would protect them from DT's lethal inhibition of protein translation.

DT kills hair cells in other vestibular organs beside the utricle. Hicks et al. [87] found that, in addition, hair cells were killed, and type II hair cells were regenerated in the anterior and lateral ampullae and in the saccule. It is not clear from studies if there are any spatial gradients in hair cell loss within the vestibular organs.

The degree and nature of hair cell loss in utricles of adult mice can be reduced by administering DT at lower overall doses. A single injection of DT at 25 ng/g induces less extensive hair cell loss in the utricle [77], seemingly inducing a comparable amount of type II hair cell death and regeneration but sparing approximately half of the type I hair cell population (J. Stone, unpublished data).

Golub et al. [60] also used DT to kill vestibular hair cells in organ cultures. They found that, in whole utricles that were explanted and incubated at 37 °C, overnight incubation with DT (dissolved in culture media at 3–333 ng/ml) destroyed 99% of hair cells in adult utricles within 5 days. These tests were intended as a demonstration of efficacy of the mouse model; systematic dosage and timing studies are needed to make this model more useful for studies in cultured inner ear organs. There are important considerations for studying hair cell damage and regeneration ex vivo. For instance, Lin et al. [89] showed there is substantial spontaneous death of hair cells in cultures even in the absence of a damaging agent presumably due to stress and malnutrition in those conditions, and hair cell regeneration is thwarted in whole cultured utricles, which can only be maintained for 3–4 weeks before tissue undergoes degeneration.

Vestibular hair cells of neonatal $Pou4f3^{DTR/+}$ mice can also be lesioned by systemic DT treatment. Much lower dosages are used in neonates. For instance, a single 5 ng/g DT injection kills ~80% of hair cells in the cristae of the semicircular canals (M. Warchol, unpublished data). However, this dose is highly toxic to the sensory epithelium of the neonatal utricle. Treatment of neonates with 5 ng/g DT at P0-1 results in a large epithelial "wound" in the central region of the utricular sensory epithelium, which is caused by the loss of *both* hair cells and supporting cells [90]. Such wounds are apparent at 7 days post-injection, and lead to the mixing of the inner ear fluids (perilymph and endolymph). Such extensive damage also results in the death of the majority of afferent neurons. Interestingly, such wounds begin to close between 7 and 14 days post-DT, probably via the contraction of a "purse-string" actin ring that surrounds the outer border of the wound [90]. These unexpected findings point to an epithelial repair process that is present in the neonatal utricle, similar to that described by Meyers and Corwin [91] after induction of small "punch" wounds in organotypic cultures.

6   Methods

6.1   Mouse Breeding

To drive expression of *hDTR* in mouse hair cells, the gene for human HB-EGF was inserted into exon 1 of *Pou4f3* in the mouse genome (Fig. 1). Mice that are heterozygous for this allele retain one functional copy of *Pou4f3*, and their auditory and vestibular hair cells appear to develop normally [58, 60, 69–71]. However, mice that possess two copies of *hDTR* lack a functional *Pou4f3* gene, and such mice exhibit a profound loss of cochlear and vestibular hair cells after they are formed during development. For this reason, an optimal breeding strategy for studies targeting hair cell damage in an otherwise physiologically normal animal is to mate $Pou4f3^{DTR+/-}$ mice to WT mice, which will yield 50% of offspring that are $Pou4f3^{DTR+/-}$ ("experimental" mice) and 50% that are WT and can be used as controls.

For genotyping, DNA is extracted from mouse tissue (typically via tail-clip), polymerase chain reaction amplifies a portion of the *Pou4f3* gene that is either WT or contains the *hDTR* allele, and gel electrophoresis is used to distinguish between these two gene segments, which differ in size. Detailed methods for genotyping these mice can be found in Tong et al. [58] and Kaur et al. [77].

Genetic background is important to consider when breeding. Most published studies have utilized C57Bl6/J mice, in which consistent lesions have been achieved. However, González-Garrido et al. [83] noted that full hair cell lesions were not reliably attained with the standard DT dose (25 ng/g) in adult $Pou4f3^{DTR+/-}$ mice on a CBA/CaJ background. This same result was experienced by the University of Washington team, and a single cross of congenic CBA-CaJ mice to C57Bl6/J mice restored the sensitivity of the mice to DT (unpublished observations).

6.2 DT Administration

Diphtheria toxin in unnicked form is injected either intramuscularly (IM) to juvenile and adult mice or intraperitoneally (IP) to neonatal mice. While some investigators report intraperitoneal (IP) injections in neonatal mice, in our hands, both IM and subcutaneous injections in neonates can lead to unreliable results likely due to leakage of the solution from the injection site. With this approach, we have attained nearly symmetric hair cell lesions in organs from the left and right sides of the mouse, and we have achieved similar lesions in all mice from a given cohort.

Monaural DT treatment, causing unilateral hair cell destruction, would be useful for both inner ear and CNS studies. Pilot studies were undertaken in the Rubel lab to determine if local injections of DT into the middle ear of mature mice will produce single-sided hair cell loss and deafness. The results were promising but optimal formulations, dosages, and timing need to be resolved.

DT powder is purchased from Sigma-Aldrich #D0564 or List Biological Labs #150 and can be stored at 2–8 °C. DT powder is readily dissolved in water. We make 1 μg/100 stock solution dissolved in saline (0.9% NaCl sterile) solution and store it in a non-defrosting (−20 °C) freezer. The drug is most dangerous in powder form or at high concentrations in solution. People using the drug should be vaccinated and should consult Material Safety Datasheets, standard operating procedures, and university resources (e.g., Occupational Nurse, Environmental Health and Safety) for instructions and assistance in handling.

In the first 3–7 days after injection, mice can react to DT by grouping together and consuming less water and food, but they usually improve by 1 week. Nutritional supplements such as the high-calorie gel Nutri-Cal (Tomlyn/Vétoquinol USA) have been added to cages to keep juvenile or adult mice healthier during the first week post-DT. If they are sick, we administer subcutaneous lactated Ringers solution. Supplements are more likely required when adult mice receive two doses of DT at 25 ng/g or higher.

A higher DT dose is required to achieve full destruction of vestibular hair cells than to induce complete loss of cochlear hair cells in adult mice [58, 60]. To reliably kill all vestibular hair cells, we inject either 25 or 50 ng/g DT (IM), once a day, for 2 days, skipping 1 day between injections. A single injection at either dose typically results in loss of most type II hair cells but only half of the type I hair cell population (J. Stone, unpublished data).

We found that DT's efficacy in hair cell killing can vary lot-by-lot, and DT solution loses its efficacy over months when stored in the freezer. Therefore, we run a dose-response test for each new lot of DT, measuring hair cell loss at 14 days post-DT, and we test the DT solution every 4–6 months, discarding it when it no longer induces a lesion at doses equal or less than 50 ng/g.

Finally, we have observed a small degree of hair cell loss in the cochleae of WT mice after treatment with high doses of DT (25–50 ng/g). This loss is usually confined to IHCs and is very minor when compared to the hair cell death that occurs in $Pou4f3^{DTR+/-}$ mice after the same DT treatment. Also, this hair cell loss is not extensive enough to cause elevated ABR thresholds in DT-injected "control" mice (e.g., see ref. 58). Still, this effect has the potential to influence the outcomes of certain types of studies and should be carefully monitored. The reason for this small degree of DT-induced cell death in WT mice (which do not possess $hDTR$) is not clear. However, genomic studies indicate that some mouse hair cells express HB-EGF (data publicly available at umgear.org). Furthermore, even though the affinity of DT for human HB-EGF is significantly greater than its affinity for the mouse form of this protein, it is still possible that a small number of DT molecules are transported into hair cells of WT mice. A complete explanation of this phenomena will require further investigation.

References

1. Von Bekesy G (1960) Experiments in hearing. McGraw-Hill, New York
2. Wever EG, Lawrence M (1954) Physiological acoustics. Princeton University Press, Princeton, NJ
3. Rubel EW (1978) Ontogeny of structure and function in the vertebrate auditory system. In: Jacobson M (ed) Development of sensory systems, vol IX. Springer-Verlag, pp 135–237

4. Born DE, Durham D, Rubel EW (1991) Afferent influences on brainstem auditory nuclei of the chick: nucleus magnocellularis neuronal activity following cochlea removal. Brain Res 557(1-2):37–47. https://doi.org/10.1016/0006-8993(91)90113-a

5. Rubel EW, Fritzsch B (2002) Auditory system development: primary auditory neurons and their targets. Annu Rev Neurosci 25:51–101. https://doi.org/10.1146/annurev.neuro.25.112701.142849

6. Rubel EW, Parks TN, Zirpel L (2004) Assembling, connecting, and maintaining the cochlear nucleus. In: Parks TN, Rubel EW, Popper AN, Fay RR (eds) Plasticity of the auditory system, Springer handbook of auditory research, vol 23. Springer-Verlag, New York, pp 8–48

7. Harris JA, Rubel EW (2006) Afferent regulation of neuron number in the cochlear nucleus: cellular and molecular analyses of a critical period. Hear Res 216-217:127–137. https://doi.org/10.1016/j.heares.2006.03.016

8. Tucci DL, Born DE, Rubel EW (1987) Changes in spontaneous activity and CNS morphology associated with conductive and sensorineural hearing loss in chickens. Ann Otol Rhinol Laryngol 96(3 Pt 1):343–350. https://doi.org/10.1177/000348948709600321

9. Kiang NY-S (1965) Discharge patterns of single fibers in the cat's auditory nerve. M.I.T Press

10. Sie KC, Rubel EW (1992) Rapid changes in protein synthesis and cell size in the cochlear nucleus following eighth nerve activity blockade or cochlea ablation. J Comp Neurol 320(4):501–508. https://doi.org/10.1002/cne.903200407

11. Pasic TR, Rubel EW (1989) Rapid changes in cochlear nucleus cell size following blockade of auditory nerve electrical activity in gerbils. J Comp Neurol 283(4):474–480. https://doi.org/10.1002/cne.902830403

12. Yuan Y, Shi F, Yin Y, Tong M, Lang H, Polley DB, Liberman MC, Edge AS (2014) Ouabain-induced cochlear nerve degeneration: synaptic loss and plasticity in a mouse model of auditory neuropathy. J Assoc Res Otolaryngol 15(1):31–43. https://doi.org/10.1007/s10162-013-0419-7

13. Cox BC, Dearman JA, Brancheck J, Zindy F, Roussel MF, Zuo J (2014) Generation of Atoh1-rtTA transgenic mice: a tool for inducible gene expression in hair cells of the inner ear. Sci Rep 4:6885. https://doi.org/10.1038/srep06885

14. Cotanche DA (1987) Regeneration of hair cell stereociliary bundles in the chick cochlea following severe acoustic trauma. Hear Res 30(2-3):181–195. https://doi.org/10.1016/0378-5955(87)90135-3

15. Cruz RM, Lambert PR, Rubel EW (1987) Light microscopic evidence of hair cell regeneration after gentamicin toxicity in chick cochlea. Arch Otolaryngol Head Neck Surg 113(10):1058–1062. https://doi.org/10.1001/archotol.1987.01860100036017

16. Corwin JT, Cotanche DA (1988) Regeneration of sensory hair cells after acoustic trauma. Science 240(4860):1772–1774. https://doi.org/10.1126/science.3381100

17. Ryals BM, Rubel EW (1988) Hair cell regeneration after acoustic trauma in adult Coturnix quail. Science 240(4860):1774–1776. https://doi.org/10.1126/science.3381101

18. Adler HJ, Raphael Y (1996) New hair cells arise from supporting cell conversion in the acoustically damaged chick inner ear. Neurosci Lett 205(1):17–20. https://doi.org/10.1016/0304-3940(96)12367-3

19. Roberson DW, Alosi JA, Cotanche DA (2004) Direct transdifferentiation gives rise to the earliest new hair cells in regenerating avian auditory epithelium. J Neurosci Res 78(4):461–471. https://doi.org/10.1002/jnr.20271

20. Warchol ME, Lambert PR, Goldstein BJ, Forge A, Corwin JT (1993) Regenerative proliferation in inner ear sensory epithelia from adult guinea pigs and humans. Science 259(5101):1619–1622. https://doi.org/10.1126/science.8456285

21. Forge A, Li L, Nevill G (1998) Hair cell recovery in the vestibular sensory epithelia of mature guinea pigs. J Comp Neurol 397(1):69–88

22. Forge A, Li L, Corwin JT, Nevill G (1993) Ultrastructural evidence for hair cell regeneration in the mammalian inner ear. Science 259(5101):1616–1619. https://doi.org/10.1126/science.8456284

23. Palmiter R (2001) Interrogation by toxin. Nat Biotechnol 19(8):731–732. https://doi.org/10.1038/90770

24. Saito M, Iwawaki T, Taya C, Yonekawa H, Noda M, Inui Y, Mekada E, Kimata Y, Tsuru A, Kohno K (2001) Diphtheria toxin receptor-mediated conditional and targeted cell ablation in transgenic mice. Nat Biotechnol 19(8):746–750. https://doi.org/10.1038/90795

25. Kujawa SG, Liberman MC (2009) Adding insult to injury: cochlear nerve degeneration

after "temporary" noise-induced hearing loss. J Neurosci 29(45):14077–14085. https://doi.org/10.1523/JNEUROSCI.2845-09.2009

26. Kujawa SG, Liberman MC (2006) Acceleration of age-related hearing loss by early noise exposure: evidence of a misspent youth. J Neurosci 26(7):2115–2123. https://doi.org/10.1523/JNEUROSCI.4985-05.2006

27. Wang Y, Hirose K, Liberman MC (2002) Dynamics of noise-induced cellular injury and repair in the mouse cochlea. J Assoc Res Otolaryngol 3(3):248–268. https://doi.org/10.1007/s101620020028

28. Hirose K, Sato E (2011) Comparative analysis of combination kanamycin-furosemide versus kanamycin alone in the mouse cochlea. Hear Res 272(1-2):108–116. https://doi.org/10.1016/j.heares.2010.10.011

29. Wu WJ, Sha SH, McLaren JD, Kawamoto K, Raphael Y, Schacht J (2001) Aminoglycoside ototoxicity in adult CBA, C57BL and BALB mice and the Sprague-Dawley rat. Hear Res 158(1-2):165–178. https://doi.org/10.1016/s0378-5955(01)00303-3

30. Brummett RE, Bendrick T, Himes D (1981) Comparative ototoxicity of bumetanide and furosemide when used in combination with kanamycin. J Clin Pharmacol 21(11):628–636. https://doi.org/10.1002/j.1552-4604.1981.tb05675.x

31. Oesterle EC, Campbell S, Taylor RR, Forge A, Hume CR (2008) Sox2 and JAGGED1 expression in normal and drug-damaged adult mouse inner ear. J Assoc Res Otolaryngol 9(1):65–89. https://doi.org/10.1007/s10162-007-0106-7

32. Taylor RR, Nevill G, Forge A (2008) Rapid hair cell loss: a mouse model for cochlear lesions. J Assoc Res Otolaryngol 9(1):44–64. https://doi.org/10.1007/s10162-007-0105-8

33. Fernandez K, Wafa T, Fitzgerald TS, Cunningham LL (2019) An optimized, clinically relevant mouse model of cisplatin-induced ototoxicity. Hear Res 375:66–74. https://doi.org/10.1016/j.heares.2019.02.006

34. Girod DA, Duckert LG, Rubel EW (1989) Possible precursors of regenerated hair cells in the avian cochlea following acoustic trauma. Hear Res 42(2-3):175–194. https://doi.org/10.1016/0378-5955(89)90143-3

35. Cotanche DA, Messana EP, Ofsie MS (1995) Migration of hyaline cells into the chick basilar papilla during severe noise damage. Hear Res 91(1-2):148–159. https://doi.org/10.1016/0378-5955(95)00185-9

36. Slattery EL, Warchol ME (2010) Cisplatin ototoxicity blocks sensory regeneration in the avian inner ear. J Neurosci 30(9):3473–3481. https://doi.org/10.1523/JNEUROSCI.4316-09.2010

37. Slattery EL, Oshima K, Heller S, Warchol ME (2014) Cisplatin exposure damages resident stem cells of the mammalian inner ear. Dev Dyn 243(10):1328–1337. https://doi.org/10.1002/dvdy.24150

38. Santi PA, Duvall AJ 3rd (1978) Stria vascularis pathology and recovery following noise exposure. Otolaryngology 86(2):ORL354–ORL361. https://doi.org/10.1177/019459987808600229

39. Sluyter S, Klis SF, de Groot JC, Smoorenburg GF (2003) Alterations in the stria vascularis in relation to cisplatin ototoxicity and recovery. Hear Res 185(1-2):49–56. https://doi.org/10.1016/s0378-5955(03)00260-0

40. Zheng JL, Stewart RR, Gao WQ (1995) Neurotrophin-4/5, brain-derived neurotrophic factor, and neurotrophin-3 promote survival of cultured vestibular ganglion neurons and protect them against neurotoxicity of ototoxins. J Neurobiol 28(3):330–340. https://doi.org/10.1002/neu.480280306

41. Calls A, Carozzi V, Navarro X, Monza L, Bruna J (2020) Pathogenesis of platinum-induced peripheral neurotoxicity: insights from preclinical studies. Exp Neurol 325:113141. https://doi.org/10.1016/j.expneurol.2019.113141

42. Mizisin AP, Powell HC (1995) Toxic neuropathies. Curr Opin Neurol 8(5):367–371. https://doi.org/10.1097/00019052-199510000-00008

43. Fujioka M, Tokano H, Fujioka KS, Okano H, Edge AS (2011) Generating mouse models of degenerative diseases using Cre/lox-mediated in vivo mosaic cell ablation. J Clin Invest 121(6):2462–2469. https://doi.org/10.1172/JCI45081

44. Cox BC, Chai R, Lenoir A, Liu Z, Zhang L, Nguyen DH, Chalasani K, Steigelman KA, Fang J, Rubel EW, Cheng AG, Zuo J (2014) Spontaneous hair cell regeneration in the neonatal mouse cochlea in vivo. Development 141(4):816–829. https://doi.org/10.1242/dev.103036

45. Burns JC, Cox BC, Thiede BR, Zuo J, Corwin JT (2012) In vivo proliferative regeneration of balance hair cells in newborn mice. J Neurosci 32(19):6570–6577. https://doi.org/10.1523/JNEUROSCI.6274-11.2012

46. Kawamoto K, Izumikawa M, Beyer LA, Atkin GM, Raphael Y (2009) Spontaneous hair cell

regeneration in the mouse utricle following gentamicin ototoxicity. Hear Res 247 (1):17–26. https://doi.org/10.1016/j.heares.2008.08.010

47. Takimoto Y, Imai T, Kondo M, Hanada Y, Uno A, Ishida Y, Kamakura T, Kitahara T, Inohara H, Shimada S (2016) Cisplatin-induced toxicity decreases the mouse vestibulo-ocular reflex. Toxicol Lett 262: 49–54. https://doi.org/10.1016/j.toxlet.2016.09.009

48. Sayyid ZN, Wang T, Chen L, Jones SM, Cheng AG (2019) Atoh1 directs regeneration and functional recovery of the mature mouse vestibular system. Cell Rep 28(2):312–324.e314. https://doi.org/10.1016/j.celrep.2019.06.028

49. Wilkerson BA, Artoni F, Lea C, Ritchie K, Ray CA, Bermingham-McDonogh O (2018) Effects of 3,3'-iminodipropionitrile on hair cell numbers in cristae of CBA/CaJ and C57BL/6J mice. J Assoc Res Otolaryngol 19 (5):483–491. https://doi.org/10.1007/s10162-018-00687-y

50. Gianutsos G, Suzdak PD (1985) Neurochemical effects of IDPN on the mouse brain. Neurotoxicology 6(3):159–164

51. Sandvig K, van Deurs B (2005) Delivery into cells: lessons learned from plant and bacterial toxins. Gene Ther 12(11):865–872. https://doi.org/10.1038/sj.gt.3302525

52. Llorens J, Crofton KM, O'Callaghan JP (1993) Administration of 3,3'-iminodipropionitrile to the rat results in region-dependent damage to the central nervous system at levels above the brain stem. J Pharmacol Exp Ther 265(3):1492–1498

53. Alwelaie MA, Al-Mutary MG, Siddiqi NJ, Arafah MM, Alhomida AS, Khan HA (2019) Time-course evaluation of iminodipropionitrile-induced liver and kidney toxicities in rats: a biochemical, molecular and histopathological study. Dose Response 17 (2):1559325819852233. https://doi.org/10.1177/1559325819852233

54. Greguske EA, Llorens J, Pyott SJ (2021) Assessment of cochlear toxicity in response to chronic 3,3'-iminodipropionitrile in mice reveals early and reversible functional loss that precedes overt histopathology. Arch Toxicol 95 (3):1003–1021. https://doi.org/10.1007/s00204-020-02962-5

55. Greguske EA, Carreres-Pons M, Cutillas B, Boadas-Vaello P, Llorens J (2019) Calyx junction dismantlement and synaptic uncoupling precede hair cell extrusion in the vestibular sensory epithelium during sub-chronic 3,3'-iminodipropionitrile ototoxicity in the

mouse. Arch Toxicol 93(2):417–434. https://doi.org/10.1007/s00204-018-2339-0

56. Sedo-Cabezon L, Jedynak P, Boadas-Vaello P, Llorens J (2015) Transient alteration of the vestibular calyceal junction and synapse in response to chronic ototoxic insult in rats. Dis Model Mech 8(10):1323–1337. https://doi.org/10.1242/dmm.021436

57. Zeng S, Ni W, Jiang H, You D, Wang J, Lu X, Liu L, Yu H, Wu J, Chen F, Li H, Wang Y, Chen Y, Li W (2020) Toxic effects of 3,3'-iminodipropionitrile on vestibular system in adult C57BL/6J mice in vivo. Neural Plast 2020:1823454. https://doi.org/10.1155/2020/1823454

58. Tong L, Strong MK, Kaur T, Juiz JM, Oesterle EC, Hume C, Warchol ME, Palmiter RD, Rubel EW (2015) Selective deletion of cochlear hair cells causes rapid age-dependent changes in spiral ganglion and cochlear nucleus neurons. J Neurosci 35(20):7878–7891. https://doi.org/10.1523/JNEUROSCI.2179-14.2015

59. Tong L, Hume C, Palmiter R, Rubel EW (2011) Ablation of mouse cochlea hair cells by activating the human diphtheria toxin receptor (DTR) gene targeted to the Pou4f3 locus. association for research in otolaryngology meeting. Abstracts 34

60. Golub JS, Tong L, Ngyuen TB, Hume CR, Palmiter RD, Rubel EW, Stone JS (2012) Hair cell replacement in adult mouse utricles after targeted ablation of hair cells with diphtheria toxin. J Neurosci 32(43):15093–15105. https://doi.org/10.1523/JNEUROSCI.1709-12.2012

61. Lord JM, Smith DC, Roberts LM (1999) Toxin entry: how bacterial proteins get into mammalian cells. Cell Microbiol 1(2):85–91. https://doi.org/10.1046/j.1462-5822.1999.00015.x

62. Holmes RK (2000) Biology and molecular epidemiology of diphtheria toxin and the tox gene. J Infect Dis 181(Suppl 1):S156–S167. https://doi.org/10.1086/315554

63. Collier RJ (2001) Understanding the mode of action of diphtheria toxin: a perspective on progress during the 20th century. Toxicon 39 (11):1793–1803. https://doi.org/10.1016/s0041-0101(01)00165-9

64. Naglich JG, Metherall JE, Russell DW, Eidels L (1992) Expression cloning of a diphtheria toxin receptor: identity with a heparin-binding EGF-like growth factor precursor. Cell 69 (6):1051–1061. https://doi.org/10.1016/0092-8674(92)90623-k

65. Dao DT, Anez-Bustillos L, Adam RM, Puder M, Bielenberg DR (2018) Heparin-binding epidermal growth factor-like growth factor as a critical mediator of tissue repair and regeneration. Am J Pathol 188 (11):2446–2456. https://doi.org/10.1016/j.ajpath.2018.07.016

66. Bektas M, Varol B, Nurten R, Bermek E (2009) Interaction of diphtheria toxin (fragment A) with actin. Cell Biochem Funct 27 (7):430–439. https://doi.org/10.1002/cbf.1590

67. Yamaizumi M, Mekada E, Uchida T, Okada Y (1978) One molecule of diphtheria toxin fragment A introduced into a cell can kill the cell. Cell 15(1):245–250. https://doi.org/10.1016/0092-8674(78)90099-5

68. Mahrt EJ, Perkel DJ, Tong L, Rubel EW, Portfors CV (2013) Engineered deafness reveals that mouse courtship vocalizations do not require auditory experience. J Neurosci 33 (13):5573–5583. https://doi.org/10.1523/JNEUROSCI.5054-12.2013

69. Erkman L, McEvilly RJ, Luo L, Ryan AK, Hooshmand F, O'Connell SM, Keithley EM, Rapaport DH, Ryan AF, Rosenfeld MG (1996) Role of transcription factors Brn-3.1 and Brn-3.2 in auditory and visual system development. Nature 381(6583):603–606. https://doi.org/10.1038/381603a0

70. Xiang M, Gao WQ, Hasson T, Shin JJ (1998) Requirement for Brn-3c in maturation and survival, but not in fate determination of inner ear hair cells. Development 125(20):3935–3946

71. Xiang M, Gan L, Li D, Chen ZY, Zhou L, O'Malley BW Jr, Klein W, Nathans J (1997) Essential role of POU-domain factor Brn-3c in auditory and vestibular hair cell development. Proc Natl Acad Sci U S A 94(17):9445–9450. https://doi.org/10.1073/pnas.94.17.9445

72. Keithley EM, Erkman L, Bennett T, Lou L, Ryan AF (1999) Effects of a hair cell transcription factor, Brn-3.1, gene deletion on homozygous and heterozygous mouse cochleas in adulthood and aging. Hear Res 134 (1-2):71–76. https://doi.org/10.1016/s0378-5955(99)00070-2

73. Palmiter RD, Behringer RR, Quaife CJ, Maxwell F, Maxwell IH, Brinster RL (1987) Cell lineage ablation in transgenic mice by cell-specific expression of a toxin gene. Cell 50 (3):435–443. https://doi.org/10.1016/0092-8674(87)90497-1

74. Breitman ML, Clapoff S, Rossant J, Tsui LC, Glode LM, Maxwell IH, Bernstein A (1987) Genetic ablation: targeted expression of a toxin gene causes microphthalmia in transgenic mice. Science 238(4833):1563–1565. https://doi.org/10.1126/science.3685993

75. Mellado Lagarde MM, Cox BC, Fang J, Taylor R, Forge A, Zuo J (2013) Selective ablation of pillar and deiters' cells severely affects cochlear postnatal development and hearing in mice. J Neurosci 33 (4):1564–1576. https://doi.org/10.1523/JNEUROSCI.3088-12.2013

76. Pan H, Song Q, Huang Y, Wang J, Chai R, Yin S, Wang J (2017) Auditory neuropathy after damage to cochlear spiral ganglion neurons in mice resulting from conditional expression of diphtheria toxin receptors. Sci Rep 7 (1):6409. https://doi.org/10.1038/s41598-017-06600-6

77. Kaur T, Zamani D, Tong L, Rubel EW, Ohlemiller KK, Hirose K, Warchol ME (2015) Fractalkine signaling regulates macrophage recruitment into the cochlea and promotes the survival of spiral ganglion neurons after selective hair cell lesion. J Neurosci 35 (45):15050–15061. https://doi.org/10.1523/JNEUROSCI.2325-15.2015

78. Wang HC, Bergles DE (2015) Spontaneous activity in the developing auditory system. Cell Tissue Res 361(1):65–75. https://doi.org/10.1007/s00441-014-2007-5

79. Kaur T, Hirose K, Rubel EW, Warchol ME (2015) Macrophage recruitment and epithelial repair following hair cell injury in the mouse utricle. Front Cell Neurosci 9:150. https://doi.org/10.3389/fncel.2015.00150

80. Weatherstone JH, Kopp-Scheinpflug C, Pilati N, Wang Y, Forsythe ID, Rubel EW, Tempel BL (2017) Maintenance of neuronal size gradient in MNTB requires sound-evoked activity. J Neurophysiol 117(2):756–766. https://doi.org/10.1152/jn.00528.2016

81. Qian ZJ, Ricci AJ (2020) Effects of cochlear hair cell ablation on spatial learning/memory. Sci Rep 10(1):20687. https://doi.org/10.1038/s41598-020-77803-7

82. Bucks SA, Cox BC, Vlosich BA, Manning JP, Nguyen TB, Stone JS (2017) Supporting cells remove and replace sensory receptor hair cells in a balance organ of adult mice. elife 6. https://doi.org/10.7554/eLife.18128

83. González-Garrido A, Pujol R, Ramirez OL, Finkbeiner C, Eatock RA, Stone JS (2021) The differentiation status of hair cells that regenerate naturally in the vestibular inner ear of the adult mouse. J Neurosci. https://doi.org/10.1523/JNEUROSCI.3127-20.2021

84. Eatock RA, Songer JE (2011) Vestibular hair cells and afferents: two channels for head motion signals. Annu Rev Neurosci 34:

501–534. https://doi.org/10.1146/annurev-neuro-061010-113710

85. Desai SS, Zeh C, Lysakowski A (2005) Comparative morphology of rodent vestibular periphery. I. Saccular and utricular maculae. J Neurophysiol 93(1):251–266. https://doi.org/10.1152/jn.00746.2003

86. Pujol R, Pickett SB, Nguyen TB, Stone JS (2014) Large basolateral processes on type II hair cells are novel processing units in mammalian vestibular organs. J Comp Neurol 522 (14):3141–3159. https://doi.org/10.1002/cne.23625

87. Hicks KL, Wisner SR, Cox BC, Stone JS (2020) Atoh1 is required in supporting cells for regeneration of vestibular hair cells in adult mice. Hear Res 385:107838. https://doi.org/10.1016/j.heares.2019.107838

88. Wu Q, Howell MP, Cowley MA, Palmiter RD (2008) Starvation after AgRP neuron ablation is independent of melanocortin signaling. Proc Natl Acad Sci U S A 105(7):2687–2692. https://doi.org/10.1073/pnas.0712062105

89. Lin V, Golub JS, Nguyen TB, Hume CR, Oesterle EC, Stone JS (2011) Inhibition of Notch activity promotes nonmitotic regeneration of hair cells in the adult mouse utricles. J Neurosci 31(43):15329–15339. https://doi.org/10.1523/JNEUROSCI.2057-11.2011

90. Borse V, Barton M, Arndt H, Kaur T, Warchol ME (2021) Dynamic patterns of YAP1 expression and cellular localization in the developing and injured utricle. Sci Rep 11(1):2140. https://doi.org/10.1038/s41598-020-77775-8

91. Meyers JR, Corwin JT (2007) Shape change controls supporting cell proliferation in lesioned mammalian balance epithelium. J Neurosci 27(16):4313–4325. https://doi.org/10.1523/JNEUROSCI.5023-06.2007

Chapter 2

Cochlear Explant Cultures: Creation and Application

Elizabeth Carroll Driver and Matthew W. Kelley

Abstract

Development of the cochlea occurs within the temporal bone of the skull making visualization or perturbation of this process challenging. One approach to address this problem is through the use of cochlear explant cultures. In this chapter, we describe the history of this technique and then provide a detailed protocol on the dissection, establishment, and maintenance of cochlear explants. We also provide details on several different ways to perturb cochlear development including electroporation and viral vectors.

Key words Cochlea, Organ of corti, Organ culture, Electroporation, Viral gene delivery, Sensory, Hair cells

1 Introduction

The development of the mammalian cochlea and organ of Corti is a fascinating process that requires the coordination of multiple processes including terminal mitosis, specification of unique cell types, and precise cellular patterning. A significant challenge in studying these events is the location of the developing cochlear duct within the temporal bone of the skull. In addition, in mice, a significant portion of cochlear development occurs during the embryonic period which makes access to this structure even more difficult. However, over the last 30 years, several laboratories have demonstrated that the cochlear duct can be dissected from mice as early as embryonic day 12 (E12) and established as an explant culture that will survive and, most importantly, develop relatively normally for 7–8 days. Explants can then be used to both visualize cochlear development and to manipulate developmental processes using pharmacologic or genetic interventions. In this chapter we will describe the dissection of the embryonic mouse cochlea, the establishment and maintenance of cochlear explants and discuss some of the methods that can be used to alter normal development.

Andrew K. Groves (ed.), *Developmental, Physiological, and Functional Neurobiology of the Inner Ear*, Neuromethods, vol. 176, https://doi.org/10.1007/978-1-0716-2022-9_2,

**1.1 Cochlear
Structure**

The bony spiral of the cochlea encapsulates three ducts, the scala vestibuli, scala media, and scala tympani. In cross section, the scala media has a triangular shape comprised of three unique structures, Reissner's membrane, which separates scala media from scala vestibuli, the stria vascularis which maintains ionic homeostasis within the scala and the sensory epithelium, the organ of Corti (OC) [1–3]. The cellular structure within the OC is truly remarkable. Two types of mechanosensory hair cells (inner hair cells and outer hair cells) are arranged in precise rows that extend along the length of the spiral. In addition, at least six unique types of associated non-sensory cells, called supporting cells, are similarly aligned in unique rows. The development of this cellular pattern is crucial for auditory perception.

**1.2 Cochlear
Development**

The majority of cells within the mammalian inner ear orginate in the otocyst, a placodally derived sphere that forms adjacent to the hindbrain between embryonic day 9 (E9) and E10 in mice. Once it has submerged beneath the surface ectoderm, a series of morphometric changes occur to transform the spherically shaped otocyst into the elaborate structures of the inner ear, including the semicircular canals and cochlear duct [4, 5]. The cochlear duct begins as an out-pocketing from the posterior-ventral region of the otocyst that is first apparent around E11. The duct continues to grow and coil through the early postnatal period.

The extending cochlear duct is lined with epithelial cells that will give rise to the different structures within the scala media of the cochlea. In particular, hair cells and supporting cells within the OC are thought to derive from a specialized population of cells referred to as prosensory cells [6–9]. These cells, which can be identified based on co-expression of Sox2 and Cdkn1b, become post-mitotic by E14 and then undergo an intricate process of cellular specification and patterning to give rise to the OC. This process occurs in a gradient that begins near the base of the cochlear spiral and extends in a graded fashion towards the apex. At birth, precise rows of inner and outer hair cells are already present along most of the cochlear spiral, with only those cells in the most apical region still undergoing cell fate specification and alignment.

By the time cochlear extension begins, the otocyst has become surrounded by the developing temporal bone of the skull which prevents in vivo imaging of cells or cellular processes within the duct. Therefore, several laboratories have developed and refined techniques for the dissection and maintenance of the developing cochlear duct from both embryonic and early postnatal mice. The development of these protocols provides a powerful technique for visualizing cochlear development as well as perturbation of that process using either pharmacological or genetic manipulations.

The method for establishment of cochlear explants was initially described by Sobkowicz [10]. Her work utilized, primarily, early postnatal animals and focused on the development of synaptic innervation between hair cells and afferent spiral ganglion neurons [11, 12]. Using an elaborate culture system that involved roller tubes and a culture media that included human serum, explants were maintained for as long as 30 days with good survival of hair cells and supporting cells [13]. The basic approach was fairly straightforward. The developing inner ear was isolated from the temporal bone and the cochlear capsule removed to expose the cochlear duct/scala media. The scala media was then separated from ductus reuniens and the modiolus severed at the cochlear base. Because the spiral extends for 1.75 turns at this time, the cochlea was separated into basal, middle, and apical portions, which are then individually adhered to glass coverslips. Each explant comprised a half-turn of the scala media including Reissner's membrane, the stria vascularis and spiral ligament, and the organ of Corti and Kölliker's organ. The explant also contained the tympanic border cells located on the contralateral side of the basilar membrane, the cell bodies of the spiral ganglion neurons (SGNs) and glial cells, and surrounding otic mesenchyme. Since the central axons of the SGNs have been severed, survival and normal development of these cells was significantly compromised, although the unique culture conditions established by Sobkowicz seemed to promote SGN survival. Finally, because Reissner's membrane and the stria vascularis were not dissected away, it seems likely that each explant contained an endolymphatic space. The dissection process led to a collapse of the lateral wall such that Reissner's membrane contacted the lumenal surface of the organ of Corti, so the normal geometric shape of this region was distorted, but the potential maintenance of endolymph in this region could have promoted hair cell survival.

By P0, the murine organ of Corti already contains a full complement of hair cells, so in order to visualize the process of hair cell development, Kelley et al. dissected cochleae from embryonic cochleae as early as embryonic day 13 [14, 15]. Explants were established in a very similar manner with the exception that the developing Reissner's membrane and stria vascularis were removed to allow better visualization of the developing organ of Corti. Because the cochlear duct is significantly shorter at embryonic time points, it was possible to simply establish the entire cochlear as a single explant. Examination of embryonic explants over the course of 5–7 days in vitro demonstrated a relatively normal recapitulation of cell fate specification and cellular patterning. However, cellular patterning does become disrupted in the apical 1/3 of these explants, in large part because the two-dimensional nature of the explants constrains the extension of the apical region of the duct (*see* **Note 1**). In fact, patterning can be improved by cutting of the apical region of the duct, which allows extension to occur [16–18]. Subsequent studies have combined this approach with time-lapse imaging and genetic labeling to visualize different processes in cochlear development including hair cell differentiation, cellular migration, spontaneous activity, and responses to injury [19–21].

2   Materials and Reagents

2.1   Equipment

Dissection microscope.

Laminar flow clean bench.

Surgical scissors.

No. 5 and No. 55 forceps.

Minutien pins.

Sylgard-coated glass petri dishes.

Electrodes.

Electroporator (ECM-830, BTX).

35 mm Mattek dishes, 10 mm glass diameter.

2.2   Reagents

HBSS/HEPES solution.

1× Hanks' balanced salt solution (HBSS), 5 mM HEPES, pH 7.2–7.3. Store at 4 °C for 1–2 months.

Cochlear explant media.

DMEM (Thermo Fisher cat. no. 12430) or DMEM/F12 (Thermo Fisher cat. no. 11039), 10% fetal bovine serum, 10 ng/ml ciproflaxin. Store at 4 °C for 1–2 weeks. Matrigel solution. 10% Matrigel Matrix Phenol Red-Free (Corning Cat. no. 356237) in DMEM. Make fresh immediately before coating Mattek dishes.

3   Methods

3.1   Dissection of the Embryonic Cochlear Duct

Timed-pregnant female mice are euthanized, and the uterine horns are exposed. Individual embryos are transferred to sterile saline. The remainder of the dissection is carried out in a clean bench. It is a good idea to stage the embryos prior to beginning the dissection. Once staging has been confirmed, embryos are decapitated, and the skin is then removed to visualize the dorsal region of the skull. The skull should be opened along the dorsal midline (Fig. 1a) and then dissected laterally to a point where the bone begins to curve ventrally. Once this process is completed on both sides, the majority of the brain can be removed intact to reveal the ventral floor of the skull (Fig. 1b). The developing external auditory meatus (EAM)

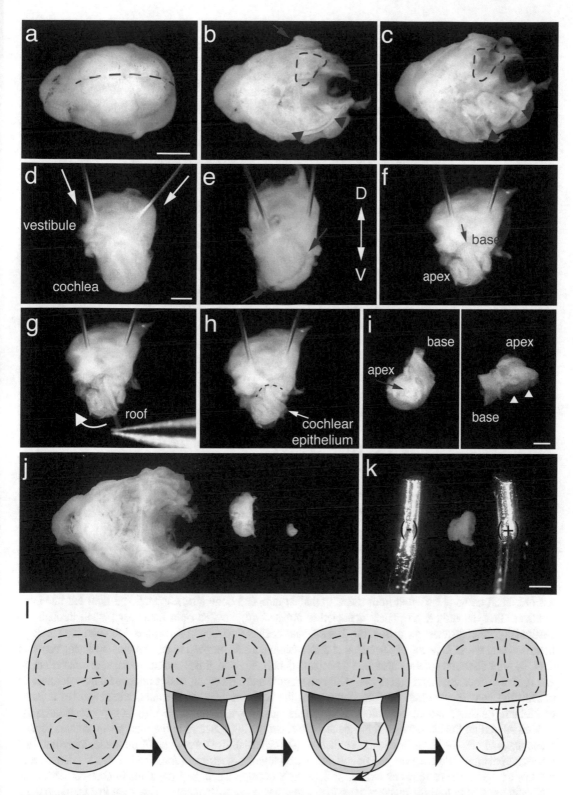

Fig. 1 Dissection of the cochlea. (**a–h**) Bright field images of the different stages of cochlear dissection. (**a**) Dorsal surface of an E14 mouse embryo head. Dotted line indicates midline dorsal suture. (**b**) Similar image as in (**a**), except the dorsal portion of the skull has been dissected away and the brain has been removed. The

and pinna can be used to locate the inner ear which are embedded in the skull directly along the lateral-to-medial axis from the EAM. The developing sigmoid sinus along the caudal side of the inner ear is also useful for determining the position of the inner ear. The inner ear has a slight flexure such that the developing semicircular canals are located in the lateral wall of the skull while the extending cochlear duct is located in the ventral region. Cut away the skull along the rostral and caudal boundaries of the inner ear and then along the ventral midline to isolate the inner ear.

At this point, transfer the inner ear to a dissecting dish containing blackened Sylgard. Flip the inner ear over, such that the concave side is facing towards the Sylgard and place dissecting pins through the vestibular structures at the wider end of the inner ear to hold the tissue in place (Fig. 1d). At this stage the entire structure is cartilaginous, and the developing cochlear duct can usually be visualized within the bony cochlea. Next, remove any remaining overlying skin or mesenchyme to expose the surface of the developing bony cochlea.

To begin to expose the cochlear duct, carefully use a pair of forceps to cut through the cartilage (arrows, Fig. 1e), starting near the base of the cochlea. Gently peel the cartilage back and away from the duct and remove it by severing the tissue near the cochlear-vestibular boundary with forceps. Once the duct is exposed, the spiral shape of the cochlea is apparent (Fig. 1f). In this orientation, the apex of the cochlea is closer to the lens of the microscope. Therefore, the region of the duct that will give rise to Reissner's membrane and the stria vascularis is on the side facing the lens of the microscope. To remove the developing Reissner's membrane and stria vascularis (the "roof" of the duct), use fine forceps to pinch the roof near the base of the cochlea and pull in the

Fig. 1 (continued) right inner ear is outlined. On the left side of the head, arrow heads indicate the developing sigmoid sinus which outlines the vestibular portion of the inner ear. (**c**) Similar view as in **b**, but the left inner ear has been partially loosened from the surrounding temporal bone (arrowheads). The right ear has been removed, and the area it previously occupied is outlined. (**d**) Isolated right inner ear pinned through the vestibular portion (arrows) for dissection. The spiraled cochlear duct can be visualized through the cartilaginous cochlear capsule. (**e**) Initial dissection of the cochlear capsule (arrows) to expose the developing cochlear duct. (**f**) After dissection of the capsule, the cochlear duct is exposed. Basal and apical turns are indicated. (**g**) The roof of the cochlear duct (future Reissner's membrane and stria vascularis) is removed (arrow, forceps) to expose the floor of the duct. (**h**) After removal of the roof, the floor of the duct, containing the sensory epithelium, Kölliker's organ, and future outer sulcus can be visualized (arrow). (**i**) Isolated cochlear spiral viewed as in **h** on the left panel and in side view to show the 3D structure in the right panel, with remaining mesenchymal and SG tissue under the epithelium (arrowheads). (**j**) Comparison of the head, inner ear, and cochlea from littermate embryos. (**k**) Orientation of cochleae for electroporation. The cochlea is oriented with the apex towards the negative electrode. (**l**) Schematic drawing illustrating the steps in cochlear dissection including removal of the cartilaginous capsule and the roof of the cochlear duct. Scale bar in **a** (same in **b**, **c**), 2 mm, Scale bar in **d** (same in **e–h**), 500 μm, Scale bar in **i**, 250 μm, Scale bar in **k**, 500 μm

direction of the apex (Fig. 1g). At embryonic stages E14–E16, the roof will usually separate easily from the remaining cochlear epithelium and can be removed in one piece. Once the roof is removed, the cochlear duct can be separated from the vestibular portion with forceps (dotted line, Fig. 1h). At E14, the epithelium is quite thick compared to later stages, and the relatively short coil around the remaining SGN and otic mesenchyme gives the dissected cochlea a fairly sturdy 3-D structure (Fig. 1i).

3.2 Cochlear Explants

Once the dissection is complete, transfer each explant to a Matrigel-coated dish. We typically use Mattek dishes which are 35 mm plastic petri dishes with a hole drilled in the center and a coverslip mounted on the outside of the hole. To coat the dish, dilute a frozen aliquot of Matrigel 1:10 in cold (4 °C) DMEM. Be sure to keep the Matrigel/DMEM cold until plated as Matrigel will come out of solution at warmer temperatures. Use about 150 μl of diluted Matrigel to cover the well created in the Mattek dish by the coverslip. Incubate the dishes at 37 °C for at least 30 min prior to use. Once the incubation period is completed, place the explant directly into the Matrigel solution. The explant should be oriented such that it is sitting on the glass with the sensory epithelium pointed "up" (Figs. 1i and 2a). Remove the excess Matrigel solution, forcing the cochlea against the glass. Allow the explant to adhere briefly but without drying out (approximately 5 min), and then add 150 μl culture media (DMEM +10% FBS) directly onto each explant. Watch the explant as you add the culture media. Most will remain adhered to the glass, but some may be detached as the media is added. In these cases, the explant is typically trapped by surface tension at the top edge of the media droplet. If this is the case, use forceps to carefully push the explant back down to the surface of the glass. Each Mattek dish is then placed in a 37 °C incubator and left overnight. By the next morning, signs of attachment and outgrowth of underlying mesenchymal cells should be evident (Fig. 2b). Over subsequent days, there will be continued mesenchymal outgrowth and flattening (Fig. 2c). By 3 or 4 days in vitro (E17–E18 equivalent), a distinct region of increased cell density will be evident in an arc around the outside of the explant (Fig. 2d, arrows). This region contains the developing organ of Corti. Explants can be maintained for as long as 7 days. We typically change the media every 3 days. Once an explant has been fixed, supporting cells, as indicated by expression of Sox2 (Fig. 2e) and hair cells, which express Myo7A, can be visualized in an arc that extends along the length of the explant (Fig. 2f).

3.3 Perturbation of Cochlear Development

At E14, the developing OC is comprised of prosensory precursor cells. Therefore, once an explant is established, the processes of cell fate specification and patterning can be perturbed in vitro. One of the simplest ways to disrupt development is by the addition of

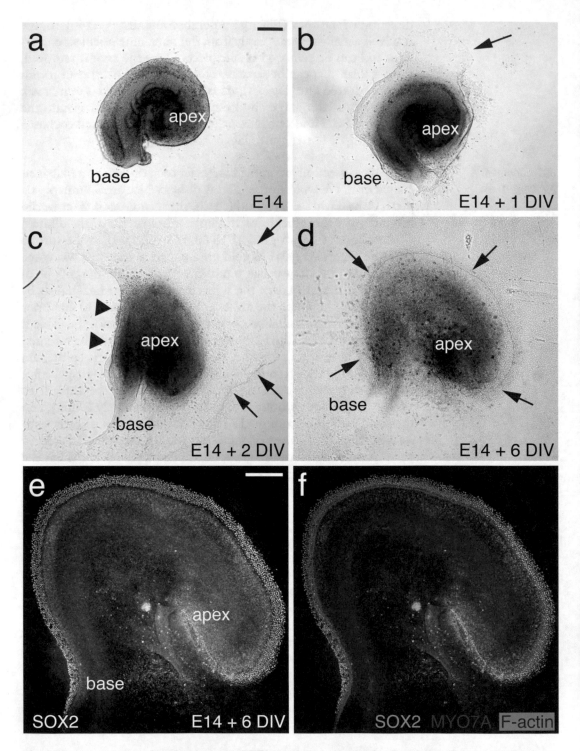

Fig. 2 Time course of explant development. (**a**) E14 cochlea immediately after plating. Basal and apical turns are indicated. (**b**) After 1 day in vitro (DIV) a region of cellular outgrowth is evident (arrow). The majority of the outgrowth cells are mesenchymal tympanic border cells. (**c**) By 2 DIV, cellular outgrowth has expanded significantly (arrows). For this explant, outgrowth along the left-hand edge (arrowheads) is limited at this time point (*see* **Note 2**). (**d**) By 6 DIV, growth of mesenchyme has extended in all directions, creating a flattening of the explant. The developing sensory epithelium can now be visualized (arrows) as an arc of higher density

pharmacological agents to the culture media. For instance, activation of the Notch pathway, which requires γ-secretase-mediated cleavage of the Notch receptor, has been shown to play a role in inhibition of hair cell fate [22–24]. Addition of the γ-secretase inhibitor LY411575 to cochlear explants beginning at E14 leads to a significant increase in the number of hair cells (Fig. 3a–d). In contrast, explants treated with FH353, a Wnt antagonist, show a significant decrease in hair cells, a result that is also consistent with in vivo studies (Fig. 3e, f) [25, 26]. Finally, treatment with blebbistatin, which inhibits non-muscle Myosin II, inhibits cochlear outgrowth and patterning, leading to larger and more poorly organized inner and outer hair cells (Fig. 3g, h) [17, 18].

A second method for modulating gene activity within a cochlear explant is by using square wave electroporation to transfect individual cells with plasmid expression constructs [27, 28]. The images in Fig. 4 illustrate explants that were electroporated on E14, immediately following dissection, with an expression construct for either GFP alone (Fig. 4a) or GFP-IRES-ATOH1 (Fig. 4b, c). The majority of transfected cells are located in Kölliker's organ, but cells located within the developing OC are typically observed as well. Forced expression of Atoh1 in cells within Kölliker's organ leads to induction of hair cell fates which are evident based on double labeling with an anti-Myosin6 antibody (Fig. 4c). The higher magnification images in Fig. 4c illustrate the small clusters of ectopic hair cells in Kölliker's organ induced by Atoh1 expression.

This approach can also be used to delete genes at a single cell level using CRISPR/Cas9 expression constructs. Our preliminary data indicates highly efficient deletion of *Atoh1* and inhibition of hair cell fate using this approach (not shown).

The protocol for electroporation is straightforward and should be carried out between completion of the cochlear dissection and adherence of each explant in a Mattek dish.

Generate plasmid expression vectors using standard cloning techniques. Resuspend the expression plasmid at a concentration of 1 μg/ml in water. In a dry, sterile Sylgard dish, make a 10 ml drop of the plasmid DNA solution. Using a microcurette, carefully transfer a single dissected cochlea into the drop. Try to minimize addition of extra solution to the drop. Orient the cochlear explant so that it is standing on its edge with the lumenal surface of the

Fig. 2 (continued) cells located along the outer edge of the original cochlear explant. (**e**) The same explant as in **d**, fixed and labeled with an antibody against SOX2 which marks supporting cells. The arc of SOX2$^+$ cells corresponds to the arc of high-density cells in **d**. (**f**) The same sample as in **e**, labeled for SOX2 (green), the hair cell marker MYOSIN7A (magenta), and F-actin (phalloidin, blue). A single row of inner hair cells and approximately three rows of outer hair cells are present along most of the explant. Scale bar in **a** (same in **b–d**), 200 μm. Scale bar in **e** (same in **f**), 200 μm

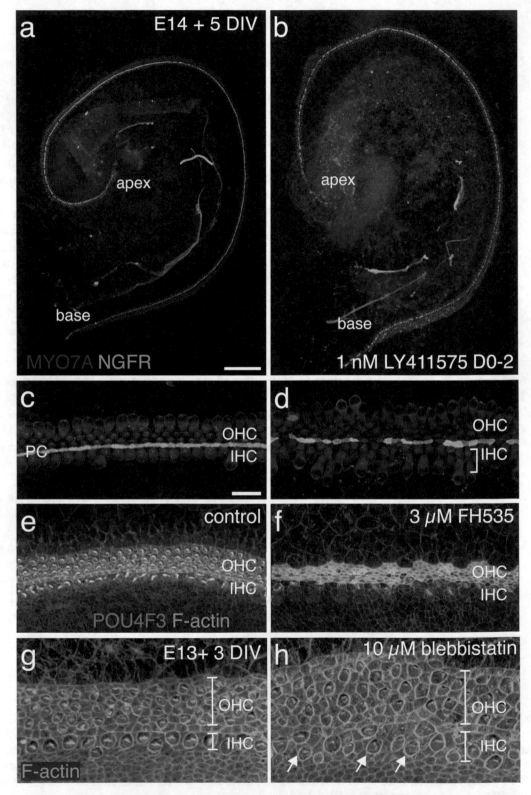

Fig. 3 Pharmacological treatments alter cell fate and patterning in cochlear explants. (**a**) Low magnification image of an entire cochlear duct explanted at E14 and maintained for 5 DIV. Hair cells are labeled with anti-

cochlear epithelial cells oriented towards one side of the droplet (Fig. 1k).

Adjust the distance between the electroporation electrodes so that their width roughly matches the width of the droplet. Then place the electrodes in the droplet such that the lumenal surfaces of the cochlear epithelium face towards the negative electrode (*see* **Note 3**).

Apply a pulse train of 27 V, 30 ms pulse duration, and nine to ten pulses across the cochlea in the droplet. Transfer the cochlea to a clean dish of HBSS and allow to recover for approximately 1 min, then mount the cochlear explant in a Mattek dish as described above.

Viral vectors can also be used to alter gene expression in cochlear explants. There has been considerable research in recent years examining the tropism of different viral serotypes within the adeno- and adeno-associated families [29–31]. Our own experience indicates that human adenovirus Type5 (dE1/E3) with RGD-fiber modification will robustly infect mouse embryonic hair cells, supporting cells, and prosensory cells, but low transfection rates are observed without the RGD modification. This is an appealing vector for in vitro studies as it is available from a commercial vendor and can be engineered to carry multiple expression constructs. The carrying capacity of an adenovirus is on the order of 20 kb. Also, since multiple explants can be cultured in a single dish, a high degree of efficiency can be obtained. Most stock adenoviruses are supplied at high titer, about 1×10^{10} PFU/ml. We have found a final concentration of 1×10^7 to 1×10^8 PFU/ml works well for overnight incubation (*see* **Note 4**). To transfect cochlear explants with adenovirus, dissect and plate explants as described. Working in a BSL2 hood, dilute the adenovirus to the desired concentration in normal culture media, including serum. Add adenovirus-

---

Fig. 3 (continued) MYO7A (magenta) and pillar cells are labeled with anti-NGFR (green). Note the single line of pillar cells located between the row of inner hair cells and the first row of outer hair cells. (**b**) A cochlear explant established as in **a** but treated with the γ-secretase inhibitor LY411575 (1 nM) for the first 2 DIV. Inhibition of γ-secretase blocks activation of the Notch-signaling pathway. Note the increased number of hair cells, particularly on the inner hair cell side of the row of pillar cells. (**c**) High magnification view of the organ of Corti from the middle region of the explant in **a**. A single row of inner hair cells and three rows of outer hair cells are present. (**d**) High magnification of a similar region of the cochlear from **b**. The number of hair cells is significantly increased, consistent with the known role of Notch signaling during cochlear development. (**e**) Control image triple labeled for hair cell cytoplasm (anti-MYO7A) and nuclei (anti-POU4f3) and F-actin (phalloidin). (**f**) A similar view as in **e**, but from an explant treated with the Wnt-inhibitor FH535 (3 μM). As has been reported previously, blocking Wnt signaling leads to a decrease in hair cell formation. (**g**) Labeling of F-actin in an E13 control explant after 3 DIV illustrates the single row of inner hair cells and three to four rows of outer hair cells. (**h**) Inhibition of non-muscle Myosin II disrupts cellular patterning, leading to changes in alignment of both inner (arrows) and outer hair cells. Scale bar in **a** (same in **b**), 200 μm, Scale bar in **c** (same in **d–h**), 25 μm

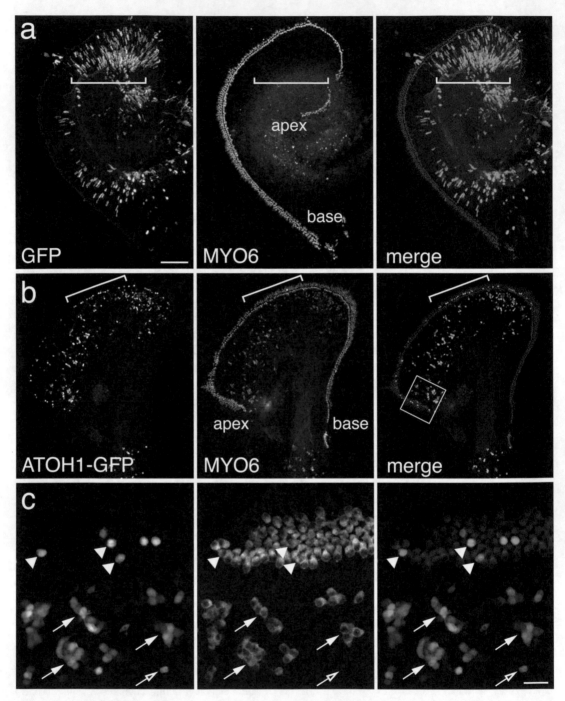

Fig. 4 Electroporation of expression constructs in cochlear explants. (**a**) E14 Cochlear explant electroporated with a GFP-expression construct after 5 DIV. GFP channel indicates regions with varying levels of electro-porated cells. In this particular explant there is region of high-density electroporation (bracket) in the apical turn. The MYO6 channel demonstrates that most of the electroporated cells are located in Kölliker's organ. Merge channel shows a limited number of electroporated cells within the organ of Corti. (**b**) E14 cochlear explant electroporated with an Atoh1-IRES-GFP expression construct. Overall level of expression is decreased relative to the GFP construct with most cells located in Kölliker's organ, although a region of transfected cells in the organ of Corti is present (bracket). (**c**) High magnification view of the boxed region in **b**. Transfected cells

containing media to the explants and incubate overnight at 37 °C. Remove and replace the adenovirus-containing media the following day with fresh culture media. Fluorescent reporter proteins can usually be seen 1–2 days after transfection. Figure 5 illustrates examples of explants transfected with high (Fig. 5a, b) or low (Fig. 5c, d) concentrations of an RGD-modified adenoviral vector expressing GFP. This explant was transfected at E14, prior to the formation of hair cells or supporting cells, and both GFP+ hair cells and supporting cells are present in the OC (Fig. 5b, d) after 5 DIV.

In summary, by embryonic day 13 the development of the organ of Corti in the mouse, at least in terms of determination of cell fate and patterning, is a largely autonomous process. As a result, the developing cochlea can be established as an in vitro explant at this time point, or any later time point, with minimal impact on the developmental process. This provides an ideal opportunity for the design of experiments to screen the effects of different pharmacological or genetic manipulations on the development of the organ of Corti. Cochlear dissections are easily mastered with practice and maintenance in vitro is not difficult. Previous work from our laboratory and others have demonstrated and/or visualized changes in cell fate or patterning using this approach. The ability to visualize different developmental processes is a particular strength of this approach given the relative inaccessibility of the cochlea during normal development.

4   Notes/Troubleshooting

1. Embryonic cochlear explant development largely recapitulates in vivo development, in terms of cell fate and patterning as determined by molecular markers, but a few differences can arise. Extra hair cells, especially outer hair cells in the apical region, are often observed, even in control samples (Fig. 2f). During dissection, slight damage to the cochlear epithelium can lead to disrupted patterning (arrow, Fig. 5a) or areas of reduced sensory development (apex, Fig. 5c). More extensive physical damage to the cochlea from dissection can cause stretches of absent hair and supporting cells, or explants with severely disrupted morphology, making analysis difficult (not shown).

2. Getting explants to adhere to the Matrigel substrate can be challenging. In our experience, if explants are still adhered to

Fig. 4 (continued) are present in both the organ of Corti (arrowheads) and Kölliker's organ (arrows). Note that expression of Atoh1 in non-sensory Kölliker's organ cells has induced expression of MYO6 in virtually all cells, with one exception (open arrow). Scale bar in **a** (same in **b**), 200 μm, Scale bar in **c**, 25 μm

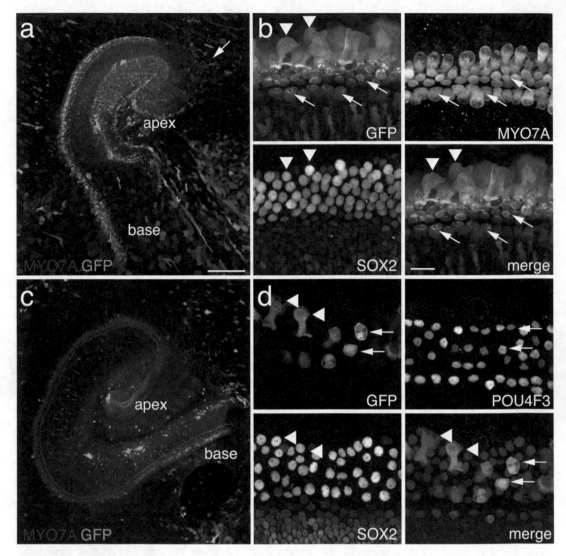

Fig. 5 Adenoviral vectors transfect cochlear prosensory cells. (**a**) Low magnification image of an E14 cochlear explant after 5 DIV. Explant was treated with an adenoviral vector expressing GFP as described. Note the extensive transfection within the organ of Corti. Arrow indicates disrupted patterning in the apical region, likely due to damage during dissection. (**b**) High magnification view of the organ of Corti from the explant in **a**. Arrowheads indicate transfected supporting cells (SOX2⁺) while arrows indicate transfected hair cells (MYO7A⁺). (**c**) Second example of viral transfection of a cochlear explant using a lower concentration of virus. (**d**) While overall number of cells transfected is reduced, individual POU4F3⁺ hair cells and SOX2⁺ supporting cells expressing GFP can be identified. Scale bar in **a** (same in **c**), 200 μm. Scale bar in **b** (same in **d**), 20 μm

the dish the day after plating, they will usually remain in place (if handled gently) and will spread out over the next few days, even if outgrowth initially appears uneven (Fig. 2b, c). Explants not adhered after 1 DIV will rarely attach to the substrate later and are best discarded.

3. Electroporation can also present some technical problems. The electrodes must be held reasonably close to the cochlea in order to get transfection but placing them too close together can lead to excessive cell death. Some evidence of cell death (missing outer hair cells) is apparent in the apical regions of the explants in Fig. 4.

4. We have not found much, if any, evidence of cell death in explants treated with adenovirus, even in areas of very high transfection (Fig. 5b). The degree of transfection is dependent on both the concentration of adenovirus added and the length of incubation time with the virus. In addition, transfection varies somewhat depending on the adenovirus construct, so it must be determined empirically to achieve the level desired for experimental analysis (broad overexpression/knock-down-vs. individual cell fate analysis).

References

1. Kelley MW (2006) Regulation of cell fate in the sensory epithelia of the inner ear. Nat Rev Neurosci 7(11):837–849

2. Groves AK, Fekete DM (2012) Shaping sound in space: the regulation of inner ear patterning. Development 139(2):245–257. https://doi.org/10.1242/dev.067074

3. Driver EC, Kelley MW (2020) Development of the cochlea. Development 147(12). https://doi.org/10.1242/dev.162263

4. Wu DK, Kelley MW (2012) Molecular mechanisms of inner ear development. Cold Spring Harb Perspect Biol 4(8):a008409. https://doi.org/10.1101/cshperspect.a008409

5. Alsina B, Whitfield TT (2017) Sculpting the labyrinth: morphogenesis of the developing inner ear. Semin Cell Dev Biol 65:47–59. https://doi.org/10.1016/j.semcdb.2016.09.015

6. Dabdoub A, Puligilla C, Jones JM, Fritzsch B, Cheah KS, Pevny LH, Kelley MW (2008) Sox2 signaling in prosensory domain specification and subsequent hair cell differentiation in the developing cochlea. Proc Natl Acad Sci U S A 105(47):18396–18401

7. Kiernan AE, Pelling AL, Leung KK, Tang AS, Bell DM, Tease C, Lovell-Badge R, Steel KP, Cheah KS (2005) Sox2 is required for sensory organ development in the mammalian inner ear. Nature 434(7036):1031–1035

8. Pan W, Jin Y, Chen J, Rottier RJ, Steel KP, Kiernan AE (2013) Ectopic expression of activated notch or SOX2 reveals similar and unique roles in the development of the sensory cell

progenitors in the mammalian inner ear. J Neurosci 33(41):16146–16157. https://doi.org/10.1523/JNEUROSCI.3150-12.2013

9. Puligilla C, Kelley MW (2017) Dual role for Sox2 in specification of sensory competence and regulation of Atoh1 function. Dev Neurobiol 77(1):3–13. https://doi.org/10.1002/dneu.22401

10. Sobkowicz HM, Bereman B, Rose JE (1975) Organotypic development of the organ of Corti in culture. J Neurocytol 4(5):543–572. https://doi.org/10.1007/BF01351537

11. Sobkowicz HM, Rose JE, Scott GE, Slapnick SM (1982) Ribbon synapses in the developing intact and cultured organ of Corti in the mouse. J Neurosci 2(7):942–957

12. Sobkowicz HM, Slapnick SM (1992) Neuronal sprouting and synapse formation in response to injury in the mouse organ of Corti in culture. Int J Dev Neurosci 10(6):545–566. https://doi.org/10.1016/0736-5748(92)90055-5

13. Sobkowicz HM, Loftus JM, Slapnick SM (1993) Tissue culture of the organ of Corti. Acta Otolaryngol Suppl 502:3–36

14. Kelley MW, Talreja DR, Corwin JT (1995) Replacement of hair cells after laser microbeam irradiation in cultured organs of corti from embryonic and neonatal mice. J Neurosci 15(4):3013–3026

15. Kelley MW, Xu XM, Wagner MA, Warchol ME, Corwin JT (1993) The developing organ of Corti contains retinoic acid and forms supernumerary hair cells in response to exogenous retinoic acid in culture. Development 119(4):1041–1053

16. Wang J, Mark S, Zhang X, Qian D, Yoo SJ, Radde-Gallwitz K, Zhang Y, Lin X, Collazo A, Wynshaw-Boris A, Chen P (2005) Regulation of polarized extension and planar cell polarity in the cochlea by the vertebrate PCP pathway. Nat Genet 37(9):980–985

17. Yamamoto N, Okano T, Ma X, Adelstein RS, Kelley MW (2009) Myosin II regulates extension, growth and patterning in the mammalian cochlear duct. Development 136(12):1977–1986

18. Driver EC, Northrop A, Kelley MW (2017) Cell migration, intercalation and growth regulate mammalian cochlear extension. Development 144(20):3766–3776. https://doi.org/10.1242/dev.151761

19. Gale JE, Piazza V, Ciubotaru CD, Mammano F (2004) A mechanism for sensing noise damage in the inner ear. Curr Biol 14(6):526–529. https://doi.org/10.1016/j.cub.2004.03.002

20. Tritsch NX, Yi E, Gale JE, Glowatzki E, Bergles DE (2007) The origin of spontaneous activity in the developing auditory system. Nature 450(7166):50–55. https://doi.org/10.1038/nature06233

21. Yamahara K, Asaka N, Kita T, Kishimoto I, Matsunaga M, Yamamoto N, Omori K, Nakagawa T (2019) Insulin-like growth factor 1 promotes cochlear synapse regeneration after excitotoxic trauma in vitro. Hear Res 374: 5–12. https://doi.org/10.1016/j.heares.2019.01.008

22. Lanford PJ, Lan Y, Jiang R, Lindsell C, Weinmaster G, Gridley T, Kelley MW (1999) Notch signalling pathway mediates hair cell development in mammalian cochlea. Nat Genet 21(3):289–292

23. Kiernan AE, Ahituv N, Fuchs H, Balling R, Avraham KB, Steel KP, Hrabe de Angelis M (2001) The notch ligand Jagged1 is required for inner ear sensory development. Proc Natl Acad Sci U S A 98(7):3873–3878

24. Kiernan AE, Cordes R, Kopan R, Gossler A, Gridley T (2005) The notch ligands DLL1 and JAG2 act synergistically to regulate hair cell development in the mammalian inner ear. Development 132(19):4353–4362

25. Shi F, Hu L, Jacques BE, Mulvaney JF, Dabdoub A, Edge AS (2014) Beta-catenin is required for hair-cell differentiation in the cochlea. J Neurosci 34(19):6470–6479. https://doi.org/10.1523/JNEUROSCI.4305-13.2014

26. Jacques BE, Montgomery WH, Uribe PM, Yatteau A, Asuncion JD, Resendiz G, Matsui JI, Dabdoub A (2014) The role of Wnt/beta-catenin signaling in proliferation and regeneration of the developing basilar papilla and lateral line. Dev Neurobiol 74(4):438–456. https://doi.org/10.1002/dneu.22134

27. Zheng JL, Gao WQ (2000) Overexpression of Math1 induces robust production of extra hair cells in postnatal rat inner ears. Nat Neurosci 3(6):580–586

28. Jones JM, Montcouquiol M, Dabdoub A, Woods C, Kelley MW (2006) Inhibitors of differentiation and DNA binding (ids) regulate Math1 and hair cell formation during the development of the organ of Corti. J Neurosci 26(2):550–558

29. Luebke AE, Rova C, Von Doersten PG, Poulsen DJ (2009) Adenoviral and AAV-mediated gene transfer to the inner ear: role of serotype, promoter, and viral load on in vivo and in vitro infection efficiencies. Adv Otorhinolaryngol 66:87–98. https://doi.org/10.1159/000218209

30. Maguire CA, Corey DP (2020) Viral vectors for gene delivery to the inner ear. Hear Res 394:107927. https://doi.org/10.1016/j.heares.2020.107927

31. Lan Y, Tao Y, Wang Y, Ke J, Yang Q, Liu X, Su B, Wu Y, Lin CP, Zhong G (2020) Recent development of AAV-based gene therapies for inner ear disorders. Gene Ther 27(7–8):329–337. https://doi.org/10.1038/s41434-020-0155-7

Immunohistochemistry and In Situ mRNA Detection Using Inner Ear Vibratome Sections

Mirko Scheibinger, Amanda Janesick, Giovanni H. Diaz, and Stefan Heller

Abstract

This chapter describes the application of immunohistochemistry and in situ mRNA detection to vibratome sections derived from chicken and mouse inner ears. The protocols portray simple strategies to investigate the cellular organization of vestibular and cochlear sensory epithelia. All inner cell types are confidently identified due to the excellent three-dimensional preservation of the inner ear tissue in vibratome sections. Shown are examples for detecting proteins that label cell subtypes as well as subcellular structures such as synapses. Vibratome sections are suitable for conventional colorimetric in situ hybridization as well as fluorescence-based hybridization chain reaction.

Key words Cryostat, Utricle, Cochlea, Basilar papilla, Synapse, Afferent, Inner hair cell, Outer hair cell

1 Introduction

Visualization of the complex cellular architecture of the inner ear is challenging. Particularly in adult mammals, where the body's hardest bone encases the inner ear, it is difficult to perform atraumatic tissue dissections and histological sections. Routine histology methods such as cryo- and paraffin-sectioning apply additional stresses to the tissue due to solvents or freezing. Vibratome sectioning is an alternative tool for the inner ear field that can be utilized to study the cellular architecture of the organ of Corti [1, 2]. Recent studies that have utilized vibratome sections of the mouse and chicken inner ear show an exceptional level of cellular detail and tissue preservation [3–7].

The vibrating microtome (vibratome) is an economical histology instrument for producing thick sections from unfixed or fixed tissue. In this chapter, we describe the generation of vibratome

---

Mirko Scheibinger and Amanda Janesick contributed equally to this work.

Andrew K. Groves (ed.), *Developmental, Physiological, and Functional Neurobiology of the Inner Ear*, Neuromethods, vol. 176, https://doi.org/10.1007/978-1-0716-2022-9_3,

sections from fixed inner ear tissue. This relatively simple procedure is the basis for the in situ mRNA detection and immunostaining methods that are also describe. Vibratome sections preserve three-dimensional structures and subcellular organization, which allows for high resolution imaging of cellular organization and details within tissues, all with a standard confocal microscope. We have not explored the possibility of using fresh vibratome sections for inner ear studies, which is an exciting perspective. Fresh vibratome sections are used to study living cells by electrophysiology, in vitro slice cultures, live-cell imaging, and other manipulations such as viral transduction and electroporation.

Here, we provide simple instructions for vibratome sectioning of the utricle and the basilar papilla (cochlea) of chicken and mouse, including postnatal day (P) 28 mouse and older animals. We provide practical instructions for examining these sections with immunohistochemistry and how to visualize mRNA in situ using Digoxigenin-labeled and fluorescent probes.

2   Materials and Methods

Using RNase-free technique and solutions preserves all options for the detection of mRNA in subsequent applications. RNase-free handling is not necessary if the primary goal of the procedure is the detection of proteins.

2.1   Preparation of the Chicken and Mouse Utricle

Split heads along the midline and remove the brain tissue to expose the temporal bone with the inner ear (Fig. 1a, b). The inner ears are carefully excised with surgical scissors (chicken) or the ear capsule is carefully dislodged and removed from surrounding tissue with forceps (mouse), and placed into 4% paraformaldehyde (PFA, Electron Microscopy Sciences, #15710) in RNase-free phosphate-buffered saline at pH 7.4 (RNase-free PBS; ThermoFisher, #AM9624) and fixed for 2 h at room temperature or overnight at 4 °C.

For embryonic or young chickens and mice younger than postnatal day 7 (P7), the utricle is dissected from the bony ear capsule in a dissection dish lined with black Sylgard® 184 silicone (World Precision Instruments, #SYLG184), allowing improved visualization. Tissue can be mounted to the black Sylgard with sterile pins, if needed. Preparation of charcoal-blackened Sylgard dishes is described in [8]. Utricles are treated with RNase-free 0.25 M EDTA (ThermoFisher, #AM9261) in RNase-free PBS overnight at 4 °C until the otoconia appear clear and invisible to the eye. The decalcification step is not necessary for in situ hybridization because the formamide in the hybridization buffer dissolves the otoconia. The utricles can be stored at 4 °C in RNase-free PBS until further use but should be processed within 2–3 days if the ultimate goal is in situ detection of mRNA.

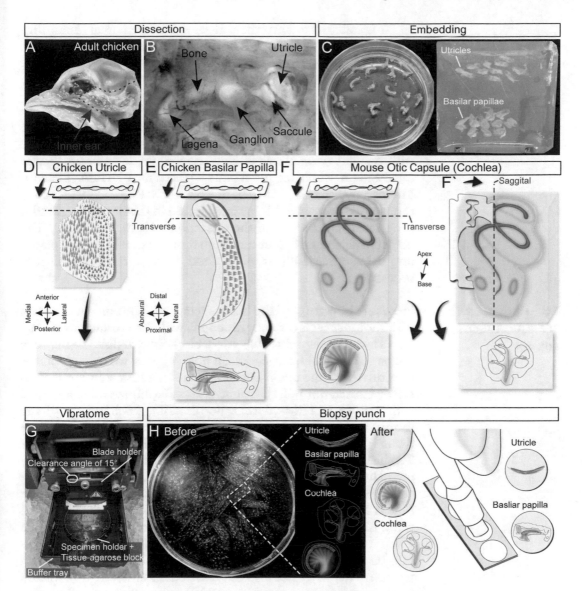

Fig. 1 Vibratome sectioning workflow. (**a**) Bisected adult chicken head. The temporal bone with the inner ear inside is indicated with red dashed line. (**b**) Anatomy of the chicken inner ear. Specific landmarks are indicated. Ganglion = cochleovestibular ganglion. (**c**) Dissected and fixed utricles and basilar papillae before and after embedding. Orientation for transverse vibratome sections of the (**d**) chicken utricle, (**e**) chicken basilar papilla, and (**f**) mouse otic capsule. (**f'**) Orientation for sagittal vibratome sections of the mouse otic capsule. (**g**) Cutting position of the Vibratome Leica VT1200/S. (**h**) Biopsy punches to obtain round specimen agarose slices for further processing

For mice older than postnatal day 7, the fixed temporal bone is decalcified in RNase-free 0.25 M EDTA in PBS at 4 °C until the bony ear capsule is soft to a light touch, which can be between 2 and 6 days depending on the age of the animal. Then, the utricle is dissected from the temporal bone. If the otoconia are not completely translucent, expose the utricle to RNase-free 0.25 M EDTA in RNase-free PBS for another night at 4 °C.

It is essential to keep the utricles submerged at all times, thus transferring the tissue should be done with a micro-spoon (F.S.T., #10360-13; Moria #1121B) with hair bundles facing down to the micro-spoon's bottom.

2.2 Preparation of the Chicken Cochlea

For dissection of the basilar papilla from P7 chicken, the temporal bone with the inner ear is removed from the bisected head with dissection scissors (F.S.T., #14002-13). The ear capsule is exposed with one cut using a scalpel between the utricle and semicircular canal and the whole temporal bone is placed into 4% PFA (Electron Microscopy Sciences, #15710) in RNase-free PBS and incubated at room temperature for 2 h. Decalcification in 0.25 M EDTA in PBS is optional, but can aid in the dissection to soften the bone. The basilar papillar duct is dissected from the temporal bone, and placed into RNase-free PBS at 4 °C, and is now ready for further processing.

2.3 Preparation of the Adult Mouse Cochlea

Heads are bisected, and the brain is removed. The bony otic capsule with the inner ear is carefully dislodged with forceps and transferred into ice-cold RNase-free PBS into a black Sylgard dish. To improve the penetration of fixative, a small hole is poked with sharp forceps into the tip of the apical cochlear bone. 4% PFA solution is then slowly infused through the round and oval windows using a syringe with a 30G 1/2 hypodermic needle (BD, #305106) until the fixative is seen flowing out the small hole made in the apex. The tissue is then placed into 4% PFA in RNase-free PBS and fixed for 2–6 h at room temperature for downstream immunohistochemical procedures. For the in situ detection of mRNA by hybridization chain reaction or conventional in situ hybridization, the fixation time is 2 h at room temperature. The ear capsule is then decalcified using RNase-free 0.25 M EDTA in RNase-free PBS until the bone is soft to the touch, which requires incubation between 2 and 6 days at 4 °C with light agitation on an orbital shaker. Extended EDTA treatment might reduce mRNA detection sensitivity in downstream in situ hybridization experiments but is usually compatible with immunolabeling. The decalcified cochlea is transferred into RNase-free PBS at 4 °C and is now ready for further processing.

2.4 Tissue Embedding

The fixed tissues are equilibrated for 5 min in RNase-free PBS in a 35 mm petri dish placed on a metal block inside a shallow 55 °C water bath, followed by two 15 min incubations, first in 2%, and then in 4% low gelling temperature (low melting point) agarose (BioRad, #1613111) in RNase-free PBS at 55 °C. Transfer of tissues is done with a micro-spoon (F.S.T., #10360-13; Moria #1121B) or wide bore transfer pipet ensuring that the specimens, particularly the hair bundles, are submerged in solution at all times.

Chicken and mouse utricles, and chicken cochleae are handled the same in subsequent steps. The tissue is transferred into disposable molds (VWR, #15160-215) filled with warm and bubble-free

4% low gelling temperature agarose (Fig. 1c). For transverse sections, orient the utricle sensory epithelium such that the anterior-posterior axis is perpendicular to the mold's floor (Fig. 1d). The posterior end of the utricle points to the floor, and the hair cells face toward you. When cutting appropriately positioned utricles, the vibratome blade will cut first through the stromal layer, then the supporting cell layer, and last through the hair cell layer. This approach ensures maximum hair bundle preservation (Fig. 1d).

For transverse chicken cochlear sections, orient the cochlea such that the distal-to-proximal axis is "standing up" perpendicular to the mold's floor with the basilar papilla's distal tip pointing upwards. The tegmentum vasculosum and the cochlear hair bundles face you to ensure that hair bundles keep their upright position during vibratome sectioning (Fig. 1e).

For the decalcified adult mouse cochlea, the vestibule can be removed with one slice near the base using a disposable scalpel. For transverse sections of the mouse cochlea, orient the capsule upright with the cochlear apex pointing upwards and the base sitting on the mold's floor. The oval and round windows should point toward you, and the first turn 22–45 kHz region away from you (Fig. 1f). For sagittal sections of the cochlea, use the same embedding orientation as for traverse sections, but when orienting the solidified agarose block for sectioning (*see* next subheading), flip the agarose block so that the apex is pointing away and the base pointing toward you (Fig. 1f).

The agarose turns opaque when solidified, which happens after 10–15 min at room temperature. Carefully monitor the specimen's orientation with a dissection microscope during this process and, if necessary, correct the orientation of the tissue with forceps when agarose is still fluid. The solidified agarose block can now be sectioned immediately or stored inside the plastic mold at 4 °C in a humidified chamber (Komax, # B083BGL3WK) for multiple days, if you wish to delay sectioning the block. However, consider mRNA stability when storing specimens for extended durations at 4 °C. Before sectioning, remove the agarose block from the plastic mold with a razor blade. Pay attention to preserve the specimen's orientation. Next, trim away unessential agarose so that the cut agarose slices are reasonably sized for further processing. Ensure that enough agarose encapsulates the specimen to ensure that the tissue is well-supported during cutting. (Fig. 1g, h).

2.5 Vibratome Sections

For mounting the tissue-agarose block, a drop of cyanoacrylate adhesive (Krazy glue, #KG585) is placed onto the flat metal specimen holder plate supplied with the vibratome. Transfer the properly oriented tissue-agarose block with forceps gently into the adhesive and let it set for 5 min at room temperature.

Please read the vibratome manufacturer's operations manual carefully for the setup and proper use of the instrument. Adjust the slice thickness to 50–200 μm for immunohistochemistry and

40–50 μm for in situ mRNA detection. Parameters are set to amplitude = 1 mm, and speed between 0.5 and 0.8 mm/s. Attach the specimen holder plate with the glued tissue-agarose block into the buffer tray. Pay attention to maintain the desired orientation of the specimen. Place the ice tray into the dovetail holder on the vibratome stage and push the buffer tray with the glued embedded specimen in as far as it will go. Fill the buffer tray with ice-cold RNase-free PBS (Fig. 1g). Rotate the blade holder 90° clockwise to insert a new single-edge razor blade (Schick Injector Blades 7) with the sharp side facing up. Return the blade holder to the cutting position by rotating the blade holder to a clearance angle of 15° (Fig. 1g). Start sectioning at the desired thickness once the blade has begun to cut into the sample. Optionally, if the intention is to document the specimens' anatomical locations, transfer each section into a well of a multi-well plate with cold RNase-free PBS using either a micro-spoon (Moria #1121B) or forceps. Otherwise, sections are decanted into a 10 cm Petri dish coated with Sylgard® 184 silicone [8]. Excess agarose is removed with 1–5 mm diameter biopsy punches (Acuderm Acu-Punch) to obtain round agarose slices with the specimen slices inside (Fig. 1h). For immunohistochemistry, sections are transferred into cold RNase-free PBS. For in situ mRNA detection, sections are transferred to glass vials, washed with 50% methanol in RNase-Free PBS for 5 min, then stored in 100% methanol at −20 °C.

2.6 Vibratome Leica VT1200/S and Compresstome VF-310-0Z

Vibratome sectioning utilizes a vibrating razor blade, which produces relatively undistorted sections of delicate tissue. The term "Vibratome" is a trade name of Leica, but other manufacturers produce equally proficient—albeit different—equipment. We routinely use the Leica VT1200/S and have also used the Compresstome VF-310-0Z (Precisionary Instruments Inc., San Jose, California). Amplitude, speed, and angle of the blades are adjustable in both types of instruments, and both produce sections from either fresh or fixed specimens with high preservation of the cellular and subcellular features of the tissue. We noticed no differences in the quality of the sections from fixed tissue between both instruments.

For cutting sections with the Compresstome, the specimen needs to be placed inside a specimen tube. Unfortunately, this way of mounting makes it difficult to properly orient the specimens. To overcome this technical hurdle when using a Compresstome, the inner ear tissue is embedded in low gelling temperature agarose in disposable molds as described for vibratome sectioning (*see* Subheading 2.4). After the agarose is solidified, the Compresstome tube is placed on top of the agarose so that the inner ear tissue is positioned in the center of the tube. Next, the tube is used similar to a biopsy punch, and the tissue is transferred into the tube by pushing and rotating firmly in one direction until the tube penetrates the agarose block down to the mold's floor. An agarose cylinder that includes the inner ear tissue is now located inside the Compresstome tube. The Compresstome blade is vertical, and the tube will be placed horizontally. Blade and the specimen are still orthogonal to each other and only the sequence how the sections are cut is reversed. For example, suppose the utricle would be cut anterior-to-posterior with the vibratome. In that case, it will be cut posterior-to-anterior with the Compresstome—if the samples are embedded as described earlier (Subheading 2.4). We did not notice differences between vibratome or Compresstome sections when analyzed with confocal microscopy.

2.7 Immunolabeling of Vibratome Sections

Vibratome sections are processed for immunolabeling using a 24-well cell culture plate (Corning Inc., #3524) and a cell-strainer cap from a 5 mL polystyrene round-bottom tube (Falcon Corning Inc., #352235). The round mesh screen of the cell-strainer cap is removed with a pair of scissors and placed into one well of the 24-well cell culture plate. Add 500 μL of PBS and transfer 6–8 vibratome sections onto the mesh using a micro-spoon (F.S.T., #10360-13) or a 3 mL disposable transfer pipet (VWR, #414004-037). All incubation steps are conducted on an orbital shaker at low rpm to prevent spilling. To avoid tissue sticking to the transfer pipet's wall, rinse the pipet with 0.1% Triton X-100 (Sigma-Aldrich, #X100) in PBS before transferring the sections. Sections stored in 100% methanol (Sigma-Aldrich, #34860) need to be rehydrated in PBS (ThermoFisher, #AM9624). Pass sections through a stepwise series of 75%, 50%, 25% methanol in PBS, and finally PBS. Sections are permeabilized at room temperature for 30 min with 1% Triton-X-100 in PBS and blocked for 3–4 h with a PBS-based blocking solution containing 1% BSA (Millipore Sigma, #9048-46-8), 5% donkey serum (Abcam, #ab138579), and 0.1% Triton-X-100. Sections are then incubated at 4 °C overnight in blocking solution with diluted primary antibodies. The primary antibodies solution's volume can be scaled down to 250 μL. On the next day, sections are rinsed three times for 15 min in 0.2% Triton-X-100 in PBS, and incubated for 1–2 h in blocking solution with diluted secondary antibodies (Alexa Fluor 488, 546, 647, ThermoFisher) at room temperature. After washing three times for 15 min with 0.2% Triton-X-100 in PBS, sections can be exposed for 15 min to 4,6-Diamidino-2-phenylindole (DAPI, 1 μg/ml PBS; ThermoFisher, #D1306) to visualize nuclei, and to fluorophore-conjugated phalloidin (0.4 μM in PBS; Thermo-Fisher) to visualize F-actin filaments. Finally, sections are rinsed three times for 15 min with 0.2% Triton-X-100 in PBS, and three times for 15 min with PBS. For mounting, a Secure-Seal™ Spacer (single well, 13 mm diameter, 0.12 mm deep; Invitrogen, #S24735) is attached to a glass slide. A drop of FluorSave™ Reagent (Calbiochem, #345789—20 ml) is administered into the center of the spacer. Sections are transferred into the drop with a micro-spoon (F.S.T., #10360-13) or plastic transfer pipet. If desired, additional washes with FluorSave™ Reagent can be conducted to reduce the transferred volume of PBS. Excessive Fluor-Save™ Reagent is removed with a pipette, and the sections are covered with an appropriately sized and typed coverslip. To avoid trapping of air bubbles under the coverslip, angle the coverslip over the sample with on edge touching the sticky Secure-Seal™ Spacer and gently lower the coverslip into place using a forceps. Seal the coverslip with nail polish and let it rest until the nail polish has set. Sealed mounted specimen can be stored at 4 °C. For best results, however, it is recommended to image the samples shortly after immunolabeling.

2.8 In Situ mRNA Detection on Vibratome Sections

Janesick and colleagues [7] introduced an in situ hybridization method specifically adjusted for vibratome sections. Compared with cryosections, this method preserves the tissue's morphology much better, and it provides an increased signal-to-noise ratio due to the thickness of the vibratome sections. In situ hybridization is adapted from Harland [9] with all steps conducted in a hybridization chamber (described below).

2.8.1 Probe Synthesis

cDNA template of inner ear tissue is typically created by homogenizing whole inner ears using BeadBug™ (Millipore Sigma, #Z763705) and extracting total RNA using TRIzol™ (ThermoFisher, #15596026). RNA is further DNase-treated (ThermoFisher, #Am2239) followed by LiCl precipitation (ThermoFisher, #AM9480) [10], and resuspension in RNase-free water. Total RNA yield is determined with a spectrophotometer (Nanodrop), and 1 μg is reverse transcribed with SuperScript™ IV (ThermoFisher, #18090050) primed with 20-mer oligo dTs (IDT, #51-01-15-01) in a 20 μL reaction volume. cDNA is diluted 1:10, and 1–5 μL are used for a 50 μL PCR reaction. Amplified cDNA is validated by gel electrophoresis and sequencing.

In situ hybridization with Digoxigenin-labeled probes and hybridization chain reaction (HCR) are conducted as described [7, 9, 11]. Digoxigenin-labeled antisense probes are prepared from PCR-amplified and validated cDNA of the gene of interest that incorporate a bacteriophage T7 promoter in the reverse amplification primer. Primers are designed using IDT's PrimerQuest searching tool and appropriate GenBank accession numbers. We typically use default settings, except for the amplicon size that is adjusted to 500 bp. We utilize the "Excluded Region List" for the exclusion of UTR sequence; the rationale is that UTR sequence could contain repetitive elements causing background. If a gene is considerably small, then UTR sequence is included. If a gene has many alternatively spliced forms, and it is unknown which form is expressed, then an area of the gene is chosen that is common across all isoforms. The T7 promoter sequence (5′–taatacgactcactataggg–3′) is appended to the reverse primer, and the primers are then re-evaluated for extensible dimers. The PCR reaction is carried out with cDNA template, 10 μM primer mix, and GoTaq® G2 master mix (Promega, #M7848) for 35 cycles, annealing at 58 °C, and extension for 1 min. 10% of PCR reaction is analyzed on a 1.2% TBE gel to ensure the presence of a single band of the correct size. The amplicon can be sequenced to confirm its identity. Gel extraction can be carried out if desired, but we usually omit this step if the gel analysis shows a clean and robust amplicon. Yield is measured spectrophotometrically (Nanodrop) and can be expected to be at least 2.5 μg.

The probe template is cleaned by phenol/chloroform extraction (ThermoFisher, #15593031), followed by ethanol/NaOAc precipitation in the presence of Pellet Paint® (Millipore Sigma, #70748). Molecular biological standard methods are conducted as described by Maniatis and colleagues [10]. Probes are transcribed using 0.2 μg DNA template with MEGAscript™ T7 Transcription Kit (ThermoFisher, #AM1334) and digoxigenin-11-UTP mix (0.875 mM DIG-11-UTP, 1.625 mM UTP, 2.5 mM ATP, 2.5 mM CTP, 2.5 mM GTP; Sigma-Aldrich, #11209256910; ThermoFisher, #R0481). Note that free-floating sections are not suitable for the high temperatures and formamide (ThermoFisher, #AM9342), which is often used in standard in situ hybridization protocols. Instead, vibratome sections need to be affixed to a slide as described above (Subheading 2.7), which makes them suitable for conventional hybridization.

2.8.2 In Situ Hybridization and Hybridization Chain Reaction

In situ hybridization is carried out using the Secure-Seal™ hybridization sealing system (9 mm diameter, Electron Microscopy Sciences #70333-40) attached to a Gold Seal™ UltraStick™ Adhesion Microscope Slide (ThermoFisher, #3039-002). The hybridization chamber and the slide are loosely aligned without attaching. Use a marker to indicate the location of the round hybridization chamber by drawing along the perimeter of the hybridization chamber onto the uncoated side of microscope slide. Remove the hybridization chamber and place six to eight vibratome sections in methanol on the coated side into the marked area. Arrange the sections with forceps to avoid overlap. Allow the methanol to evaporate, which dries and bonds the sections onto the slide. Align the marked area with the hybridization chamber and firmly attach the chamber to the slide. The chambers hold approximately 50 μL of liquid. The following steps are carried out in a chamber built from a sealed plastic container (Tupperware or equivalent) with paper towels soaked in water at the bottom for humidification.

Sections are rehydrated in 50% methanol, 50% 1× PTw (RNase-Free PBS + 0.1% Tween-20 (Sigma-Aldrich, #P9416)) for 5 min, followed by two washes with 1× PTw at room temperature. Sections are permeabilized with 10 µg/mL Proteinase K (New England BioLabs, #P8107S; in PTw) for 8–10 min at room temperature. We then proceed with two washes in 0.1 M Triethanolamine pH 7.8 (Sigma-Aldrich, #90279), followed by 10 min incubation in 0.1 M Triethanolamine pH 7.8 + 0.25% Acetic Anhydride (Sigma-Aldrich, #320102), which reportedly reduces electrostatic binding of the probe. Sections are washed with 1× PTw, then incubated for 10 min in 50% 1× PTw and 50% hybridization buffer (50% formamide (ThermoFisher, #AM9342), 5× SSC (ThermoFisher, #AM9770), 1 mg/mL Torula Yeast (Sigma-Aldrich, #R6625), 100 µg/mL Heparin (Sigma-Aldrich, #R6625), 1× Denharts (ThermoFisher, #750018), 0.1% CHAPS (Sigma-Aldrich, #C3023), 10 mM EDTA (ThermoFisher, #AM9261), 0.1% Tween-20 (Sigma-Aldrich, #P9416)). The solution is replaced with hybridization buffer that includes specific probes at a concentration of 0.5 ng/µL. Chamber ports are sealed with 3 M VHB Adhesive Seal Tabs (Electron Microscopy Sciences, #70328-00), and then incubated overnight at 55 °C.

The following day, adhesive tabs are removed and the sections are washed three times for 20 min in 2× SSC at 55 °C. Sections are then exposed to 2× SSC and 1:1000 RNase Cocktail for 30 min at 37 °C, and washed three times for 20 min in 0.2× SSC at 55 °C. To prepare for anti-digoxigenin antibody incubation, sections are equilibrated in MAB (100 mM maleic acid, 150 mM NaCl, pH 7.5) followed by 1 h incubation in blocking solution (20% heat-inactivated lamb serum (ThermoFisher, #16070096) and 2% blocking reagent (Millipore Sigma 11096176001) in MAB) at room temperature. Digoxigenin-labeled mRNA is detected with Anti-Digoxigenin-AP (Millipore Sigma, #11093274910), dilutes 1:3000 in blocking solution for 2–4 h at room temperature, followed by four to five washes in MABT (MAP with 0.1% TWEEN-20). The first washes can be quick, but then make sure that the main washes are considerably long (totaling ~2 h, or alternatively overnight). Sections are then equilibrated in AP reaction buffer (100 mM Tris-HCl pH 9.5, 50 mM $MgCl_2$, 100 mM NaCl, 0.1% TWEEN-20) for 5 min at room temperature. BM-Purple (Millipore Sigma, #11442074001) is added for the colorimetric detection, and sections are incubated until staining is satisfactory (3 h to 3 days). The reaction is stopped with AP Stop buffer (100 mM Tris-HCl pH 7.5, 100 mM EDTA pH 8.0), followed by 5–10 min fixation in 4% PFA. Sections are mounted in 50% Glycerol (Sigma-Aldrich, #G5516) in PBS, coverslipped, and sealed with clear nail polish. Images are acquired with a Leica DM2000 microscope, HCX Plan Apochromatic objective (40×/1.30, oil immersion), Nikon D7000 color camera, 1.6× projection lens/adapter (Best Scientific), and digiCamControl software (V2.1.1.0). Sections are manually tiled and merged using Microsoft Image Composite Editor (V2.0.3.0).

Hybridization chain reaction (HCR) is conducted as described [6, 11] with distinct modifications: 40–50 µm vibratome sections are used and attached to slides (*see* Subheading 2.8). After denaturing at 95 °C, the hairpins are snap cooled on ice for 2 min, warmed to room temperature, and added to the hybridization buffer. The amplification time is probe-specific and should be determined empirically. We found that robust HCR detection depends on empirical adjustment of the duration of the earlier Proteinase K treatment and that some probes work better with longer Proteinase K incubation times. On the other hand, HCR on mouse cochlear specimens requires shorter Proteinase K incubation times. After washing off the hairpins, the vibratome sections are fixed again with 4% PFA for 10 min. Sections are exposed for 15 min to 4,6-Diamidino-2-phenylindole (DAPI, 1 µg/mL in PBS; Thermo-Fisher, #D1306) to visualize nuclei. *OTOF* probes used the B2 initiator and B2-Alexa 546 amplification hairpins. *TECTA* probes used the B1 initiator and B1-Alexa 488 amplification hairpins. Sections are mounted with a Secure-Seal™ Spacer (single well, 13 mm diameter, 0.12 mm deep, Invitrogen, #S24735) in a drop of FluorSave™ Reagent (Calbiochem, #345789—20 mL). For more details, *see* Subheading 2.7.

2.9 Confocal Microscopy

For immunocytochemistry and HCR, chicken and mouse vibratome sections are imaged with a Plan-Apochromat 20×/0.8M27 or a Plan-Apochromat 40×/1.3 NA oil DIC UV-IRM27 objective. Digital zoom is adjusted between 0.6 and 0.9. We use a Zeiss LSM 880 Airyscan laser scanning confocal microscope and Zen Black software. Confocal z-stacks are collected with 6% overlap, tiling is set to bounding grid mode, and z-stacks are stitched together with the Zen Blue software.

For high resolution imaging for synapse and hair cell counts, the adult mouse cochlea is imaged with an oil-immersion Plan-Apochromat 63×/1.4 oil DIC M27 and 0.9× digital zoom. For each image z-stack, the thickness was adjusted to include the CTBP2-positive area of the synaptic apparatus of inner and outer hair cells. Each z-stack of the shown example included 12–16 adjacent inner hair cells.

3 Results

3.1 Chicken Utricle

To visualize hair and supporting cells in 80 µm-thick vibratome sections of P7 utricles, we use antibodies to MYO7A (Proteus Biosciences, # 25-6790, 1:1000) and SOX2 (Santa Cruz Biotechnology, #sc-17320, 1:100). Alexa Fluor 647-conjugated phalloidin (ThermoFisher, #A22287, 1:100) is used to highlight F-ACTIN in hair bundles and DAPI (1 µg/mL PBS; ThermoFisher, #D1306) is used to visualize nuclei. MYO7A and SOX2-labeled type II hair cells were identified in extrastriolar regions, whereas MYO7A-positive and SOX2-negative type I hair cells were located in striolar regions, respectively (Fig. 2a). Moreover, this methodology permits to study hair bundle morphology demonstrated by many upright hair bundles (insets, Fig. 2a). To visualize extrastriolar type II hair cells in 50 μm-thick vibratome sections by in situ hybridization, we used a probe against the type II hair cell marker *CALB2* (Primers: Forward 5′–CGAAGGCAAAGAGCTGGAAA–3′; Reverse 5′–taatacgactcactataggg CGCTGCCATCCTTGTCA TAAA–3′). Labeling for CALB2 mRNA was strong in hair cells in the extrastriolar regions and absent in striolar type I hair cells (Fig. 2b). Conversely, a probe against OCM mRNA (probe primers in [7], a known striolar hair cell marker, labeled hair cells in the striolar region (Fig. 2c). Hair bundles of the crista ampullaris of the anterior semicircular canal, adjacent to the utricle, are well preserved using low melting point agarose embedding and vibratome sectioning (Fig. 2d).

3.2 Chicken Basilar Papilla

In 50 μm-thick vibratome sections of P7 basilar papillae, hair cells and supporting cells were identified using hybridization chain reaction (HCR) probes for OTOF and TECTA mRNA (Fig. 2e). Visualization of mRNA in supporting cells of post-hatch inner ear sensory epithelia is challenging. Supporting cells are long and slender, and span from the basilar membrane up to the apical surface of sensory epithelia. Most of the space is occupied by hair cells, thus the supporting cell cytoplasm only expands at the nuclei levels and at the very top of the apical surface. In thin cryostat sections, this cell morphology can cause issues because of low cytoplasmic signal strength. Better tissue preservation and thickness allow for a higher signal-to-noise ratio with vibratome sections, which overcomes the stated limitations of conventional cryosectioning. Detection of MTCL1 mRNA (probe primers in [7]), a recently discovered supporting cell marker, with a DIG-labeled in situ probe shows a specific and strong signal in P7 basilar papillae vibratome sections (Fig. 2f).

Next, we want to showcase that vibratome sectioned inner ear tissue is not only useful for analysis of hair cells and supporting cells but can also be applied for visualization of cochlear innervation. Projections of the spiral ganglion were visualized using a TUBB3 antibody (Millipore Sigma, #MAB5564, 1:1000) (Fig. 2g). To highlight presynaptic ribbons in chicken basilar papilla hair cells, we used a CTBP2 antibody (BD Transduction Labs, #612044, 1:200). Characteristic punctae of CTBP2-labeled presynaptic ribbons were discernable at the basal regions of cochlear hair cells (Fig. 2h). If the spiral ganglion is not removed during the dissection, it is possible to highlight the somata of spiral ganglion neurons. To label neurons including their fibers as well as hair cells, we use an antibody against AC-TUB (Cell Signaling Technology, #D20G3,

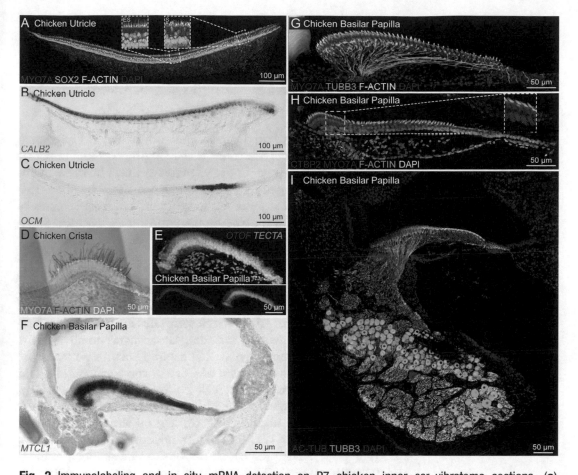

Fig. 2 Immunolabeling and in situ mRNA detection on P7 chicken inner ear vibratome sections. (**a**) Immunolabeling of hair cells and supporting cells on 80 μm-thick vibratome sections of P7 chicken utricle. Hair cells are labeled with antibodies against MYO7A (red). Supporting cells are labeled with antibodies against SOX2 (green). F-ACTIN is visualized with phalloidin (white). Nuclei are labeled with DAPI (blue). Insets show magnified views of the extrastriolar (ES) and striolar (S) regions indicated with dashed lines. (**b**) In situ hybridization for the extrastriolar marker *CALB2* in a 40 μm-thick vibratome section of the chicken utricle. (**c**) In situ hybridization for the striolar marker *OCM* in a 40 μm-thick vibratome section of the chicken utricle. (**d**) Immunolabeling of hair cells on an 80 μm-thick vibratome section of the crista ampullaris of the anterior semicircular canal. Hair cells are labeled with antibodies against MYO7A (turquoise). F-ACTIN is labeled with phalloidin (magenta). Nuclei are labeled with DAPI (blue). Sensory epithelium and the stroma are visible in the transmitted-light section (tPMT). (**e**) Hybridization chain reaction provides individual cell resolution of the hair cell marker *OTOF* (magenta) and the supporting cell marker *TECTA* (green) on a 40 μm-thick vibratome section of the chicken basilar papilla. (**f**) In situ hybridization for the supporting cell marker *MTCL1* on a 40 μm-thick vibratome section of the chicken basilar papilla. (**g**) Immunolabeling of hair cells and spiral ganglion neurites on 80 μm-thick vibratome sections of P7 chicken basilar papilla. Hair cells were labeled with antibodies against MYO7A (magenta). Neurites are labeled with antibodies against TUBB3 (green). F-ACTIN in hair bundles was labeled with phalloidin (white). Nuclei were labeled with DAPI (blue). (**h**) Immunolabeling of presynaptic ribbons in hair cells on 80 μm-thick vibratome sections of P7 chicken basilar papilla. Hair cells were labeled with antibodies against MYO7A (blue). Presynaptic ribbons were labeled with antibodies against CTBP2 (magenta). F-ACTIN in hair bundles was labeled with phalloidin (green). Nuclei were labeled with DAPI (white). Inset shows a magnified view of CTBP2 immunoreactivity in tall hair cells indicated with dashed lines. (**i**) Immunolabeling of supporting cells, hair cells, and the spiral ganglion on 80 μm-thick vibratome sections of P7 chicken basilar papilla. Supporting and hair cells were stained with antibodies against acetylated-tubulin AC-TUB (magenta). Spiral ganglion neurons and their processes are visualized with antibodies to TUBB3 (green). Nuclei were labeled with DAPI (white)

1:1000) in combination with anti-TUBB3 (Millipore Sigma, #MAB5564, 1:1000) (Fig. 2i). We postulate that excellent preservation of the tissue in vibratome sections enables detailed investigation of innervation in the chicken inner ear (Fig. 2i).

3.3 Mature Mouse Cochlea

Preserving the sophisticated cytoarchitecture of the adult mammalian organ of Corti in cryosections is challenging. Paraffin or plastic embedding usually results in better preservation, but these methods are not well compatible with fluorescence-based imaging. It is particularly difficult to preserve the different cellular processes of various supporting cell subtypes in the adult mouse cochlea. Sagittal vibratome sections at 80 μm thickness were collected from a decalcified P28 mouse cochlea. Inner and outer hair cells were labeled with a MYO6 antibody (Proteus Biosciences, # 25-6791, 1:1000). AC-TUB (Sigma-Aldrich, #AB_477585, 1:1000) and TUBB3 (BioLegend, #801202, 1:250) antibodies were combined in the green channel to label the microtubules in cochlear supporting cells and also the neurons including the somata and fibers of the spiral ganglion neurons (Fig. 3a). A sagittal section provides the option to zoom in on the organ of Corti at the basal, middle, and apical turns as well as into the cell bodies of the spiral ganglion (Fig. 3b–d). To demonstrate that it is possible to image subcellular specializations of mature hair cells on sagittal vibratome sections, we used antibodies to CTBP2 (BD Transduction Labs, #612044, 1:200) (Fig. 3e). The presynaptic ribbons can be assigned to inner hair cells and a few CTBP2-positive puncta are associated with outer hair cells. We also noticed that MYO6 immunoreactivity is higher in inner hair cells than in outer hair cells (Fig. 3a–e). Next, we tested whether in situ hybridization for mRNA encoding the outer hair cell marker *OCM* would work on 50 μm-thick vibratome sections and found that signal strength is comparable to in situ hybridization on whole mounts (Primers: Forward 5′–TCTGAGC GCTGATGACATTG –3′; Reverse 5′–taatacgactcactataggg ATC CACCACCACCAAAGAAG –3′) (Fig. 3f, g). When imaging 80 μm-thick vibratome sections from a P2 mouse expressing Lfng-promoter driven eGFP, we noted excellent tissue preservation (Fig. 3h).

Synapse counts often utilize immunolabeling of dissected fragments from the different cochlear turns of the adult rodent cochlea. Top-down view high resolution confocal microscopy is performed on these fragments stained with antibodies for hair cell markers, and pre- and postsynaptic markers. Synapse numbers per inner hair cell are usually quantified and correlated to hearing loss or recovery from synaptopathy [12]. Whereas dissecting cochleae for whole mount processing is easily achievable from young animals, cochlea dissection from older mice remains a challenging endeavor. The margin of error when dissecting the adult cochlea is small: the

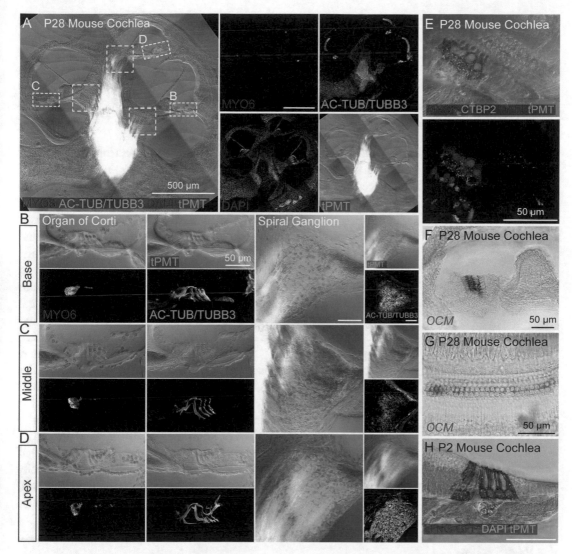

Fig. 3 Immunolabeling and in situ *mRNA* detection on mouse inner ear vibratome sections. (**a**) Sagittal 80 μm-thick vibratome section of the P28 mouse cochlea. Inner and outer hair cell are immunolabeled with antibodies against MYO6 (magenta). AC-TUB and TUBB3 antibodies were combined (green) to visualize microtubules in supporting cells of the Organ of Corti, and the neuronal somata and fibers of the spiral ganglion. (**b–d**) Confocal z-stacks for magnified organ of Corti and spiral ganglion sections from the apex, middle, and base. Z-stacks series were used for maximum intensity projections. (**e**) Presynaptic ribbons (CTBP2, green) in inner and outer hair cells (MYO6, magenta) on an 80 μm-thick vibratome section of the P28 mouse cochlea. (**f**) In situ hybridization for the outer hair cell marker *OCM* on a 40 μm-thick transverse vibratome section, and (**g**), on a whole mount fragment of P28 mouse cochlea. (**h**) Lfng-promoter driven eGFP expression (shown in magenta) in 80 μm-thick vibratome sections from a P2 mouse

organ of Corti and surrounding tissues can be easily damaged. The duration of a single dissection of an adult cochlea can initially last up to an hour and with practicing can be reduced to 20–30 min [8]. This is owed to the delicate nature of the adult cochlea and the fragility increases when dissections are performed on pathologically affected cochlear tissue. Here, we provide an alternative approach for a complete adult mouse cochlear dissection and whole mount-based detection of proteins and mRNA.

We suggest utilizing transverse vibratome sections of the adult cochlea because the organ of Corti and surrounding tissue is well preserved. These sections can be obtained without extensive training. We envision that this approach can also be translated to other species. In our example, a decalcified P28 mouse cochlea was embedded as described in Subheading 2.4 and in Fig. 1f so that the apex-to-base axis is orientated perpendicular to the razor blade plane of the vibratome. Round and oval window should point toward and the base-mid turn of the cochlea away from the embedder. Excessive agarose is removed until the inner ear capsule of the apex is reached, and then 150 μm-thick vibratome sections are collected. Sections were collected individually to maintain a record of the apex-to-base tonotopic positions. Specimens from the apical, middle, and basal turns were immunostained with antibodies against MYO7A (Proteus Biosciences, # 25-6790, 1:1000) to hair cells and CTBP2 (BD Transduction Labs, #612044, 1:200) for detection of presynaptic ribbons (Fig. 4a–c). Preservation of the tissue was assessed using DAPI staining and brightfield imaging. The tissue sections did not show tears or other dissection-related disfigurations. Synapse numbers in inner hair cells could be easily quantified using maximum intensity projections of MYO7A and CTBP2.

4   Conclusions

Excellent preservation of inner ear tissue across different species can be achieved with vibratome sectioning. Relatively little training is necessary to successfully execute the methods described in this chapter. Because vibratome sections are free-floating, we offer a protocol for immunostaining without affixing to slides. Moreover, when attachment to slides is required such as for in situ hybridization and HCR, we provide a detailed and rigorous method. Transverse vibratome sections from the adult cochlea can serve as an alternative to whole mounts dissection.

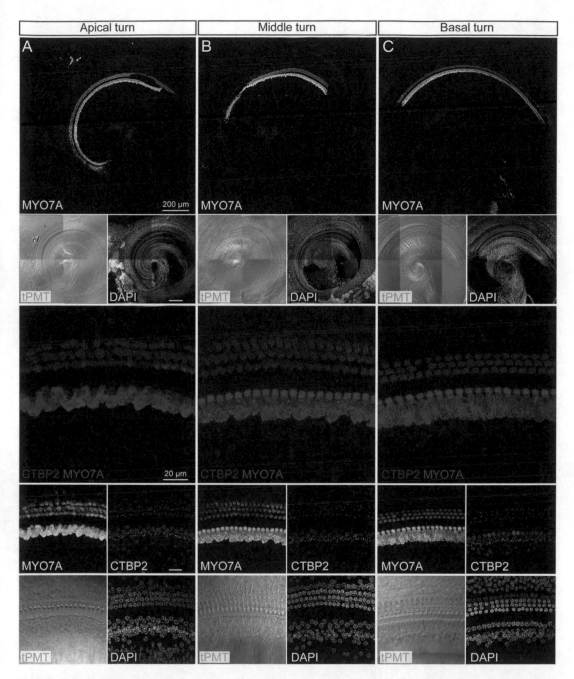

Fig. 4 Transverse vibratome sections from the adult cochlea. (**a**–**c**) Z-stacks of 150 μm-thick sections of P28 mouse cochlea for apical, middle, and basal regions were acquired using the tiling function, and stitched together; maximum intensity projections are shown. Hair cells are labeled with antibodies to MYO7A (blue). Presynaptic ribbons in inner and outer hair cells were stained using antibodies against CTBP2 (magenta). DAPI labels nuclei and transmitted-light sections (tPMT) are used to inspect the sensory epithelium and surrounding tissue of the P28 mouse cochlea

References

1. Shim K (2011) Vibratome sectioning for enhanced preservation of the cytoarchitecture of the mammalian organ of Corti. J Vis Exp 52

2. Shim K et al (2005) Sprouty2, a mouse deafness gene, regulates cell fate decisions in the auditory sensory epithelium by antagonizing FGF signaling. Dev Cell 8(4):553–564

3. Zhu Y et al (2019) Single-cell proteomics reveals changes in expression during hair-cell development. elife 8

4. Hartman BH et al (2018) Fbxo2(VHC) mouse and embryonic stem cell reporter lines delineate in vitro-generated inner ear sensory epithelia cells and enable otic lineage selection and Cre-recombination. Dev Biol 443 (1):64–77

5. Kubota M et al (2021) Greater epithelial ridge cells are the principal organoid-forming progenitors of the mouse cochlea. Cell Rep 34 (3):108646

6. Benkafadar NJ, Janesick A, Scheibinger M, Ling AH, Jan TA, Heller S (2021) Transcriptomic characterization of dying hair cells in the avian cochlea. Cell Rep 34(12):108902

7. Janesick AS, Scheibinger M, Benkafadar N, Kirti S, Ellwanger DC, Heller S (2021) Cell-type identity of the avian cochlea. Cell Rep 34 (12):108900

8. Montgomery SC, Cox BC (2016) Whole mount dissection and immunofluorescence of the adult mouse cochlea. J Vis Exp 107

9. Harland RM (1991) In situ hybridization: an improved whole-mount method for Xenopus embryos. Methods Cell Biol 36:685–695

10. Green MR, Sambrook J (2014) Molecular cloning: a laboratory manual, 4th edn. Cold Spring Harbor Laboratory Press, Cold Spring Harbor, NY

11. Choi HMT et al (2018) Third-generation in situ hybridization chain reaction: multiplexed, quantitative, sensitive, versatile, robust. Development 145(12)

12. Suzuki J, Corfas G, Liberman MC (2016) Round-window delivery of neurotrophin 3 regenerates cochlear synapses after acoustic overexposure. Sci Rep 6:24907

Chapter 4

Genetic Manipulation of the Embryonic Chicken Inner Ear

Nicolas Daudet, Magdalena Żak, Thea Stole, and Stephen Terry

Abstract

The chicken embryo has historically been a key animal model to investigate the mechanisms of inner ear development. With the ongoing progress in methods for genetic manipulation of avian embryos, and in particular the advent of CRISPR/Cas9 technology, a wide range of approaches are now available to investigate gene function in the chicken inner ear. In this chapter, we provide a standard protocol for in ovo electroporation of the inner ear and discuss the advantages and limitations of the genetic methods available for gain- and loss-of-function studies in the embryonic chicken inner ear.

Key words Inner ear, Development, Otocyst, Avian, Chicken embryo, In ovo electroporation, Transposon, Retrovirus, CRISPR/Cas9

1   Introduction

For centuries, the chicken embryo has been one of the favorite animal models of developmental biologists. It is large and easily accessible for surgical manipulation such as the "cut-and-paste" grafting experiments that have paved the way to our modern understanding of cell lineages and inductive interactions between developing tissues. It is fair to say that its popularity has decreased in the functional genomic era, due to the availability of more convenient and tractable vertebrate models for genetic studies, such as the mouse or the zebrafish. Although it is possible nowadays to generate transgenic chicken (or quails), the technology is not trivial and setting up a dedicated facility for their production and maintenance is a hurdle that most laboratories will find difficult to overcome.

On the other hand, it is very easy to induce mosaic genetic modifications in avian embryos using transfection of plasmid DNA

Supplementary Information The online version of this chapter (https://doi.org/10.1007/978-1-0716-2022-9_4) contains supplementary material, which is available to authorized users.

Andrew K. Groves (ed.), *Developmental, Physiological, and Functional Neurobiology of the Inner Ear*, Neuromethods, vol. 176, https://doi.org/10.1007/978-1-0716-2022-9_4,

or infection with viral vectors. These techniques have been used with great success to investigate the mechanisms of inner ear development. For example, the avian replication-competent RCAS retroviruses [1, 2], able to induce widespread and sustained expression of a transgene after in ovo infection, were used to investigate the roles of Notch [3, 4], BMP [5, 6], FGF [7], or Wnt signaling [8] in the developing inner ear. Lineage studies performed with replication-defective version of the RCAS provided key insights into the clonal relationships between distinct otic cell types and direct evidence that inner ear hair cells and supporting cells are derived from a common progenitor [9–11]. One of the major advantages of viruses is that once a highly concentrated viral stock is produced, the in ovo injection of the otic vesicle is relatively easy (at early developmental stages at least) and does not require much equipment beyond a stereomicroscope, a manipulator, and an injector. There are however two significant limitations with RCAS-derived vectors: they cannot accommodate transgenes larger than 2.5 kilobases and virus-infected cells become refractory to secondary infection. Hence, the co-expression of several transgenes can only be achieved through infection by retroviruses harboring different envelope proteins.

Two major advances helped to lift these limitations. The first one was the introduction of in ovo electroporation to transfect avian cells with plasmid DNA [12, 13]. The principle of the method is to cover or inject the targeted tissue with a concentrated plasmid solution then apply square-wave pulses of current to "push" the negatively charged DNA through the cell membranes temporarily disrupted by the electric pulses. The second advance was the introduction of transposon-based vectors for genetic manipulation. Transposons are mobile genetic elements that can stably integrate into the cell genome when transposase activity is present. Vectors derived from the Tol2, Sleeping Beauty or PiggyBac transposons have been successfully used for stable integration of transgenes into a variety of animal cells [14, 15], including avian cells [16, 17].

In comparison to viral infection, in ovo electroporation of plasmid DNA is more time-consuming and requires slightly more equipment, but it is much more versatile: any type of eukaryotic expression plasmid (including conditional and inducible ones) can be electroporated, alone or in combination, greatly expanding the field of experimental possibilities. Another important advantage is that electroporation of plasmid DNA offers a better temporal control of transgene expression: all electroporated cells start to express the transgene at the same time (typically within 1–2 h post-electroporation, unless an inducible system is used) and for a duration that depends on the type of vector, while cells initially infected with replication-competent viruses act as virus factories, propagating the original infection to "naïve" cells for a long period of time. This is advantageous if one aims to produce large patches of modified cells. However, the uncertainty regarding the onset of transgene expression can complicate the interpretation of some experiments in which the timing of the genetic manipulation is critical.

In our lab, transposons are currently the favorite vectors for stable integration of transgenes in the developing chicken inner ear (Fig. 1). We have successfully used Tol2 and PiggyBac vectors to overexpress transgenes for extended period of time (up to pre-hatching stages) in a conditional or constitutive manner [18–20], to monitor the activities of the Notch and Wnt signaling pathways with genetically encoded fluorescent reporters [4, 18, 21], or to drive the expression of transgenes in a doxycycline-inducible manner [4, 18, 22].

1.1 Loss-of-Function in Avian Embryos: The CRISPR/Cas9 Revolution

The overexpression of a gene of interest, or "gain-of-function," is easily achieved in chicken embryos. It can be very informative, but it may potentially lead to spurious results due to non-physiological levels of gene expression. The gold-standard in functional genomics remains the "loss-of-function" experiment, whereby the expression or function of a gene of interest is reduced or knocked-out entirely. Until very recently, this was relatively difficult to achieve in a consistent manner in chicken embryos. Morpholinos, which are small modified oligonucleotides able to interfere with protein translation, have been used successfully in some studies but their effects are only temporary and they require extensive controls for cellular uptake, specificity, and absence of toxicity [23]. The RNA interference (RNAi) approach, which relies on the ability of double-stranded RNAs to silence gene expression, has also been used for loss-of-function in avian embryos. Avian-tailored vectors have been successfully developed for transient or long-term expression of short interfering RNA (siRNA) after in ovo electroporation or viral infection [24]. However, RNAi does not guarantee a complete and sustained gene knock-down and applying this technique requires specific antibodies and careful controls to monitor the extent of gene silencing.

With the advent of CRISPR-Cas9 genome editing tools, a new range of genetic and epigenetic manipulations targeting both coding and non-coding DNA is now possible in avian cells. The CRISPR/Cas9 system is a two-component system based on a complex between a guide RNA (gRNA) and an endonuclease enzyme derived from *Streptococcus pyogenes* (Cas9). Once this complex is bound to genomic DNA, cleavage will occur, causing double-stranded DNA breaks (DSBs) that will be imperfectly repaired by non-homologous end joining repair enzymes. This repair leads to indels formation (insertions or deletions) resulting in frameshifts and subsequent truncation or deletion of the transcript. High specificity of the system is conferred by base pairing of the protospacer domain located in the gRNA, a region of typically

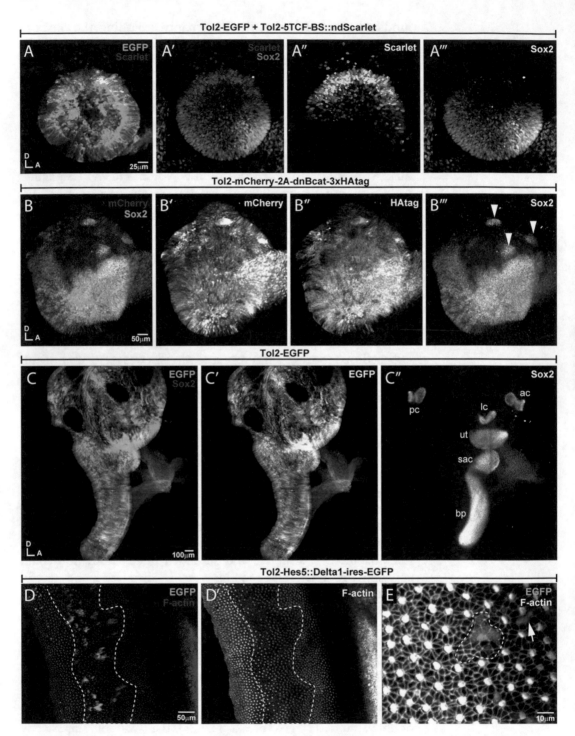

Fig. 1 Whole-mount views of chicken inner ear samples electroporated at E2 with different types of Tol2 transposon constructs and a Tol2 transposase expression plasmid for long-term integration of the transgenes. The images are adapted from [18, 21]. (**a–a′′′**) An E3 otocyst co-electroporated with a constitutive (control) Tol2-EGFP and a Tol2 Wnt reporter (Tol2-5xTCF::BS-ndScarlet), in which five binding sites for TCF transcription factors regulate the expression of a nuclear and destabilized Scarlet protein. The co-electroporation of a constitutively expressed EGFP is crucial to demonstrate the specific activation of the Wnt reporter in the dorsal region of the otocyst. (**b–b′′′**) An E4 otocyst electroporated with a Tol2-mCherry-2A-dnBcat-3xHAtag plasmid.

~20 base pairs homologous to the target genomic DNA together with the Protospacer Adjacent Motif (PAM), typically NGG for *S. pyogenes* Cas9 [25, 26]. However, the DSBs induced by the wild-type form of Cas9 leads to indels that are repaired imperfectly, making the gene editing not precisely controlled. To circumvent this problem, modified forms of Cas9 with single "base-editing" (BE) capabilities were engineered. These consist of a nickase form of Cas9 inducing single-strand break fused to the cytidine deaminase (APOBEC1) and uracil glycosylase inhibitor (UGI), able to mediate direct conversion of C to T (or G to A) and introduce premature "stop" codons in the coding sequence of targeted genes [27]. This CRISPR-stop method [28, 29] was for example used to engineer, by genome editing of blastocyst-stage mouse zygotes, single (*Atoh1*) and triple (*vGlut3, Otoferlin, Prestin*) knock-out mice for genes essential for auditory hair cell development and function [30]. Another type of CRISPR/Cas9 applications is the epigenetic silencing or activation of regulatory elements, which relies on the use of catalytically "dead" forms of Cas9 (dCas9) coupled to chromatin modifiers. This allows the investigation of the function of specific non-coding regions such as promotors, enhancers, and insulators [31–33].

A range of CRISPR/Cas9 tools have already been combined with in ovo electroporation in chicken embryos [33–35], with some modifications to improve their efficiency in avian cells. Firstly, the fusion of Cas9 to nuclear localization signals (NLS-Cas9-NLS) [36]. Secondly, a modification of the gRNA scaffold to reduce premature termination of the gRNA transcription and to improve the assembly of Cas9/gRNA complexes via extension of the Cas9-binding hairpin structure [36, 37]. Finally, the use of a chicken U6.3 promotor to increase the expression of the gRNA guide [33, 34, 38].

Fig. 1 (continued) In this vector, the same constitutive promoter drives expression of a dominant negative form of Beta-catenin tagged with human influenza hemagglutinin (HA) and a mCherry reporter (separated by a cleavable 2A peptide sequence) to easily trace transfected cells. Note the induction of ectopic Sox2-positive neurosensory patches (arrowheads in **b'''**) in the dorsal aspect of the inner ear. (**c–c''**) An E7 inner ear electroporated with a constitutive Tol2-EGFP plasmid and immunostained for Sox2 to visualize the distinct sensory patches (*ac, lc,* and *pc* anterior, lateral, and posterior crista, *ut* utricle, *sac* saccule, *bp* basilar papilla). Note the variation in EGFP fluorescence intensity, due to the mosaicism in transfection. (**d-e**). Surface view of an E12 basilar papilla transfected with Tol2-Hes5::Delta1-ires-EGFP, in which the mouse Hes5 promoter regulates the expression of the Notch ligand Delta1 and EGFP (separated by an IRES-sequence). With this vector, the transgenes are expressed in a conditional manner within Notch-active cells. Regions containing EGFP-positive cells exhibit a reduced density of hair cells (in between the dotted lines in **d-d'**), recognized by their F-actin rich bundle of stereocilia. At higher magnification (**e**), the mosaicism in transfection and expression of the transgene makes it possible to compare the developmental fate of individual (arrow) versus clusters of EGFP-positive cells (dotted line)

To date, there has been no published report of CRISPR/Cas9 genetic manipulation of the avian inner ear, but preliminary results from our group and others are very promising. Given that the development of the inner ear proceeds a relatively long period of time, additional modifications of the above components and their incorporation in transposon vectors would be beneficial to allow the long-term tracing of edited cells and a conditional or temporal control of gene editing. Once such tools are available, there is no doubt that this transformative technology will stimulate a renewed interest in using chicken embryos for investigating the molecular mechanisms of inner ear development and regeneration.

1.2 Final Recommendations

One important consideration with either viral infection or plasmid transfection is the fact that the genetic modification is mosaic: a mixture of transfected and untransfected cells is generated. While this may be perceived as a problem, mosaicism can be highly beneficial in some experiments: it bypasses the potential lethality that could result from an "animal-wide" manipulation of a signaling pathway, and it allows the experimenter to compare directly, within the same tissue, the properties of modified versus non-modified cells. This is particularly advantageous to study the function of intercellular signaling pathways, such as Notch or Wnt signaling, and to decipher their cell-autonomous and non-cell-autonomous effects (Fig. 1).

The key to circumvent the potential problem of mosaicism is to use vectors enabling an unambiguous identification of transfected cells. For gain-of-function studies, this can easily be achieved by expressing either a tagged-form of the transgene, or driving from the same DNA construct the transgene with a fluorescent reporter protein using either an internal ribosomal entry site (IRES) for a bi-cistronic construct, a 2A cleavable peptide sequence for a 1:1 ratio of transgene/reporter expression, or a bi-directional promotor. Solving the "trackability" issue is more challenging for approaches such as CRISPR/Cas9, which requires multiple components (one or more gRNA guides, Cas9, and a reporter) to be expressed within the same cell. Fortunately, the large cargo capacity of transposons should enable the design of standalone vectors containing all of these components.

Finally, one significant limitation of the in ovo electroporation technique is the relatively poor survival of embryos at late (E11 and older) embryonic stages. In our hands at least, the survival rate of electroporated embryos decreases approximately from 80–100% at 1–4 days post-electroporation to 10–25% at 16 days post-electroporation, with maximum embryonic death occurring at about 5–6 days post-electroporation. Therefore, it is not a high-throughput method for investigating some of the final steps of inner ear development occurring between mid-incubation and hatching stages.

Having introduced the molecular tools available for genetic manipulation of avian cells, we now turn to a description of the material and methods used in our laboratory for in ovo electroporation of the embryonic chicken inner ear. We use the "microelectroporation" approach, in which a fine tungsten needle is used as a cathode in order to transfect more precisely the target cells while applying low voltage conditions, which improve embryo survival [12]. A video of this protocol can be accessed at https://mediacentral.ucl.ac.uk/Play/46572.

2 Material

2.1 Electroporation Setup

- One bench and a metal plate for positioning the magnetic stands (Fig. 2).

- One stereomicroscope (e.g., PZMTIV Trinocular Body With 10× Eyepieces, 1× Objective) on a boom-stand. For training purposes, we use a stereomicroscope with a C-mount camera and tablet.

- One cold-light source with a flexible light guide (e.g., Schott KL-300 LED).

- One small magnifying lens (e.g., 10× 18 mm geology hand-held pocket lens) attached to a suitable holder to focus the incident light on the surface of the embryo (*see* Fig. 2c and Video 1). This drastically improves contrast and visualization of the embryonic structures.

- One electroporator able to deliver square-wave pulses of current (e.g., BTX ECM830, Sonidel CUY-21, Intracel TSS20 Ovodyne), equipped with a foot pedal and a pair of banana plug to crocodile clip leads.

- Two three-dimensional micromanipulators (e.g., MM-33 Rechts) on vertical magnetic stands to hold the positive and negative electrodes.

- One "chopstick" (e.g., Sonidel CUY611 and CUY613 series) platinum, gold-plated electrode as an anode on the left-hand side micromanipulator. With time, the gold-plating will degrade but this does not prevent successful electroporation.

- One tungsten electrode with a sharp tip as a cathode on the right-hand side micromanipulator. Electrodes can be purchased from electroporation device suppliers or custom made (*see* **Note 1**).

- Borosilicate glass needles (e.g., Harvard Apparatus ref. 30-0050 with OD—1.2 mm and ID—0.94 mm) for injecting the plasmid DNA solution. Use heating/pulling parameters that produce

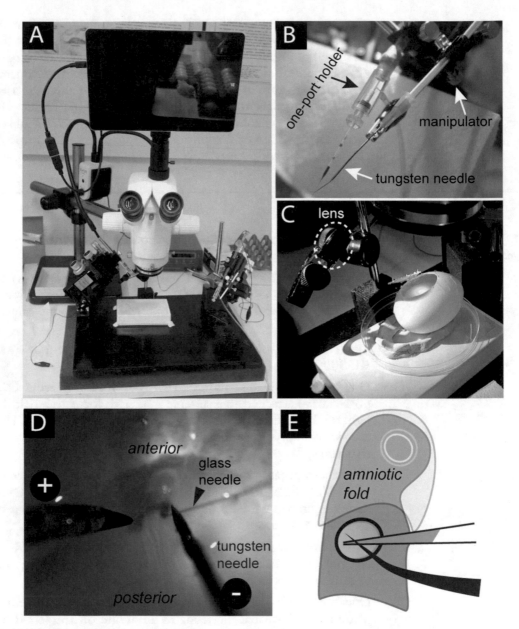

Fig. 2 The electroporation setup. (**a**) Boom-stand stereomicroscope and small equipment for electroporation. (**b**) The right manipulator with the one-port holder with the injection glass needle; the tungsten electrode is mounted on a small manipulator to facilitate its positioning. (**c**) The magnifying lens is used to focus the light on the surface of the embryo. (**d**) An E2 chicken embryo with the electrodes and injection needle in place for electroporation. (**e**) Schematic representation of the relative position of the otic cup, glass needle, and tungsten needle (on top of the injection needle)

needles with a long and fine tip, which can then be easily trimmed down in case the needle would get clogged.

- A one-port holder (e.g., Harvard Apparatus MP series Microinjection Electrode Holder) for the glass needle. The port of the holder is connected to a ~50 cm long piece of tubing ended by a P200 pipette tip through which air is mouth-blown to force the DNA solution out of the glass needle.

- A one-dimensional manipulator with a pivot and tilt mechanism (U-12C, Narishige) mounted onto the (glass needle) one-port holder handle and holding the tungsten electrode. This manipulator greatly facilitates fine adjustment of the position of the tungsten electrode relative to the glass needle.

- Some "Blue tack" to build a rudimentary egg holder at the bottom of a 50 mm petri dish and restrict the movement of the electrode cables when using the manipulators.

2.2 Fertilized Eggs and Incubators

- Fertilized chicken eggs obtained from a commercial source. We store our eggs at 14 °C for up to 1 week (in a wine cooler cabinet) before starting the incubation.

- One egg incubator (set at 38 °C) preferably equipped with an automatic humidity control system (e.g., Brinsea Ovation-EX). The use of an automatic egg-turning system is not necessary for short incubation times (up to 7 days), but seems to improve survival rates for embryos kept past this stage.

2.3 Other Material and Reagents

- One pairs of fine forceps.
- One pair of small curved scissors.
- 20 ml syringe with 18G needle.
- 1 ml syringe with 25G needle.
- Sterile Phosphate Buffer Saline (PBS) pH 7.4.
- Disposable plastic pipette (5 ml).
- Plasmid DNA, purified using anion exchange columns and ethanol-precipitated before resuspension in distilled nuclease-free water at a final concentration of 1–3 μg/μl (*see* **Note 2**).
- Injection solution 10×: distilled water containing 20% sucrose and 1–2% Fast Green (Sigma) for easy visualization of the DNA solution. Prepare 10 ml of the solution, filter using a Nalgene 0.4 μm syringe filter, and aliquot in 0.5 ml tubes that can be stored at −20 °C for an extended period of time.
- Eppendorf 0.5–20 μl GELoader tips.
- Cellulose tape (e.g., Sellotape), 15 and 50 mm wide.

3 Methods

The inner ear derives from the otic placode, an ectodermal structure located on both sides of the hindbrain. The otic placode is originally flat (stage HH10-11, 35–45 h of incubation), but it rapidly transforms into a "cup" (stage HH12-13, 45–52 h of incubation). These stages are ideal for in ovo electroporation of the otic progenitors, since the late placode/cup is easy to visualize and provides a natural "container" for the DNA solution. At HH12, the head of the embryo starts to turn on its left side; it is therefore a lot easier to electroporate the right otic cup than the left one. If late otic cup stages are used, it may become necessary to partially disrupt the amniotic membrane (using for example a small needle) extending from the head region to expose the otic cup before the electroporation. Over the following hours of incubation, the cup invaginates into the underlying mesoderm to form a closed otic vesicle (or otocyst), which will subsequently grow and transforms into the inner ear—with the vestibular system located dorsally and the cochlear duct ventrally, hosting the auditory epithelium (the basilar papilla). While some labs have successfully performed in ovo electroporation of the otocyst at 3–4 days of incubation [39–41], it requires additional manipulation of the embryo to push the injection needle within the otocyst. In our hands at least, the frequent damage to surrounding tissues and blood vessels results in poor embryo survival. We therefore recommend that the electroporation procedure is first practiced at the cup stage before adapting as necessary the method (injection and voltage parameters) for later developmental stages.

3.1 Windowing of the Eggs and Preparation of the Embryo

Windowing of the eggs is a critical step that should not be overlooked, in particular if one wants to achieve long-term survival of electroporated embryos. Sellotape is useful to prevent shell debris from falling inside the egg during the opening. If any damage to the yolk has occurred, discard the egg. Try to minimize the time the eggs are left open to the air before proceeding to electroporation or the embryos will dry and die. An ideal arrangement is for two people to work as a team, one electroporating the embryos and the other windowing and closing the eggs. If working alone, do not open more than a dozen eggs at a time and try to process these within 30 min.

- Incubate fertilized chicken eggs horizontally in a humidified incubator at 38 °C for approximately 48 h, or until stage Hamburger-Hamilton (HH) 12-13. A line is drawn on top of the egg shell to easily locate the position of the embryo in subsequent steps. We use relatively small-capacity incubators connected to plug-in timers to ensure maximum flexibility for incubation times of each lab user.

- At the end of the incubation, place the eggs in a tray and spray them with 70% ethanol. Let them dry for approximately 5 min horizontally. Apply a band of 15 mm wide Sellotape on the surface of the eggs along its long axis (marked by the permanent pen line) and using a pair of forceps, make one hole on the top of the shell, towards the pointed end of the egg.

- Using a 20 ml syringe equipped with an 18G needle inserted through the hole (approximately 2 cm depth, and angled towards the pointed end of the egg), remove carefully 3–5 ml of albumen.

- Let the egg sit for another 5 min, then open an oval (approximately 15 × 30 mm) window on the top of the egg with a pair of fine scissors. Make sure to leave a bit of the egg shell uncut so that the shell can be easily sealed back in place after electroporation.

- Place the egg within a suitable holder (such as a ring of blue tack inside a petri dish, which allows easy positioning and rotation of the egg) and under the stereomicroscope.

- Rotate the egg as necessary to orient the posterior end of the embryo towards the experimenter and the right otic cup towards the right-hand micromanipulator.

- Adjust lighting by changing the position of the light guide and the magnifying lens (positioned on the left-hand side of the embryo in our setup) so that a small spot of light is focused on the surface of the embryo and the right otic placode/cup is visible.

- Using a fine tungsten (or gauge 25G) sterile needle, poke a small hole through the vitelline membrane lateral to the embryo and close to the hindbrain level. Apply one or two drops of sterile PBS. The PBS will infiltrate through the hole and the vitelline membrane should be lifted from the surface of the embryo. It is then possible to enlarge the hole and to access easily the hindbrain/otic region with the electrodes and injection glass needle.

- If the amnion starts to cover the otic cup, make a small cut at its most caudal end until you can gain easy to the cup with the injection needle.

3.2 In Ovo Electroporation

- Clean the electrodes with 70% Ethanol and mount them on the micromanipulators. Connect the crocodile clips to the electrodes.

- Prepare the DNA solution. For 10 μl of DNA solution, pipet 1 μl of the 10× injection solution (thaw at room temperature one of the 10× aliquot; it can be re-frozen and re-used at least 10 times) into a 0.5 ml eppendorf tube and add the required amount of purified plasmid DNA (see Note 2) and water up to 10 μl, then pipet up and down several times to mix well. Store the DNA mix on ice.

- Using a p10 pipette and a microloader tip, back-fill the glass needle with 5–10 μl of DNA solution (avoid forming bubbles) and mount it onto the injection needle holder.

- Cut the tip of the glass needle with a fine pair of tweezers and adjust the position of the tungsten needle so that its tip is located slightly posterior to and on top of the tip of the glass needle.

- Bring the injection glass needle and the tungsten cathode on top of the right otic placode/cup and place the anode on the opposite (left) side of the embryo, making sure that the needle and electrodes do not touch any embryonic tissue. Readjust the position of the tungsten needle relative to that of the glass needle if needed (*see* **Note 3**).

- Inject the DNA solution into the otic cup that should become blue and clearly visible. If the otic cup is not clearly filled or the solution fails to remain in the cup, check that the membranes that may be covering the otic region have been removed properly.

- Apply the train of electric pulses (*see* **Note 4**) and keep gently injecting fresh DNA solution into the otic cup during the pulses. Bubbles should form at the tip of the tungsten needle. If not, check that electric connections are correct.

- Remove carefully the electrodes and apply two to three drops of PBS on top of the embryo, then close the egg shell window with 50 mm wide Sellotape (*see* **Note 5**).

- Return the egg to the incubator until required for analysis or further experimental work.

3.3 Common Issues and Their Solution

If performed incorrectly (poor placement of the electrodes, too high voltage, damage during egg windowing, dissection/DNA injection), electroporation will result in very poor embryo survival or abnormal development of the inner ear. It is essential to practice the technique sufficiently and determine the optimal voltage conditions for your setup with a control plasmid construct (e.g., CMV-driven EGFP expression vector) before embarking on further experiments. For long-term survival, we recommend to (a) use fresh eggs, and be careful during the windowing procedure; (b) clean all instruments, electrodes, and working space with 70% ethanol; (c) if possible, place the electroporation setup under a horizontal flow hood to minimize the risks of infection; and (d) maintain egg incubators and water trays clean.

We list in Table 1 below the common technical issues that can be encountered and their potential remedies.

Table 1
Common technical issues and their potential remedies

Issue	Possible cause	Remedy
1. Injection needle repetitively clogged.	• The concentration of DNA used is too high. • Purity of DNA is not optimal.	• Do not use DNA concentrated higher than 2–3 mg/ml. • Ensure high quality DNA preparations. Try sodium acetate/ethanol precipitation of your DNA, or alternative clean-up methods.
2. No bubbles on electrodes during electroporation.	• No electric current between the electrodes	• Check the connectivity of the electrodes and the crocodile clips. • Clean the electrodes thoroughly and remove any albumen residue. Do a few pulses in a dish containing a PBS solution.
3. Few cells transfected.	• Low DNA concentration. • Low quality DNA. • Insufficient voltage.	• Ensure high concentration and good quality DNA preparations are used (run quality control with restriction enzyme digestion and transfecting cell lines). • Consider changing/adjusting the DNA isolation kit/protocol (always use molecular biology grade reagents). • Modify voltage/pulse parameters.
4. No expression of transgene at 24–48 h post-EP.	• No transfection of cells. • Error in the design of the construct. • Transgene causing cell death.	• Sequence the construct to double check it has all required elements. • Co-transfect a control plasmid to determine the source of the problem. • Check expression at earlier time points (e.g., after 6 h).
5. No long-term (e.g., E7) expression of Tol2 transgene.	• Transgene causing cell death. • No transposase.	• Check the expression of the transgene at an earlier time point. • Check that the transposase is working with a control transgene.
6. Poor embryo survival.	• Contamination. • Damage to the embryo during the electroporation procedure. • Excessive humidity levels in the incubator.	• Clean and disinfect the work area before electroporation. Use sterile tools and PBS. • Clean the incubator regularly with soap and water, followed by veterinary disinfectant. • Clean the eggs thoroughly before incubation using veterinary disinfectant. • During electroporation, take care not to touch the embryo with the electrodes. • Practice egg windowing and incubation without electroporation to check that this aspect of the technique is mastered. • For long-term survival, use fresh eggs within the first week from delivery. • Ensure the humidity in the incubator is ca. 50%. • For long-term incubation, consider using the automatic egg-turning system.

(continued)

Table 1
(continued)

Issue	Possible cause	Remedy
7. Defects in inner ear morphology (with control transgene).	• Excessive voltage or poor placement of the electrodes. • Bad quality DNA preparation.	• Avoid touching the embryo with electrodes. Lower the voltage parameters. • Consider sodium acetate/ethanol precipitation of your DNA, or alternative clean-up methods.
8. Condensation inside the incubator.	• Excessive humidity levels in the incubator	• Avoid incubating too many windowed eggs in the same incubator. The number of eggs that can be incubated at once will depend on the size of the incubator and has to be tested empirically. In case of excessive condensation, switch off the incubator and wipe the excess of water from the lid and the container. Assemble it back, and continue incubating with the lid slightly lifted to allow the moisture to escape while keeping an appropriate temperature.

4   Notes

1. We use a "chopstick" CUY611P3-1 platinum electrode with a 1 mm tip as an anode and a custom-made tungsten electrode as a cathode. To make a tungsten electrode, cut a 3–4 cm long segment of 0.5 mm diameter tungsten then sharpen its tip by electrolysis in a bath of 1 M sodium hydroxide. Connect the tungsten rod to the positive cable and a metal paper-clip to the negative cable using banana-clips, and apply a 12 V current between them until a sharp tip is obtained. The tungsten needle can be fitted inside an injection needle to which the negative electric wire is connected via a crocodile clip.

2. The final concentration of plasmid DNA used for electroporation is adjusted according to the type of construct and promoter and the desired molar ratio when combining multiple plasmids. For most CMV/CAGGS promoter-based plasmids we have used, 0.2–0.5 µg/µl is sufficient for strong expression. When co-electroporating several plasmids, try to maintain a final DNA concentration < 3 µg/µl or the solution may become very viscous and difficult to inject through the glass needle. A centrifugal vacuum concentrator may be used for a few minutes to evaporate excess water if necessary. For a typical electroporation run of 24 eggs, 5–10 µl of DNA solution is enough.

3. The electrical field will form in between the two electrodes. Depending on the developmental stage, the otic cup will be

more or less closed and the position of the tungsten needle relative to that of the injection needle should be modified accordingly to maintain the two electrodes and injection site along a roughly linear axis. In all cases, do not allow the electrodes to touch any tissue.

4. We are using the following parameters for electroporation: 7–9 V, 3 × 50 ms with 100 ms time interval between each pulse (BTX ECM830 or Intracell TSS20 Ovodyne). These values are provided as guidelines only and should be adapted to each individual setup and developmental stage to reach a good compromise between efficiency of transfection and embryo survival.

5. Provided that the concentration of Fast Green was high enough, you should still be able to see some traces of the dye in the otic cup region. This can be very helpful when initially practicing the method.

References

1. Hughes SH (2004) The RCAS vector system. Folia Biol 50:107–119

2. Kiernan AE, Fekete DM (1997) In vivo gene transfer into the embryonic inner ear using retroviral vectors. Audiol Neurootol 2:12–24. https://doi.org/10.1159/000259226

3. Eddison M, Le Roux I, Lewis J (2000) Notch signaling in the development of the inner ear: lessons from Drosophila. Proc Natl Acad Sci 97:11692–11699

4. Eddison M, Weber SJ, Ariza-McNaughton L, Lewis J, Daudet N (2015) Numb is not a critical regulator of Notch-mediated cell fate decisions in the developing chick inner ear. Front Cell Neurosci 9:74. https://doi.org/10.3389/fncel.2015.00074

5. Chang W, Nunes FD, De Jesus-Escobar JM, Harland R, Wu DK (1999) Ectopic noggin blocks sensory and nonsensory organ morphogenesis in the chicken inner ear. Dev Biol 216:369–381. https://doi.org/10.1006/dbio.1999.9457

6. Chang W, ten Dijke P, Wu DK (2002) BMP pathways are involved in otic capsule formation and epithelial–mesenchymal signaling in the developing chicken inner ear. Dev Biol 251:380–394. https://doi.org/10.1006/dbio.2002.0822

7. Chang W, Brigande JV, Fekete DM, Wu DK (2004) The development of semicircular canals in the inner ear: role of FGFs in sensory cristae. Development 131:4201–4211. https://doi.org/10.1242/dev.01292

8. Stevens CB, Davies AL, Battista S, Lewis JH, Fekete DM (2003) Forced activation of Wnt signaling alters morphogenesis and sensory organ identity in the chicken inner ear. Dev Biol 261:149–164. https://doi.org/10.1016/S0012-1606(03)00297-5

9. Fekete DM, Muthukumar S, Karagogeos D (1998) Hair cells and supporting cells share a common progenitor in the avian inner ear. J Neurosci 18:7811–7821. https://doi.org/10.1523/JNEUROSCI.18-19-07811.1998

10. Lang H, Fekete DM (2001) Lineage analysis in the chicken inner ear shows differences in clonal dispersion for epithelial, neuronal, and mesenchymal cells. Dev Biol 234:120–137. https://doi.org/10.1006/dbio.2001.0248

11. Satoh T, Fekete DM (2005) Clonal analysis of the relationships between mechanosensory cells and the neurons that innervate them in the chicken ear. Development 132:1687–1697. https://doi.org/10.1242/dev.01730

12. Momose T, Takeuchi J, Ogawa H, Umesono K, Yasuda K, others (1999) Efficient targeting of gene expression in chick embryos by microelectroporation. Develop Growth Differ 41:335–344

13. Nakamura H, Funahashi J (2001) Introduction of DNA into chick embryos by in ovo electroporation. Methods 24:43–48. https://doi.org/10.1006/meth.2001.1155

14. Kawakami K, Largaespada DA, Ivics Z (2017) Transposons as tools for functional genomics in vertebrate models. Trends Genet 33:

784–801.   https://doi.org/10.1016/j.tig.
2017.07.006

15. Kim A, Pyykko I (2011) Size matters: versatile use of PiggyBac transposons as a genetic manipulation tool. Mol Cell Biochem 354: 301–309. https://doi.org/10.1007/s11010-011-0832-3

16. Sato, Y., Kasai, T., Nakagawa, S., Tanabe, K., Watanabe, T., Kawakami, K. and Takahashi, Y. (2007). Stable integration and conditional expression of electroporated transgenes in chicken embryos. Developmental Biology 305, 616–624

17. Watanabe, T., Saito, D., Tanabe, K., Suetsugu, R., Nakaya, Y., Nakagawa, S. and Takahashi, Y. (2007). Tet-on inducible system combined with in ovo electroporation dissects multiple roles of genes in somitogenesis of chicken embryos. Developmental Biology 305, 625–636

18. Chrysostomou E, Gale JE, Daudet N (2012) Delta-like 1 and lateral inhibition during hair cell formation in the chicken inner ear: evidence against cis-inhibition. Development 139:3764–3774. https://doi.org/10.1242/dev.074476

19. Freeman S, Chrysostomou E, Kawakami K, Takahashi Y, Daudet N (2012) Tol2-mediated gene transfer and in ovo electroporation of the otic placode: a powerful and versatile approach for investigating embryonic development and regeneration of the chicken inner ear. In: Mace KA, Braun KM (eds) Progenitor cells. Humana Press, pp. 127–139

20. Mann ZF, Gálvez H, Pedreno D, Chen Z, Chrysostomou E, Żak M, Kang M, Canden E, Daudet N (2017) Shaping of inner ear sensory organs through antagonistic interactions between Notch signalling and Lmx1a. elife 6:e33323. https://doi.org/10.7554/eLife.33323

21. Żak M, Plagnol V, Daudet N (2020) A gradient of Wnt activity positions the neurosensory domains of the inner ear. bioRxiv 2020.05.04.071035. https://doi.org/10.1101/2020.05.04.071035

22. Freeman SD, Daudet N (2012) Artificial induction of Sox21 regulates sensory cell formation in the embryonic chicken inner ear. PLoS One 7:e46387. https://doi.org/10.1371/journal.pone.0046387

23. Kos R, Tucker RP, Hall R, Duong TD, Erickson CA (2003) Methods for introducing morpholinos into the chicken embryo. Dev Dyn 226:470–477. https://doi.org/10.1002/dvdy.10254

24. Das RM, Van Hateren NJ, Howell GR, Farrell ER, Bangs FK, Porteous VC, Manning EM, McGrew MJ, Ohyama K, Sacco MA, Halley PA, Sang HM, Storey KG, Placzek M, Tickle C, Nair VK, Wilson SA (2006) A robust system for RNA interference in the chicken using a modified microRNA operon. Dev Biol 294:554–563.   https://doi.org/10.1016/j.ydbio.2006.02.020

25. Doudna JA, Charpentier E (2014) The new frontier of genome engineering with CRISPR-Cas9. Science 346:1258096. https://doi.org/10.1126/science.1258096

26. Hsu PD, Lander ES, Zhang F (2014) Development and applications of CRISPR-Cas9 for genome engineering. Cell 157:1262–1278. https://doi.org/10.1016/j.cell.2014.05.010

27. Komor AC, Kim YB, Packer MS, Zuris JA, Liu DR (2016) Programmable editing of a target base in genomic DNA without double-stranded DNA cleavage. Nature 533: 420–424.   https://doi.org/10.1038/nature17946

28. Billon P, Bryant EE, Joseph SA, Nambiar TS, Hayward SB, Rothstein R, Ciccia A (2017) CRISPR-mediated base editing enables efficient disruption of eukaryotic genes through induction of STOP codons. Mol Cell 67: 1068–1079.e4.   https://doi.org/10.1016/j.molcel.2017.08.008

29. Kuscu C, Parlak M, Tufan T, Yang J, Szlachta K, Wei X, Mammadov R, Adli M (2017) CRISPR-STOP: gene silencing through base-editing-induced nonsense mutations. Nat Methods 14:710–712. https://doi.org/10.1038/nmeth.4327

30. Zhang H, Pan H, Zhou C, Wei Y, Ying W, Li S, Wang G, Li C, Ren Y, Li G, Ding X, Sun Y, Li G-L, Song L, Li Y, Yang H, Liu Z (2018) Simultaneous zygotic inactivation of multiple genes in mouse through CRISPR/Cas9-mediated base editing. Development 145: dev168906.   https://doi.org/10.1242/dev.168906

31. Diao Y, Li B, Meng Z, Jung I, Lee AY, Dixon J, Maliskova L, Guan K, Shen Y, Ren B (2016) A new class of temporarily phenotypic enhancers identified by CRISPR/Cas9-mediated genetic screening. Genome Res 26:397–405. https://doi.org/10.1101/gr.197152.115

32. Thakore PI, D'Ippolito AM, Song L, Safi A, Shivakumar NK, Kabadi AM, Reddy TE, Crawford GE, Gersbach CA (2015) Highly specific epigenome editing by CRISPR-Cas9 repressors for silencing of distal regulatory elements. Nat Methods 12:1143–1149. https://doi.org/10.1038/nmeth.3630

33. Williams RM, Senanayake U, Artibani M, Taylor G, Wells D, Ahmed AA, Sauka-Spengler T (2018) Genome and epigenome engineering CRISPR toolkit for in vivo modulation of cis-regulatory interactions and gene expression in the chicken embryo. Dev Camb Engl 145: dev160333. https://doi.org/10.1242/dev. 160333

34. Gandhi S, Piacentino ML, Vieceli FM, Bronner ME (2017) Optimization of CRISPR/Cas9 genome editing for loss-of-function in the early chick embryo. Dev Biol 432:86–97. https://doi.org/10.1016/j.ydbio.2017. 08.036

35. Véron N, Qu Z, Kipen PAS, Hirst CE, Marcelle C (2015) CRISPR mediated somatic cell genome engineering in the chicken. Dev Biol 407:68–74. https://doi.org/10.1016/j. ydbio.2015.08.007

36. Chen B, Gilbert LA, Cimini BA, Schnitzbauer J, Zhang W, Li G-W, Park J, Blackburn EH, Weissman JS, Qi LS, Huang B (2013) Dynamic imaging of genomic loci in living human cells by an optimized CRISPR/ Cas system. Cell 155:1479–1491. https://doi. org/10.1016/j.cell.2013.12.001

37. Orioli A, Pascali C, Quartararo J, Diebel KW, Praz V, Romascano D, Percudani R, van Dyk LF, Hernandez N, Teichmann M, Dieci G (2011) Widespread occurrence of non-canonical transcription termination by human RNA polymerase III. Nucleic Acids Res 39:5499–5512. https://doi.org/10. 1093/nar/gkr074

38. Kudo T, Sutou S (2005) Usage of putative chicken U6 promoters for vector-based RNA interference. J Reprod Dev 51:411–417. https://doi.org/10.1262/jrd.16094

39. Chang W, Lin Z, Kulessa H, Hebert J, Hogan BLM, Wu DK (2008) Bmp4 is essential for the formation of the vestibular apparatus that detects angular head movements. PLoS Genet 4:e1000050. https://doi.org/10.1371/jour nal.pgen.1000050

40. Kamaid A, Neves J, Giraldez F (2010) Id gene regulation and function in the prosensory domains of the chicken inner ear: a link between Bmp signaling and Atoh1. J Neurosci 30:11426–11434. https://doi.org/10.1523/ JNEUROSCI.2570-10.2010

41. Neves J, Parada C, Chamizo M, Giraldez F (2011) Jagged 1 regulates the restriction of Sox2 expression in the developing chicken inner ear: a mechanism for sensory organ specification. Development 138:735–744. https:// doi.org/10.1242/dev.060657

Chapter 5

Molecular Tools to Study Regeneration of the Avian Cochlea and Utricle

Amanda Janesick, Mirko Scheibinger, and Stefan Heller

Abstract

The avian inner ear can regenerate sensory hair cells after damage and has served as a model for the study of hearing regeneration for more than 30 years. Here we present a detailed surgical protocol to induce rapid apoptosis of all hair cells in the chicken cochlea and utricle with a single, local infusion of the aminoglycoside sisomicin. S-phase entry of supporting cells engaged in proliferative regeneration peaks at 48 h and newly regenerated hair cells emerge as early as 4–5 days post-sisomicin. We provide reliable read-outs for hair cell loss, such as overt manifestations of vestibular deficiencies, and quick validation of regeneration using reliable markers that can be detected with commercial antibodies. Titrating down the dose of sisomicin reveals differential susceptibilities of hair cell subtypes: cochlea versus utricle, cochlear tall versus cochlear short hair cells, vestibular type I versus type II hair cells, and proximal versus distal location along the cochlea. We provide a method to quantitate cells within the sensory epithelium in 3D, leveraging vibratome sectioning and imaging methods that are presented in a companion chapter. Finally, we present the technique of cold-peeling the cochlear sensory epithelium for the purposes of RNA or protein extraction, and single-cell dissociation in preparation for RNA-seq.

Key words In vivo hair cell damage, Surgical model, Aminoglycoside, Supporting cell proliferation, EdU, Microscopy, 3D virtual reality quantitation

1 Introduction: A Damage Model to Eliminate All Cochlear Hair Cells

Complete regeneration of adult sensory hair cells happens in non-mammalian species, with chickens being the long-coveted model organism to study this phenomenon which is lacking in mammals. Many regeneration studies on avian basilar papilla (cochlea) or utricle hair cells have been conducted in vitro. However, regeneration in vitro is incomplete; thus, a culture strategy is an imperfect proxy for the full restoration of mature hair cells with properly oriented hair bundles and neuronal connections. Existing in vivo models are limited because they typically utilize partially

---

Amanda Janesick and Mirko Scheibinger contributed equally to this work.

Andrew K. Groves (ed.), *Developmental, Physiological, and Functional Neurobiology of the Inner Ear*, Neuromethods, vol. 176, https://doi.org/10.1007/978-1-0716-2022-9_5,

damaged cochlear sensory epithelia and usually require bioaccumulation of the damage-inducing drug, leading to an asynchronous regenerative response. After incomplete ablation, the presence of surviving hair cells makes it challenging to characterize newly regenerated hair cells in the absence of reliable transgenic fate markers. Research aimed to reconstruct the temporal sequence of the critical events that govern hair cell regeneration in birds benefits from a precisely initiated damage paradigm leading to complete hair cell loss. Here, we describe a damage model that uses sisomicin, a chemically defined ototoxin. Sisomicin is directly infused into the chicken inner ear via the posterior semicircular canal. This strategy results in rapid and synchronized apoptosis and extrusion of cochlear and utricular hair cells while preserving the surrounding supporting cells.

We recently utilized our in vivo damage model to characterize transcriptional changes in dying cochlear hair cells [1]. Here, we provide additional details and expand upon the damage model's technical advantages and limitations, leaving to the readers' imagination future applications for regenerative studies. Because of its reliability and relatively short procedure time of ≈15 min per surgery, we envision that our protocol provides a useful alternative to existing models. Spurring our interest in conveying our damage model to the broader community is to minimize experiment-to-experiment and laboratory-to-laboratory variability. To this end, we first discuss our choice of using sisomicin over gentamicin, which eliminates batch differences inherent to different gentamicin stock preparations and the resulting variability in ototoxicity. We hypothesize that the use of sisomicin, a chemically defined ototoxic component of gentamicin, reduces experimental variability. We further provide the reader with many checkpoints, including surgical landmarks, chicken behavior, and useful markers of newly regenerated hair cells that can be visualized with commercial antibodies. We discuss specific dosage requirements for studying the chicken utricle versus cochlea, and how lower sisomicin doses can be informative toward examining different cell types (e.g., tonotopic and medial-lateral axes of the cochlea, or striolar/extrastriolar regions of the utricle). We present a time course of regeneration focusing on hair cell morphology, whereby new hair cells progress from a supporting cell-like shape to one of a mature hair cell with hair bundles. We also provide a time window after hair cell loss when supporting cells enter S-phase, as measured by 5-ethynyl-2-′-deoxyuridine (EdU) incorporation.

Much of the data presented in this article relies on vibratome sectioning, immunochemistry, and imaging techniques presented in a companion chapter [2]. We direct the reader to this chapter for more detailed information on acquiring high-quality images of the chicken cochlea and utricle. Furthermore, we primarily utilize reputable hair cell and supporting cell markers, well-known to the inner ear field. However, it should be noted that for elucidating the mechanisms of regeneration, we rely heavily on gene discovery to characterize dying hair cells, proliferating and differentiating supporting cells, or newly regenerated hair cells. For this, we use computational methods [3]. For validation, we employ in situ hybridization and hybridization chain reaction (HCR) to detect mRNA within the sensory epithelium [1, 4], the details of which are described in the companion chapter [2]. Finally, we describe a novel cold peeling method for rapid acquisition of pure cochlea sensory epithelia for RNA-sequencing, Western blotting, proteomics, and more. All materials required for the methods outlined in this chapter are presented in tabular form in Subheadings 2.1, 2.2, and 2.3.

2 Materials

2.1 Chicken Husbandry

Name	Company	Catalog number
Fertilized Rhode Island Red eggs	AA Laboratory Eggs Inc.	NA
Egg incubator	GQF Manufacturing Company	1502
Brinsea Octagon 20 advance	Amazon	B00DPMZ37K
Chick starter feed	Farmers Warehouse Company	NA
Galvanized chick feeder	My Pet Chicken	NA
Bantam waterer	My Pet Chicken	NA
Infrared heat lamp 75 W	Amazon	B01BIBMY9E
Small animal cage	Circle K Industries	NA
Digital hygro-thermometer	Amazon	B000NI4AQY
Egg candler	Amazon	B07FQNX9WF
Newspaper	J Weekly	jweeklyusa.com

2.2 Materials: Surgical

Name	Company	Catalog number
Oxygen regulator (CGA870)	Cramer Decker Medical Inc	AREG8725-B2D
Oxygen	Praxair	OX M-AE
Isoflurane w/ anti-spill adapter	Stanford VSC Pharmacy	NA
V-1 Tabletop Anesthesia System	VetEquip Incorporated	901806
Activated Charcoal Filter VaporGuard	VWR	89012-608

(continued)

Heated Small Animal Pad	K&H Pet Products	NA
Carprofen	Stanford VSC Pharmacy	NA
Scissors	Roboz	RS-5610
Surgical microscope	Leica	M320
Eye Spears (Beaver-Visitec)	Fisher Scientific	NC0972725
UMP3 Microinjection Syringe Pump	World Precision Instruments	UMP3T-1
Nanofil needle	World Precision Instruments	NF35BL-2
SilFlex tubing	World Precision Instruments	SILFLEX-2
Nanofil Injection Holder	World Precision Instruments	NFINHLD
Nanofil Syringe (10 µL)	World Precision Instruments	NANOFIL
FrameWorks V-Base Kit	World Precision Instruments	503207
Sisomicin	Xi'an Health Biochem Technology Co	32385-11-8
Methylene blue	Sigma-Aldrich	M9140
Bone Wax (Ethicon)	Stanford VSC Pharmacy	NA
Surgical Superglue (VetClose)	Stanford VSC Pharmacy	NA
Leg Bands (1/4 in.)	Amazon (Chicken Hill)	B078BHHLSH
5-ethynyl-2′-deoxyuridine (EdU)	ThermoFisher	A10044
Digital Scale (Ozeri)	Amazon	B004164SRA
Kaytee Exact formula	Amazon	B0002DGJH8
5-bromo-2′-deoxyuridine (BrdU)	ThermoFisher	B23151
PBS, PTw, PBST	*See* recipe	*See* recipe
DMSO	Sigma-Aldrich	D8418
Nose Cone	Stanford VSC	NA
Nair	Amazon	B000TJT65M
Surgical Scissors	Fine Science Tools	14002-13
Disposable Scalpel (size 11)	VWR	89176-380
16% paraformaldehyde	Electron Microscopy Sciences	15710
EGTA	Sigma-Aldrich	E3889
Thimerosal	Sigma-Aldrich	T5125

2.3 Materials: Antibodies, EdU and BrdU Staining, and syGlass Quantitation

Name	Company	Catalog number
Rabbit Anti-Myosin VIIa (1: 1000)	Proteus	25-6790
Mouse Anti-SOX2-488 (1:200)	Santa Cruz Biotechnology	sc-365,823 AF488
Mouse Anti-Tubulin β III (1: 1000)	Millipore Sigma	MAB5564
Rabbit Anti-CALB2 (1:1000)	Swant	CR 7697
Mouse Otoferlin (HCS-1) (1: 100)	A gift from Dr. Jeff Corwin	Reference [5]
Mouse Anti-BrdU (MoBU-1) (1: 100)	ThermoFisher	B35128
Dapi (1:2500)	ThermoFisher	D1306
Donkey anti-Rabbit IgG-546 (1: 500)	ThermoFisher	A10040
Alexa Fluor™ 633 Phalloidin (1: 50)	ThermoFisher	A22284
Citifluor™ CFM-3	Electron Microscopy Sciences	17979-20
Click-iT™ EdU Kit 647	ThermoFisher	C10340
LSM880 Confocal Microscope	Zeiss	NA
Zen Black Software v.14	Zeiss	NA
syGlass Software v.1.4	IstoVisio, Inc	NA
SteamVR tracking	Valve	NA
VR headset, controllers, sensors[a]	Amazon (Oculus Rift)	B073X8N1YW

[a]Note that VR technology is changing rapidly. Check www.syglass.io for the latest headset and controller information compatible with syGlass

2.4 Materials: Cold Peeling and Lysis

Name	Company	Catalog number
Medium 199, HEPES[a]	ThermoFisher	12340030
Ciprofloxacin	Sigma-Aldrich	17850
Eyebrow Hair Knife	Self-manufactured according to [6]	NA
Sylgard® 184 silicone plate	Fisher Scientific	NC9285739; FB0875713
Micropipet & Aspirator (50 μL)	Fisher Scientific	21-180-16
RNAqueous™-Micro Kit	ThermoFisher	AM1912

(continued)

Name	Company	Catalog number
Bioanalyzer Instrument 2100	Agilent	G2939BA
Accutase®	Sigma-Aldrich	A6964
Protease Inhibitor Cocktail III	Research Products Intl.	P50700
BCA Protein Assay	ThermoFisher (Pierce™)	23227
OTOF qPCR Primers	Forward: GGCTCTCCTTCT ACACCCGA	Reverse: GGATCTCCAA GATCTCGTT CCCA
TECTA qPCR Primers	Forward: TGGATGATG GGAGCTCTCCTG	Reverse: GAAGGCTTCA GGTGTGAA CTGG
ACTB qPCR Primers	Forward: TCATGTTTGAGACC TTCAACACCC	Reverse: GTGTGGGTAACA CCATCACCAG

[a]For applications requiring Sytox Red staining such as FACS, media without phenol red is required

3   Methods

3.1   Chicken Husbandry

Fertilized chicken eggs (AA Laboratory Eggs Inc.; Westminster, CA) are placed into an egg incubator set to 38 °C, humidified, with egg rocking switched on. Developed embryos are identified by candling at embryonic day 18 or 19 and moved into a Brinsea Octagon incubator for hatching. Hatchlings are transferred to a brooder cage equipped with an infrared heat lamp, digital thermometer, chick starter feed, water, and newspaper. Chickens are raised in a calm environment, with an artificial (12 h) day-night cycle and controlled, low ambient sound levels. Daily logbook records temperature, food, water, newspaper change (every 2 days), and animal number. Animal procedures for the examples shown were approved by the Stanford University Institutional Animal Care and Use Committee.

3.2   Surgical Method: Pre-operation

Adapting from existing mouse protocols [7, 8], we established an effective method for the local infusion of aminoglycoside antibiotics via the posterior semicircular canal into the post-hatch day 7 (P7) chicken inner ear. The posterior semicircular canal (Fig. 1a, [9]) is the most accessible canal from the head's surface. Our surgical landmark is the junction of the V-shaped subarcuate artery with the mastoid air cells [8, 10] (Fig. 1b). Caudal to the subarcuate artery and below the mastoid cells is the posterior canal near the surface. The posterior semicircular canal in P7 chickens is about 200 μm in diameter (Fig. 1c). The infusion is delivered into the perilymphatic space between the membranous canal (Fig. 1c) and the bony labyrinth and not into the endolymphatic lumen of the membranous canal. We found that this route of delivery is efficient for the delivery of drugs to the chicken cochlea. We recommend first identifying the subarcuate artery in fixed tissue where the blood vessels are readily visible and bright red (Fig. 1d–f), prior to surgery in live animals.

We infuse the aminoglycoside sisomicin because gentamicin typically contains undefined and varying impurities [11] and ultimately carries the risk of batch-to-batch variability [12]. Sisomicin is the most ototoxic component of gentamicin [12, 13], is chemically defined, and therefore produces consistency and efficacy across experiments.

3.3 Surgical Method: Infusion and Post-operation

The 35-gauge Nanofil needle, SilFlex tubing, and syringe are front-loaded with sisomicin dissolved at 50–100 μg/μL (*see* Subheading 3.4 for sisomicin dose considerations) in 1× sterile PBS containing ≈100 mM methylene blue, which allows for visual feedback during the injection. Clogging of injection needles is exacerbated with dye and we recommend stopping of dye use when the surgeon is comfortable with the procedure. P7 chickens are anesthetized by inducing lateral recumbency with 4 L/min oxygen and 4% isoflurane with a nose cone. Anesthesia is maintained with constant administration of 2 L/min oxygen and 2% isoflurane. The chicken is placed on a heating pad controlled thermostatically to the chicken's standard body temperature of 102 °F, and the outer ear canal is visualized with a surgical microscope. Figure 1g, h shows the surgical setup along with a close-up of the syringe apparatus. The pump is attached to a V-base, and the Nanofil injection holder is hand-held for optimum control. If a micromanipulator or stereotaxic device is desired, it could certainly be implemented.

100 μL carprofen (1 mg/kg) is injected subcutaneously under the wing near the breast tissue. The postauricular area is prepped with Nair (to remove feathers), followed by three rotations of diluted betadine and 70% isopropanol (alternating). The respiratory rate is monitored and documented in the surgical record. An incision is made 6 mm dorsal to the ear canal with microdissection scissors, and the underlying muscle is gently pushed aside to expose the bony labyrinth (Fig. 1i). Eye spears are used to soak up any minimal blood. The subarcuate artery should be immediately visible, serving as a landmark to identify the posterior semicircular canal (Fig. 1j). The bony labyrinth is punctured adjacent to the mastoid air cell junction using a 30 1/2 gauge needle. Using the UMP3 Micropump and the blunt Nanofil needle, sisomicin or control vehicle (sterile PBS) is slowly infused (5–8 nL/s) into the perilymphatic space between the membranous posterior semicircular canal and the bone (Fig. 1j). The bone wound is sealed with bone wax, and the skin/muscle wound is closed with surgical superglue.

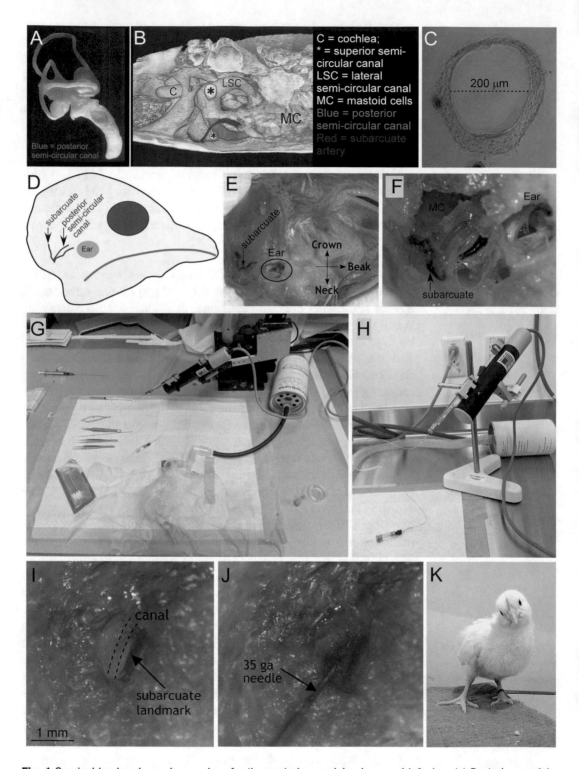

Fig. 1 Surgical landmarks and procedure for the posterior semicircular canal infusion. (**a**) Posterior semicircular canal (blue) in relation to the avian inner ear (image from [9], modified). (**b**) Posterior semicircular canal (blue) and subarcuate artery within the human temporal bone (image from [10], modified). (**c**) Cross-section through the membranous posterior semicircular canal from a P7 chicken. (**d**) Shown is a simple schematic of a chicken head and the posterior semicircular canal and its relative location to the subarcuate artery, the

Once the surgical wound is sealed, isoflurane is reduced to 0%, leaving pure oxygen flowing at 4 L/min. When the chicken recovers from lateral recumbency, it is placed in a small plastic container and closely monitored before returning to the brooder cage. Chickens are distinguished from each other using colored 1/4 in. leg bands and placed in a separate cage from chickens not receiving surgery. Chickens are weighed daily and compared to their natural growth curve, ensuring they are eating and drinking. A quick and reliable read-out for hair cell loss is the appearance of head-tilting towards the surgical ear starting as early as 12–16 h (Fig. 1k).

3.4 Dose Considerations and Partial Damage Phenotypes

Different doses of sisomicin are required for eliciting complete hair cell loss in the cochlea versus the utricle. We typically use 75 μg/μL in a total infusion volume of 2 μL for the cochlea [1] and 50 μg/μL for a total of 0.8 μL for the utricle. Note, however, that variability in the needle seal and position, and differences in surgeons' approaches do influence the results. Furthermore, although Nano-fil needles can be cleaned, they function best when they are new. We have noticed a decrease in ototoxicity with reused needles, presumably due to clogging. We advise determining the ultimate dose a specific surgeon is satisfied with by using our stated concentrations as a starting point. Although the exact dosing might vary, we are confident in the generalizable statement that the dosage of sisomicin required to reliably kill hair cells along the full tonotopic range of the cochlea (proximal-distal) does inevitably destroy the utricle (Fig. 2a–c). Indeed, a high enough sisomicin dose is also capable of damaging the cochlea beyond repair (Fig. 2d, e). This is easily identified, as the SOX2-positive supporting cells will disappear, leaving a thin layer of cells covering the basement membrane. Homogene cells will also disappear at these highly destructive doses.

Suppose the goal is to eliminate all cochlear hair cells and keep the animal alive for more than ~5 days. In that case, the chicken needs to be closely monitored for vestibular defects (Fig. 2c), which can progressively develop into a condition whereby the animal cannot feed itself. If the vestibular deficiencies worsen and body weight plateaus or decreases, the chicken needs to be fed 3× per day by oral gavage using Kaytee formula and injected subcutaneously with 5 mL lactated Ringers solution [14]. The chicken will eventually recover from vestibular defects around 2.5 weeks post-

Fig. 1 (continued) surgical landmark. "Ear" marks the external ear canal. (**e**) Fixed P7 chicken head demonstrating the location of the V-shaped subarcuate artery relative to the external ear canal. (**f**) Fixed P7 chicken head with the posterior semicircular canal false-colored in blue, and its relationship to the subarcuate artery, external ear canal, and mastoid cells (MC, dissected away). (**g**) Overview of the surgical setup with an anesthetized P7 chicken. (**h**) Close-up of hand-held injection holder with nanofil needle, syringe pump, syringe, SilFlex tubing, and V-base clamp. (**i**) After the initial incision, the subarcuate artery is visible and the posterior semicircular canal (dotted line) is located below the bony labyrinth. (**j**) Infusion of sisomicin plus methylene blue dye using a 35 gauge blunt-tip needle. (**k**) Confirmation of sisomicin-induced hair cell damage when chickens tilt their heads towards the surgical ear (a result of vestibular hair cell loss)

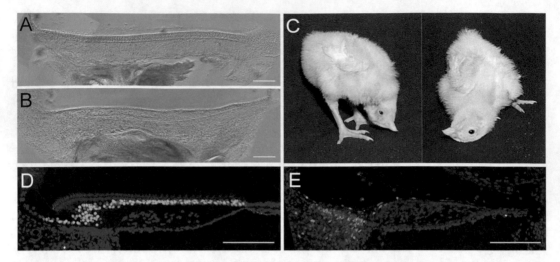

Fig. 2 High-dose sisomicin effects. (**a**, **b**) Transmitted light (tPMT) images of transverse vibratome sections through the P9 chicken utricle. (**a**) Contralateral control. (**b**) Two days post-sisomicin at doses high enough (75–100 µg/µL) to destroy the sensory epithelium. (**c**) Vestibular defects were observed in chickens with flat utricular epithelium. Scale bar (yellow line) = 100 µm. (**d**, **e**) Transverse vibratome sections through the middle part of the P11 chicken cochlea immunolabeled with antibodies against MYO7A (red) for hair cells, SOX2 (green) for supporting cells' nuclei. DAPI labels nuclear DNA (blue). Scale bar (yellow line) = 100 µm. (**d**) Contralateral control basilar papilla. (**e**) Two days post-sisomicin, at a dose high enough (100 µg/µL) to destroy the sensory epithelium, including homogene cells

sisomicin, and manual feeding can be discontinued. Based on the veterinary staff's recommendations, chickens exhibiting weight loss greater than 25% should be euthanized.

At the 75 µg/µL dose of sisomicin, our damage paradigm yields *complete* hair cell loss in the cochlea without affecting the supporting cells [1]. This includes the distal (low frequency) region. Contralateral control ears and inner ears injected with vehicle controls are unaffected (Fig. 3a), demonstrating that sisomicin does not cross the body via the cerebrospinal fluid, and that the surgery itself does not damage the sensory epithelium. MYO7A-positive young hair cells[1] appear after 4–5 days post-sisomicin (Fig. 3b). These regenerated hair cells are SOX2-positive, which contrasts with hair cells in the undamaged cochlea. The new hair cells also feature cytocaud processes, reminiscent of their supporting cell origin. We also note that debris from extruded hair cells, intensely labeled with an antibody to MYO7A, is detectable in the scala media for at least 3 weeks, the latest timepoint we have collected. By 7–9 days, small, disoriented bundles appear, SOX2 is downregulated, and cytocauds are less prevalent (Fig. 3c). By

---

[1] Note that new hair cells can also be specifically marked by strong expression of CALB2 and TUBB3, which are not expressed in controls. By contrast, antibodies to OTOF do not label new hair cells as intensely as controls (Fig. 4c).

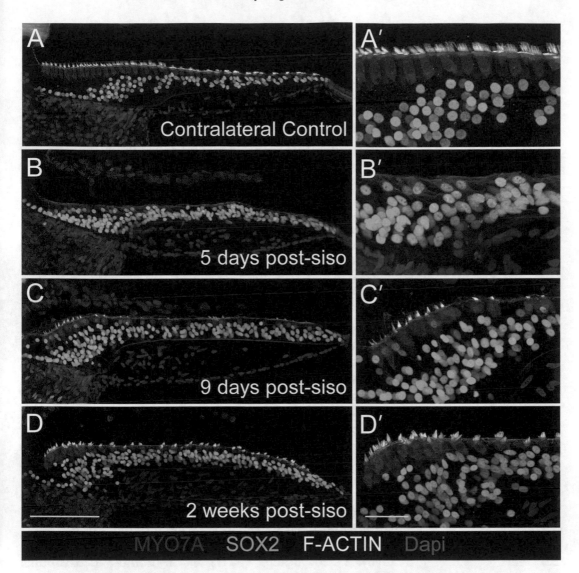

Fig. 3 Time course of hair cell regeneration in the chicken basilar papilla. Transverse vibratome sections through the middle part of the chicken basilar papilla immunolabeled with antibodies against MYO7A (red) for hair cells, SOX2 (green) for supporting cells' nuclei. F-ACTIN was visualized with phalloidin (white) and is enriched in hair cell bundles. DAPI labels nuclear DNA (blue). (**a, a'**) The contralateral control cochlea is unaffected by sisomicin and shows normal morphology. (**b, b'**) New hair cells in the regenerating cochlea are SOX2 and MYO7A positive, and exhibit cytocaud morphology 5 days after sisomicin damage. (**c, d, c', d'**) New hair cells in the regenerating cochlea have downregulated SOX2 and display bundles 9 days and 2 weeks post-sisomicin. Scale bar (yellow line) in **a–d** = 100 μm. Scale bar (yellow line) in **a'–d'** = 25 μm

2 weeks, tall hair cells are abundantly restored, with the short hair cells lagging behind (Fig. 3d).

Titrating down the dose of sisomicin to a threshold where utricle supporting cells survive (25–50 μg/μL, 1.5–2 μL infusion) yields a partial damage phenotype in the basilar papilla. Preservation of the sensory ganglia cell bodies during the dissection allows for an approximate alignment of the tonotopic location along the basilar papilla: the ganglia region's size is small in distal sections and increasingly larger in more proximal sections (Fig. 4a). In this partial damage phenotype, we consistently observe that proximal sections have complete damage, middle sections are damaged only on the abneural (inferior) and far neural (superior) side, and that distal sections are unaffected (Fig. 4b). Proximal hair cells are either more susceptible due to their closer proximity to the infusion (and receive a higher concentration of sisomicin), or more sisomicin can enter the cells, or both. Higher frequency hair cells have more mechanoelectrical transduction channels than their low-frequency counterparts. The unique susceptibilities across the neural-abneural axis (Fig. 4b, "Middle") resonates with single-cell RNA-seq analysis results which show that superior tall hair cells and short hair cells share more expressed genes compared to tall hair cells [4]. Surviving tall and distal hair cells that are not extruded from the epithelium have clearly been exposed to sisomicin and indeed have a distinct transcriptomic profile, marked by the expression of *TRIM35*, a gene strongly and rapidly upregulated in chicken hair cells after sisomicin exposure [1].

In the utricle, the dose of sisomicin required to kill hair cells in the extrastriolar region is higher than for the striolar area, which is more sensitive to sisomicin. A significant decrease in the number of extrastriolar hair cells was observed 2 and 3 days post-sisomicin at 50, 25, and 16 μg/μL doses, with 0.8 μL infusion volume (Fig. 5a, b). In contrast, striolar hair cells died at the aforementioned doses but also as low as 5 μg/μL sisomicin (Fig. 5a, c). We conclude that partially damaged specimens reveal differential sensitivity of hair cell subtypes and a concentration-dependent susceptibility across the axes of the chicken cochlea (neural-abneural, proximal-distal), as well as extrastriolar versus striolar regions of the utricle.

3.5 EdU Bioavailability and Proliferative Window: 3D Quantification

5-ethynyl-2′-deoxyuridine (EdU) and 5-bromo-2′-deoxyuridine (BrdU) permanently label cells when supplied during S-phase. These thymidine analogs have largely replaced tritiated thymidine. We use EdU labeling to monitor the S-phase entry and proliferation of supporting cells, predominantly on the medial/neural side of the sensory epithelium [15]. The detection of EdU-positive new hair cells demonstrates that they originate from supporting cells that replicated their DNA before mitosis. An EdU-negative, MYO7A-positive new hair cell has either phenotypically converted from a supporting cell or has not incorporated EdU because the thymidine analog was not available. We typically inject 50 mg/kg EdU (or BrdU) in 50:50 PBS:DMSO in a volume of ≈200 μL subcutaneously under the wing near the breast tissue after sterilizing the area with 70% isopropanol.

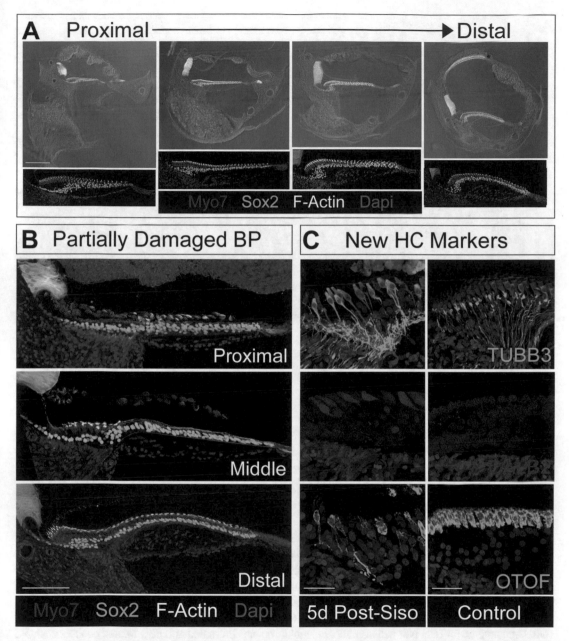

Fig. 4 Partial damage phenotype and markers for newly regenerated cochlear hair cells. (**a, b**) Transverse vibratome sections immunolabeled with antibodies against MYO7A (red) for hair cells, SOX2 (green) for supporting cells' nuclei. F-ACTIN was visualized with phalloidin (white) and is enriched in hair cell bundles. DAPI labels nuclear DNA (blue). (**a**) Vibratome sections of the basilar papilla across the tonotopic axis. Scale bar (yellow line) = 200 μm. (**b**) Low dose sisomicin (25–50 μg/μL, 2 μL infusion) yields a partial damage phenotype. Scale bar (yellow line) = 100 μm. (**c**) New hair cells in the regenerating basilar papilla, 5 days post-sisomicin. Scale bar (yellow line) = 25 μm

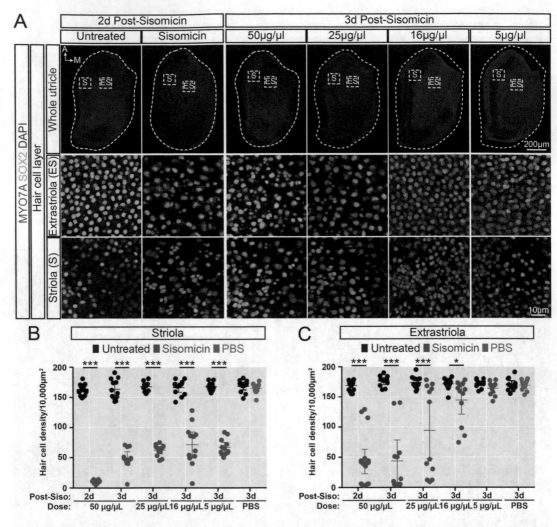

Fig. 5 Regionally specific hair cell death is dose-dependent in the utricle. (**a**) Representative images of the hair cell layer of control and damaged utricles 2 and 3 days after sisomicin infusion (for corresponding supporting cell layer, *see* Fig. 8). Whole utricles and *xy* projections (10,000 μm²) of representative striolar (S) and extrastriolar (ES) regions are shown. MYO7A (red) and SOX2 (green) immunostaining marks hair cells (type I and II) and supporting cells, respectively. Dapi (blue) labels nuclear DNA. (**b**) Total hair cell numbers in extrastriolar regions of sisomicin-treated inner ears (magenta) compared to untreated (black) and PBS-treated (blue) specimens. (**c**) Total hair cell numbers in striolar regions. Untreated and PBS-treated controls analyzed 3 days post-sisomicin showed no significant changes in the number of hair cells in striolar and extrastriolar regions. In **b** and **c**, error bars represent the standard deviation. *$q \leq 0.05$, **$q \leq 0.01$, ***$q \leq 0.001$. Post-Siso = days post-sisomicin infusion. Dose = concentration of sisomicin; 0.8 μL was infused

Systemically injected EdU passes the blood-cochlea barrier and reaches the cochlea. Its pharmacokinetic availability to cochlear cells is transient as it is ultimately absorbed and secreted. Therefore, unlike in vitro culture experiments, where EdU is available for days, in vivo experiments require that EdU is re-administered before its bioavailability diminishes. We determined EdU's bioavailability period in P7-P9 chickens by injecting the thymidine analog BrdU at different times after EdU injection. The EdU Click-iT™ reaction is conducted after permeabilization and before blocking for immunohistochemistry [2]. BrdU cannot be detected with Click-iT™ chemistry but rather by a selective antibody that does not cross-react with EdU. If EdU is injected 2 h prior to BrdU, significant co-labeling is observed. If EdU is injected 4 h prior to BrdU, co-labeling is still observed; however, EdU is weaker in the BrdU-positive cells detected (Fig. 6a). No double-positive EdU and BrdU cells are found at 6 h. Therefore, we conclude that the bioavailability of EdU to cochlear cells, when systemically injected into chickens, is between 4 and 6 h.

Quantitation of hair cells and supporting cell numbers, as well as proliferative cells in the inner ear, is conducted with two different strategies (Fig. 7a). The whole-mount approach utilizes multiple representative squares of the sensory epithelium that are imaged with a top view of the sensory epithelia's apical surface. Cells are counted within a fixed z-depth that focuses on the hair cell layer and the supporting cell layer, respectively. Such an approach is shown in Figs. 5 and 7 for the utricle, and more details are presented in [16]. A second strategy uses transverse vibratome sections, whereby a whole medial-to-lateral representation of the sensory epithelium is quantified at a specific tonotopic location (Fig. 4a). Cells are virtually tagged in 3D-space in a defined volume at distinct locations along the medial-lateral axis using the virtual reality syGlass software package. For syGlass quantitation, our preferred mounting medium is Citifluor™ CFM-3 which also clears tissue sections for confocal microscopy to a depth of ≈100 μm. Note that this mounting medium is not compatible with fluorophore-conjugated phalloidin.

Sections are imaged at 1.0× zoom with a confocal microscope at 40× magnification (Zeiss LSM880; Plan-Apochromat 1.3 numerical aperture, oil immersion) using Zen Black acquisition software at a voxel size of 0.208 × 0.208 × 0.371 μm, and z-depth of 80 μm. Raw data series are exported as tiff files and imported into syGlass. syGlass software interfaces with SteamVR tracking and the Oculus Rift virtual reality headset, touch controllers, and constellation sensors (Fig. 7b). Hair cells and supporting cells are identified by their morphologies and by the expression of MYO7A and SOX2, respectively. Nuclear EdU is visualized as outlined above. Using the "count" function on the right trigger, and the grab/manipulate/rotation function on the left trigger (Fig. 7c), individual cells are manually annotated in 3D virtual reality.

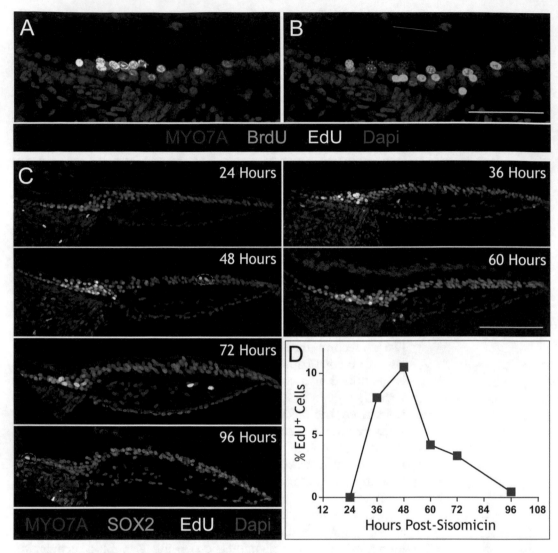

Fig. 6 S-phase re-entry peaks at 48 h post-sisomicin damage. (a–c) Transverse vibratome sections through the middle part of the chicken cochlea immunolabeled with antibodies against MYO7A (red), SOX2 (green, in c), and BrdU (green, in a, b). EdU was visualized by Click-iT™ EdU Kit 647 and marks nuclei (white) that have undergone DNA replication. Dapi labels nuclear DNA (blue). Sisomicin was infused via the posterior semicircular canal at P7 (time = 0 h). (a, b) Bioavailability experiment. EdU was administered subcutaneously 4 h prior to BrdU on day 2 post-sisomicin infusion. Co-labeling of white (EdU) and green (BrdU) nuclei is observed. Scale bar (yellow line) = 50 μm (c) Subcutaneous injection of EdU was delivered 2× surrounding the chosen timepoint (*see* Subheading 2). Maximum intensity projections of transverse vibratome sections through the cochlea ($z = 20$ mm depth) at six timepoints post-sisomicin damage. $N = 3$, with one representative experiment shown. Forty-eight hours circle marks rare abneural EdU+ cells; 96 h circle marks the one EdU-positive cell observed. Scale bar (yellow line) = 100 μm. (d) syGlass quantitation of EdU-positive cells expressed as a percentage of total supporting cell number in the z-stack

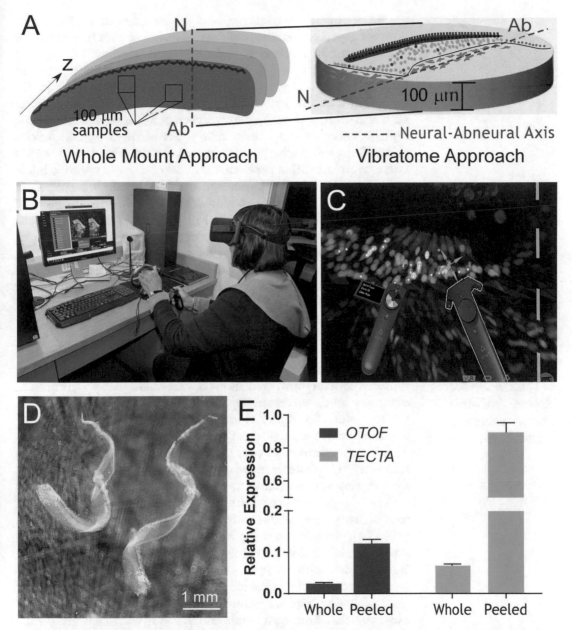

Fig. 7 syGlass quantitation and characterization of cold-peeled epithelia. (**a**) Diagram comparing the whole-mount versus vibratome imaging approach. The red color represents the hair cell layer, and the green color represents the supporting cell layer. (**b**) syGlass station showing the Oculus Rift virtual reality headset, touch controllers, constellation sensors, and computer. (**c**) Screenshot from syGlass where EdU is counted with the right-hand trigger, while the section is manipulated and rotated with the left-hand trigger. (**d**) Cold-peeled sensory epithelium from basilar papillae of P7 chickens. (**e**) QPCR analysis (relative to ß-actin) of whole versus peeled basilar papillae shows the enrichment of hair cell gene *OTOF* and supporting cell gene *TECTA* in peeled specimens

Since our in vivo damage paradigm is a single-dose, local infusion, we expected that the proliferative response after hair cell death would occur in a relatively tight temporal window. We injected EdU twice, 6 h apart, and the timing of those injections would flank the timepoint we would want to observe. For example, for collecting data for 48 h post-sisomicin, we would inject EdU at 42 h and 48 h, and sacrifice the chicken at 54 h. We found that the S-phase entry of supporting cells peaks at 48-h post-sisomicin, primarily in the basilar papilla's neural region (Fig. 6c, d), which is in agreement with a previous study using a single systemic injection of gentamicin [17]. For the utricle, we administered EdU 1 day before the analysis shown in Fig. 5. An increase of EdU-labeled nuclei was observed in sisomicin-treated utricles at 2 and 3 days post-sisomicin when compared with untreated utricles (Fig. 8a). In the striola, this increase was significant at all doses from 5 μg to 50 μg sisomicin (Fig. 8b). In contrast, in the extrastriolar region, a significant increase of EdU-labeled cells was only observed at the higher doses of 25 and 50 μg (Fig. 8c). Different doses of sisomicin therefore reveal region and subtype-specific aminoglycoside susceptibilities of utricle hair cells.

3.6 Cold Peeling of Chicken Cochlear Sensory Epithelium and Downstream Applications

Peeling sensory epithelia from the underlying basement membrane is used to obtain pure populations of hair cells and supporting cells, especially for downstream applications such as RNA isolation and Western blots. For the P7 chicken cochlea, we found that traditionally used enzymatic treatment (e.g., thermolysin) can be omitted. This improvement skips over the typical 30 min to 1 h incubation step at 37 °C. We determined empirically that keeping the cochlear duct during the dissection in ice-cold HEPES-buffered Medium 199 (+10 μg/mL ciprofloxacin) allows for an effortless peeling of the sensory epithelium without prior enzymatic treatment (Fig. 7d). Sensory epithelia are loosened with an eyebrow hair knife [6] while holding the basilar papilla's proximal superior cartilaginous plate with forceps.

We recently characterized peeled sensory epithelium by sectioning basilar papillae before and after peeling [4]. Here, we observed that homogene cells and some hyaline cells are peeled along with hair cells and supporting cells, which was confirmed by RNA-seq. Comparative analysis of whole basilar papillae and peeled sensory epithelia by qPCR shows that mRNA for supporting cell marker TECTA and hair cell marker OTOF are significantly enriched in peeled specimens (Fig. 7e). The cold peeling method refines the source material for downstream applications such as culturing, lysis for bulk RNA preparation, Western blot or proteomics, dissociation for FACS analysis, and single-cell RNA-sequencing [1, 4].

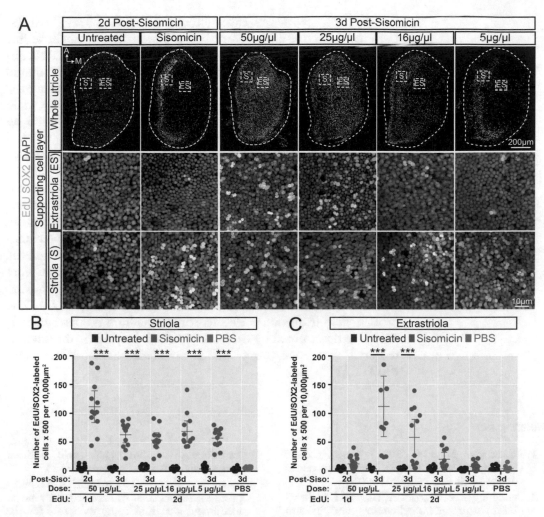

Fig. 8 Cell cycle re-entry in supporting cells corresponds with regions of utricular hair cell death post-sisomicin. To label DNA replication in cells, a single subcutaneous injection of EdU was administrated 1 day before analysis. (**a**) Representative images of the supporting cell layers of control and damaged utricles 2 and 3 days after sisomicin infusion (for corresponding hair cell layer, *see* Fig. 5). Whole utricles and *xy* projections (10,000 μm²) of representative striolar (S) and extrastriolar (ES) regions are shown. SOX2 (green) labels supporting cells. DAPI (blue) labels nuclear DNA. EdU was visualized by Click-iT™ EdU Kit 647 and marks nuclei (white) that have undergone DNA replication. (**b**) The number of EdU-labeled nuclei divided by total SOX2+ supporting cell count in striolar regions of sisomicin-treated inner ears (magenta) compared to untreated (black) and PBS-treated (blue) specimens. (**c**) The number of EdU-labeled nuclei divided by total SOX2+ supporting cell count in extrastriolar regions. Untreated and PBS-treated controls of 3 days post-sisomicin showed no significant changes in EdU-labeled nuclei in striolar and extrastriolar regions. In **b** and **c**, error bars represent the standard deviation. *$q \leq 0.05$, **$q \leq 0.01$, ***$q \leq 0.001$. Post-Siso = days post-sisomicin infusion. Dose = concentration of sisomicin; 0.8 μL was infused

For *RNA extraction*, each epithelium is washed 3× in HEPES-buffered Medium 199 before being placed into a nuclease-free tube, and excess liquid is removed with a pipet. 100 μL RNAqueous™ Lysis Solution is added, and the tissue is quickly triturated and vortexed, then frozen at −80 °C. RNA is extracted using the RNAqueous™-Micro Total RNA Isolation Kit. The RNA yield from one peeled sensory epithelium is approximately 20–30 ng with RNA integrity numbers ranging from 8.2 to 9.5 (as determined by bioanalyzer analysis). For *single-cell dissociation* [4], the epithelium is placed into Accutase® for 30 min at 37 °C, triturated, then spun and washed for FACS, or directed submitted for sequencing (e.g., 10× Genomics protocol). For *protein extraction*, each epithelium is placed into a small ultra-low absorption tube, and excess liquid is removed. 15 μL radioimmunoprecipitation assay buffer (RIPA) + 1× Protease Inhibitor is added, and the epithelium is quickly triturated using ultra-low absorption pipet tips. RIPA buffer consists of 1% Triton X-100, 0.5% deoxycholate, 0.5% SDS, 150 mM NaCl in 30 mM Tris-HCl at pH 7.4. The protein yield from one peeled sensory epithelium is approximately 15 μg (as determined by BCA assay), which is typically the amount required for one lane of a standard SDS-PAGE mini-PROTEAN® gel.

References

1. Benkafadar N et al (2021) Transcriptomic characterization of dying hair cells in the avian cochlea. Cell Rep 34(12):108902

2. Scheibinger M, Janesick A, Diaz GH, Heller S (2021) Immunohistochemistry and in situ mRNA detection using inner ear vibratome sections. Neuromethods (submitted)

3. Ellwanger DC et al (2018) Transcriptional dynamics of hair-bundle morphogenesis revealed with CellTrails. Cell Rep 23(10):2901–2914.e14

4. Janesick AS et al (2021) Cell type identity of the avian cochlea. Cell Rep 34(12):108900

5. Goodyear RJ et al (2010) Identification of the hair cell soma-1 antigen, HCS-1, as otoferlin. J Assoc Res Otolaryngol 11(4):573–586

6. Peng HB (1991) Xenopus laevis: practical uses in cell and molecular biology. Solutions and protocols. Methods Cell Biol 36:102

7. Suzuki J et al (2017) Cochlear gene therapy with ancestral AAV in adult mice: complete transduction of inner hair cells without cochlear dysfunction. Sci Rep 7:45524

8. Talaei S et al (2019) Dye tracking following posterior semicircular canal or round window membrane injections suggests a role for the cochlea aqueduct in modulating distribution. Front Cell Neurosci 13:471

9. Bissonnette JP, Fekete DM (1996) Standard atlas of the gross anatomy of the developing inner ear of the chicken. J Comp Neurol 368(4):620–630

10. Skrzat J, Wrobel A, Walocha J (2013) A preliminary study of three-dimensional reconstruction of the human osseous labyrinth from micro-computed tomography scans. Folia Morphol (Warsz) 72(1):17–21

11. Grahek R, Zupancic-Kralj L (2009) Identification of gentamicin impurities by liquid chromatography tandem mass spectrometry. J Pharm Biomed Anal 50(5):1037–1043

12. O'Sullivan ME et al (2020) Dissociating antibacterial from ototoxic effects of

gentamicin C-subtypes. Proc Natl Acad Sci U S A 117(51):32423–32432

13. Kitasato I et al (1990) Comparative ototoxicity of ribostamycin, dactimicin, dibekacin, kanamycin, amikacin, tobramycin, gentamicin, sisomicin and netilmicin in the inner ear of guinea pigs. Chemotherapy 36(2):155–168

14. Zakir M, Dickman JD (2006) Regeneration of vestibular otolith afferents after ototoxic damage. J Neurosci 26(11):2881–2893

15. Cafaro J, Lee GS, Stone JS (2007) Atoh1 expression defines activated progenitors and differentiating hair cells during avian hair cell regeneration. Dev Dyn 236(1):156–170

16. Scheibinger M et al (2018) Aminoglycoside damage and hair cell regeneration in the chicken utricle. J Assoc Res Otolaryngol 19(1):17–29

17. Bhave SA et al (1995) Cell cycle progression in gentamicin-damaged avian cochleas. J Neurosci 15(6):4618–4628

Part II

Molecular Analysis of the Inner Ear

An Efficient Method to Detect Messenger RNA (mRNA) in the Inner Ear by RNAscope In Situ Hybridization

Sumana Ghosh, Graham Casey, Kendra L. Stansak, Punam Thapa, and Bradley J. Walters

Abstract

Biological processes are largely governed by the RNA molecules and resulting peptides that are encoded by an organism's DNA. For decades, our understanding of biology has been vastly enhanced through study of the distribution and abundance of RNA molecules. Studies of the inner ear are no exception, and approaches like qPCR, RNA-seq, and in situ hybridization (ISH) have contributed greatly to our understanding of inner ear development and function. While qPCR and RNA-seq provide sensitive and broad measures of RNA quantity, they can be limited in their ability to resolve RNA localization. Thus, ISH remains a vital technique for inner ear studies. However, traditional ISH approaches can be technically challenging, time-consuming, suffer from high background, and are generally limited to the investigation of only a single RNA of interest. Recent advances in ISH approaches have overcome many of these limitations allowing for speed, high signal-to-noise, and the ability to perform multiplexed ISH where several transcripts of interest can be visualized in the same tissue or section. One such approach is RNAscope which is a commercially available option that allows for ease of use and, for many transcripts, the ability to achieve absolute quantification of RNA molecules per cell. Here we outline RNAscope methods that have been optimized for inner ear (and related) tissues and allow for relatively rapid labeling of RNA transcripts of interest in fixed tissues. Furthermore, these methods elucidate how RNAscope labeling can be imaged with brightfield or fluorescence microscopy, how it allows for quantification as well as localization, how it can be multiplexed to visualize multiple transcripts simultaneously, and how it can be combined with immunocytochemistry so that RNA and proteins may be visualized in the same sample.

Key words Cochlea, Utricle, Saccule, Vestibular, Development, In situ hybridization, RNAscope, RNA-seq, Transcriptomic profiling

1 Introduction

The central tenet of genetics states that gene function arises from the code present in deoxyribonucleic acid (DNA) which is transcribed to ribonucleic acid (RNA), which is then subsequently translated into functional proteins [1]. As transcription is a critical first step in gene expression it is often necessary to find effective

Andrew K. Groves (ed.), *Developmental, Physiological, and Functional Neurobiology of the Inner Ear*, Neuromethods, vol. 176, https://doi.org/10.1007/978-1-0716-2022-9_6,

tools to quantify and spatially map RNA molecules when one is interested in a given biological process. This has been demonstrated by numerous researchers whose work over the past several decades has helped us to gain deeper insights into gene expression changes that occur during complex biological processes such as tissue development, cell–cell signaling, cell death, neurodegeneration, cancer, and a host of other biological phenomena [2–4]. Indeed, techniques that allow for quantification and/or visualization of RNA have perhaps gained even greater importance as our understanding of RNA molecules and their functions has grown beyond the preliminary transcription-and-translation model to include not only the transcription of messenger RNA (mRNA), but the splicing and editing thereof and also the generation of other RNA molecules such as micro-RNAs (miRNA) and long non-coding RNAs (lncRNA), all of which can play key roles in regulating chromatin structure and dynamics, transcript expression and degradation, and translational and post-translational processes [5, 6]. In response to this, the advancement of next generation sequencing technologies like RNA-seq, at both the bulk and single cell (scRNA-seq) levels, has significantly improved our ability to profile RNA expression levels across the entire transcriptome in both normative and pathological conditions [7]. These RNA-seq approaches have several advantages such as quantitative analysis of differential gene expression, coverage of all or most of the RNA molecules in a given cell type or tissue, and the potential to discover novel roles for RNA molecules. However, the sheer quantity of RNA molecules that are assayed simultaneously and technical limitations such as low numbers of biological replicates can lead to insufficient statistical power and/or insufficient resolution in the analysis of RNA-seq data. In particular, bulk RNA-seq approaches are generally limited in resolving only mean expression levels across a given tissue type or large, potentially heterogeneous group of cells. While scRNA-seq can overcome these limitations to a degree, such an approach can sacrifice coverage and sensitivity, may suffer from bias in the cell sorting process, and still does not provide complete spatial representations of the RNA expression data, failing not only to distinguish exactly where each cell comes from in a given tissue, but also being indeterminate of RNA subcellular localization. More recently, in situ sequencing and similar approaches have been developed to try and improve spatial resolution compared to bulk or scRNA-seq, but these techniques, at current, can suffer from a lack of sensitivity and in some cases can still have limited resolution [8–10]. Despite these limitations, there are clear advantages to these sequencing-based approaches and as such they have been rapidly adopted by many labs and have accelerated our knowledge of RNA expression patterns in a number of different contexts. Indeed the success of RNA-seq based approaches has only further enhanced the need for techniques with greater spatial resolution and equal or greater sensitivity to validate and extend salient findings from RNA-seq data. As many recent reports demonstrate, the use of immunocytochemistry (ICC), in situ hybridization (ISH), fluorescent in situ hybridization (FISH), or single-molecule FISH (smFISH) have been used for this purpose of validating RNA seq experiments and to map spatial expression of genes of interest in intact tissue with preserved cytoarchitecture [4, 11–14]. Furthermore, approaches such as ICC and, in particular, ISH may provide advantages over RNA-seq based approaches for investigators who may not have the resources for large RNA-seq based experiments, or whose research questions require high sensitivity and spatial specificity and are focused on only one or a limited number of RNA molecules of interest.

In situ hybridization has long been an effective tool for identifying RNA transcripts in tissue sections and allows for cellular and subcellular localization as well as relative quantitation of RNA molecules. However, several limitations of traditional ISH are that it can be technically challenging, time-consuming, lacks the sensitivity to detect RNA molecules of low abundance, may provide only relative rather than absolute quantification, can sometimes result in high background or non-specific labeling, and only allows for the interrogation of a single RNA molecule of interest at a time. However, since its development significant improvements have been made which allow for higher throughput, enhanced sensitivity, absolute quantification, as well as simultaneous detection of multiple transcripts in the same tissue slice with resolution at the single cell and subcellular levels [11–13, 15–17]. One such approach that embodies several of these advantages over traditional ISH is RNA-scope (www.ACDBio.com) which uses several successive, proximity dependent, hybridization steps to vastly improve signal-to-noise and thus improve specificity and sensitivity. Additionally RNAscope can be performed for either fluorescent or brightfield microscopy and was introduced commercially in 2012 [4]. RNAscope, like other single-molecule ISH approaches, is designed to generate a single chromogenic or fluorescent point for each individual RNA molecule that is hybridized by the probes, thus allowing not only for greater specificity, sensitivity, and visualization, but also absolute quantification of RNA molecules of interest within individual cells. While these advantages are true of most single-molecule ISH approaches, the established protocols, organized "kits," and company led design and production of probes allow for the easy adoption of RNAscope into research laboratories even if they have not previously performed traditional ISH experiments [17–19]. Furthermore, RNAscope can be combined with immunohistochemistry allowing for the visualization of RNA and protein in the same tissue at the same time. This combined approach can also provide greater resolution of individual cells and subcellular compartments which can be delineated via immunostaining.

The general principle of RNAscope relies on the usage of many tandem "Z"-shaped probes that hybridize to successive, neighboring regions of a single RNA molecule. Each "Z"-shaped probe contains three domains: the antisense lower portion (typically an 18 to 25 base sequence) that hybridizes to a region on the RNA of interest; an upper portion composed of a 14-base tail sequence that provides half of the binding site for additional preamplifier probes; and a spacer region in between the upper and the lower region (*see* Fig. 1). Two of these "Z" probes must bind to the target RNA adjacent to one another for the upper regions to create a 28 base stretch of sequence which can be recognized and hybridized by a

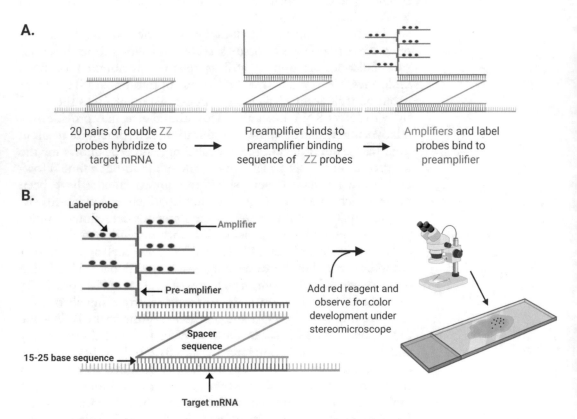

Fig. 1 Schematic of RNAscope in situ hybridization. (**a**) Z-shaped target probes bind in tandem by hybridizing to the complementary sequence on the RNA of interest. In subsequent steps, a preamplifier probe binds where two adjacent Z probes create a longer stretch of nucleotide sequence that is complementary to the binding region on the preamplifier. Binding of only a single Z probe results in decreased efficiency of hybridization of the preamplifier as it is designed to span the two probes. After the preamplifier is added, several subsequent steps add amplifier and label probes to the preamplifier exponentially increasing the overall signal. (**b**) The completed RNAscope label complex consists of the target mRNA bound by the Z probes which in turn are bound by the preamplifier which in turn is bound by the amplifier and label probes. Once these steps are complete, a color reaction can be performed for Red or other chromogenic kits and the signal can be observed under the microscope. For fluorescent kits, no chromogenic reaction is needed and the samples can proceed directly to imaging unless immunocytochemistry or other labeling is also to be performed. (This figure was created with Biorender)

preamplifier probe in a subsequent hybridization step. After hybridization with the preamplifier probes, additional rounds of hybridization allow for labeled probes to then be hybridized as branches to the preamplifier trunks to achieve signal amplification (Fig. 1) [4]. These final probes are labeled in that they are either pre-conjugated to fluorescent molecules or to chromogenic enzymes. Each of these steps contributes to greater amplification of signal and reduced background/non-specific labeling. For example, no target signal amplification occurs when only a single Z probe binds to any transcript, thus background from non-specific binding of probes is significantly reduced. According to the manufacturer, six "Z" probes must bind in tandem in order for a clearly visible punctum to be generated. Typical probe cocktails for mRNA contain 20 or more probes ensuring bright signal and further enhancing specificity. For investigators interested in detecting micro-RNAs or splice variants, probes and other kits (miRNAscope, Basescope) have been designed to detect shorter sequences while preserving sufficient signal amplification [20].

With the increasing popularity of RNAscope, multiple protocols have been developed and optimized for various tissue types and various methods of tissue preparation such as fixed frozen sections, fresh frozen sections, formalin fixed paraffin embedded (FFPE) sections, and whole-mounted tissue preparations [21–23]. Other differences across protocols may include the type of dehydrating agent used (methanol versus ethanol), the temperature at which different tissues are treated, duration of various incubation steps, and the types of solution used to dilute the probes or wash or permeabilize the tissue [12, 23, 24]. In particular, protease treatment and permeabilization steps can vary widely and are the most likely steps one needs to optimize depending on the tissue type, thickness, and method of preservation. In the inner ear research field, RNAscope ISH has been employed in the validation of RNA-seq data from a variety of contexts including spiral ganglion neurons (SGN) in developing and adult mice, sensory hair cells and supporting cells in the murine cochlea and vestibular organs, and in human fetal cochlear progenitor cells [21–23, 25–27]. Here we outline procedures for performing RNAscope that have been optimized for our work with rodent inner ear and related tissues (e.g. zebrafish). To promote relatively easy adoption into labs that may not have experience with traditional ISH, our procedures are largely based on the manufacturer's protocols and reagents with two primary areas of deviation: (1) tissue preparation and pretreatment processes must generally be optimized for the tissue type and the method by which tissues are preserved and so here we present multiple examples and (2) the wash buffer for all washes in between all of the amplification steps has been varied to better suit the use of these kits with fixed tissues. To the first point, we have outlined steps (Fig. 2) for the preparation of both sections and whole-mount

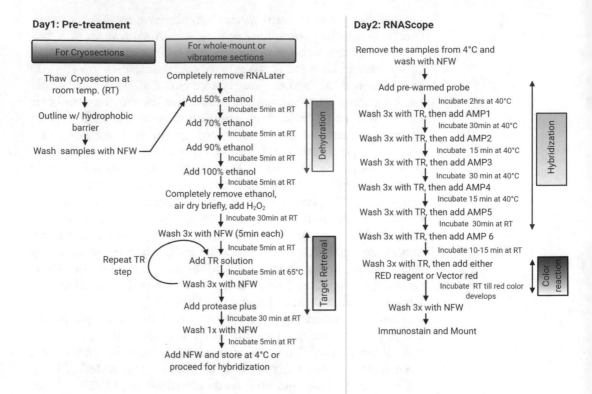

Fig. 2 RNAscope workflow

inner ear specimens from mice (Figs. 3, 4, 5, and 6) and have added a discussion of steps which may be useful when using older mice and decalcification of the temporal bone may be required. With regard to the second point, we replaced the wash buffer with a sodium citrate (pH 6.0) antigen unmasking buffer, which in our experience leads to better adherence of tissue sections to charged slides (due to lower detergent concentration) and leads to more consistent labeling results, perhaps due to greater RNA stability at lower pH [28, 29] and/or greater probe penetrance due to reverse cross-linking or other beneficial effects of the higher sodium citrate concentration [30] (Fig. 3).

Here we will provide examples of RNAscope in paraformaldehyde (PFA) fixed mouse cochlear and vestibular tissues using different ages and different preparations (whole mount, frozen section, etc.). We will also demonstrate RNAscope labeling in fixed zebrafish embryos. Our focus in the methods will be on the preparation of tissues that will allow for combined RNAscope and immunocytochemical labeling, followed by detailed protocols for single channel and duplex RNAscope labeling which can then be followed by traditional immunostaining. After, we provide a brief discussion of multiplex and HiPlex fluorescent kits for which tissue preparation is similar and protocols provided by the supplier can be used with some listed modifications on tissue that is fixed by

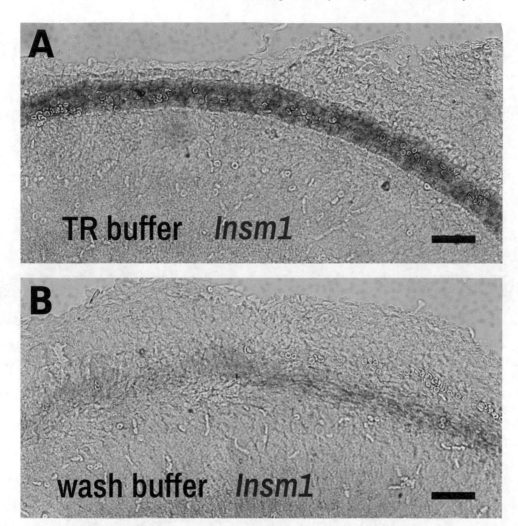

Fig. 3 Comparison of results with target retrieval buffer versus provided wash buffer. (**a**) RNAscope for *Insm1* in P0 mouse cochleae consistently results in robust labeling of outer hair cells with good contrast using the fast red label and performing all washes with pH 6.0 sodium citrate target retrieval (TR) buffer. (**b**) Results are generally not as consistent nor consistently robust when the wash buffer provided by the manufacturer is used. RNAscope for *Insm1* in P0 mouse cochlea is shown after procedure was performed identically to and in tandem with the sample shown in (**a**), the only exception being that the provided wash buffer was used in (**b**). Scale bars = 50 μm. (This figure was created with Biorender)

aldehyde cross-linking. In general, we recommend that RNAscope be performed using PFA-fixed tissues as previous studies have demonstrated that PFA fixation helps to preserve RNA integrity and maintain cytoarchitecture [31, 32], and such fixation also better allows for the use of immunostaining following the ISH procedure. However, some studies have demonstrated that the unmodified protocols provided by the manufacturer may be suitable if unfixed, ethanol fixed, or isopentane fixed, frozen tissues are used in short order and not intended to be combined with immunostaining [26].

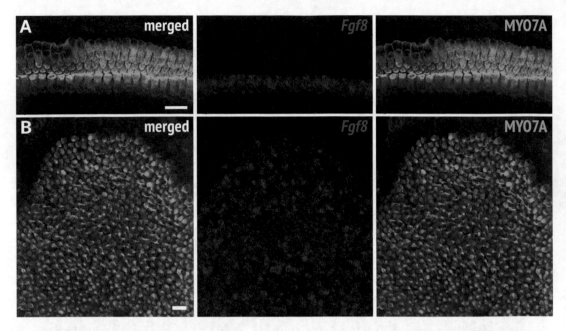

Fig. 4 RNAscope, combined with immunofluorescence, in whole-mounted inner ear tissues. (**a**) RNAscope for *Fgf8* (magenta) in P0 mouse cochlea is highly enriched in inner hair cells. In the same tissue, an antibody against myosin VIIa (MYO7A) was used to label both inner and outer hair cells (green). (**b**) RNAscope for Fgf8 (magenta) in P0 mouse utricle similarly demonstrates effective labeling of many hair cells in a whole-mount sample. Again, immunostaining for MYO7A (green) was performed in the same tissue to label all of the hair cells. Both sets of images also demonstrate the ability to image the red chromogen (Fgf8) using fluorescence microscopy. Scale bars = 20 μm

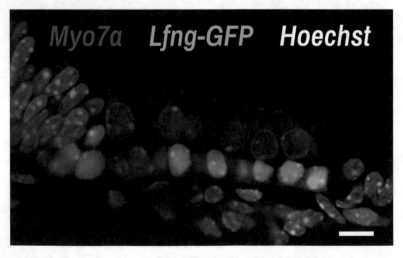

Fig. 5 RNAscope, combined with a fluorescent reporter, in vibratome sections. The image shows RNAscope labeling of myosin VIIa mRNA (*Myo7a*, magenta) in cochlear hair cells in a vibratome section from the cochlea of a transgenic mouse that expresses green fluorescent protein driven by *Lfng* regulatory sequences (green). Cell nuclei were counterstained with Hoechst dye (white). Scale bar = 10 μm

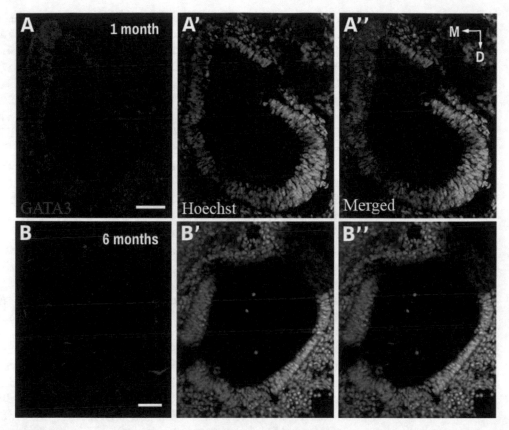

Fig. 6 Storage time, even of frozen sections, can influence efficacy of RNAscope ISH. (**a, a', a"**) When cryosections are stored at −80 °C for 1 month or less, RNAscope for *Gata3* (red) in the otic vesicle of an E10.5 mouse embryo shows robust labeling in the medial portion of the vesicle, consistent with previous reports. (**b, b', b"**) When cryosections are stored at −80 °C for 6 months or longer, RNAscope labeling is significantly diminished. Again an E10.5 mouse otic vesicle is shown. Scale bars = 50 μm

2 Materials

2.1 Animal Care and Housing

All mice used for this study were housed in the Center for Comparative Research at the University of Mississippi Medical Center and treated according to approved IACUC protocols. The mice were maintained at 12:12 light:dark cycle and they were provided food and water ad libitum. Also, the authors would like to thank the laboratory of Nathaniel Heintz (Rockefeller University) and the GENSAT project for generating the *Lfng-GFP* (MGI:4847192) mice and the laboratory of Andrew Groves (Baylor College of Medicine) for providing tissue from these mice.

2.2 Reagents

DNAse- and RNAse-free ultrapure distilled water (#10977-015) and Hoechst were purchased from Invitrogen. Electron microscope grade 16% paraformaldehyde (#15710) and Fluorogel mounting media (# 17983-100) were bought from Electron Microscopy Sciences. Basix gel-loading tips (#13-611-116), Tissue-Plus

O.C.T. Compound (#23-730-571), RNAlater (#AM7021), molecular biology grade ethylenediaminetetraacetic acid (EDTA, 0.5 M) (# 6381-92-6), nuclease free microfuge tubes (# 3453), SuperFrost positively charged slides (#12-550-17), cover slips (#12542B), sucrose (# #5212), agarose (CAS 9012-36-6), and 4-(2-hydroxyethyl)-1-piperazineethanesulfonic acid (HEPES) buffer 100 ml (#16924AE) were purchased from Thermo Fisher Scientific. Hank's Balanced Salt Solution (HBSS) (#14065-056) was bought from Gibco and molecular biology grade ethanol-CAS 64-17-5, #2701G was procured from Decon Labs. Citrate-based antigen unmasking solution (# H-3300) and vector red substrate kit alkaline phosphatase (#sk-5100) were purchased from Vector Laboratories. Glycerol (CAS No 56-81-5) was purchased from VWR Analytical. Opal 520 reagent pack (#FP1487001KT) was purchased from Perkin Elmer. All antibodies used herein are listed in Table 1.

2.3 Dissection Tools

Dissection of whole-mount tissues from murine inner ears was performed under a 7-45× zoom stereomicroscope (Fisherbrand #03-000-014) with added illumination (AmScope #NC0772783). All the dissection tools were purchased from Fine Science tools including Dumont #5 forceps (#11252-40), Dumont AA-epoxy coated forceps (#11210-10), Vannas Spring Scissor 2 mm cutting edge straight (#15000-03) and curved (#1500-04), Vannas-Tübingen spring scissors (#15003-08). MX35 premier disposable low-profile microtome blades (# 3052835), Electron Microscopy Sciences razor blade injector type (#71990), surgical design no.10 carbon scalpel blade (#22-079-690), and scalpel blade handles (#12-000-163) were purchased from Fisher Scientific. 96-well round bottom microplate (#229590) was purchased from Celltreat Scientific Products.

2.4 Instruments

For cryosectioning, a Leica CM3050S cryostat was used and vibratome sections were cut using Vibratome 1500 sectioning system. All the incubations were conducted in an Isotemp oven (Fisher Scientific). Black polypropylene storage boxes (#14-100-F) from Fisher Scientific were used for creating humidifying chambers. Liquid blocker PAP pen (#NC9827128) was purchased from Fisher Scientific. For shaking or mixing either an orbital shaker (Corning #6780FP) or a paddle tube rotator (Fisherbrand #88-861-051) was used.

2.5 RNAScope Reagents

RNAScope 2.5 HD Detection Reagent-Red Kit (#322360). RNA-Scope® HiPlex AssayRNAScope HD Duplex Assay (#322500), RNAScope LS Multiplex TSA Buffer Pack (#322810), RNAScope H2O2, and Protease Plus Reagents (#322330) were purchased from Advanced Cell Diagnostics (ACDBio). All reagents were stored at 4 °C. All of the probes used herein are listed in Table 2.

Table 1
Examples of antibodies that do and do not efficiently co-label inner ear tissues after RNAscope procedure

	Antibody	Host species	Dilution	Vendor (catalogue #)	RRID#
Antibodies that work well with RNAScope	BCl11b	Rat	1:400	Millipore (MABE1045)	N/A
	Cx29	Rabbit	1:350	Thermo Fisher (#34-4200)	AB_2533169
	HuD	Mouse	1:350	Santa Cruz Biotechnology, (# sc-28299)	AB_627765
	MYO6	Rabbit	1:200	Proteus Biosciences (#25-6791)	AB_10013626
	MYO7A	Rabbit	1:200	Proteus Biosciences (25-6790)	AB_10015251
	Neuroplastin	Sheep	1:100	R&D Systems (AF7818)	AB_2298877
	NF-H	Chicken	1:1000	Millipore (#AB5539)	AB_11212161
	Oncomodulin	Goat	1:500	Santa Cruz Biotechnology (sc-7446)	AB_2267583
	Prestin	Goat	1:500	Santa Cruz Biotechnology (sc-22692)	AB_2502038
	SOX2	Goat	1:250	Santa Cruz Biotechnology (#sc-17320)	AB_2286684
	VGlut3	Rabbit	1:500	Synaptic systems (#135203)	AB_887886
Antibodies that require further optimization or may not work with RNAScope	Gata3	Goat	1:100	R&D Systems (#AF2605)	AB_2108571
	Parvalbumin	Mouse	1:1000	Sigma-Aldrich (#P3088)	AB_477329
	SOX2	Rabbit	1:500	Millipore (#AB5603)	AB_2286686

Table 2
List of probes used to generate figures for this chapter

Probes	Species	Accession no.	Catalog no.
Dio2	Mus musc.	NM_010050.3	479331-T7
Fgf8	Mus musc.	NM_010205.2	313411
Gata3	Mus musc.	NM_008091.3	403321
Insm1	Mus musc.	NM_016889.3	430621
Myo7a	Mus musc.	NM_001256081.1	462771
notch3	Danio rerio	NM_131549.2	431951
Pou4f3	Mus musc.	NM_138945.2	546731-T1
Prox1	Mus musc.	NM_008937.2	488591-T3
Sema3c	Mus musc.	NM_013657.5	441441-T5
Sox10	Mus musc.	NM_011437.1	435931-T2

3   Methods

RNAscope can be performed on different types of inner ear sample preparations including fixed frozen sections, fresh frozen sections, formalin fixed paraffin embedded (FFPE) sections, and whole-mounted fixed tissue. In all these cases, the hybridization, pream-plification, and amplification steps do not vary to a large degree. However, the initial sample permeabilization and pretreatments may be distinct in some cases. In this section we will describe the protocol standardized for paraformaldehyde (PFA) fixed samples, which can then be frozen and cryosectioned, agarose-embedded and vibratome sectioned, or whole-mounted. By fixing the tissue in PFA, mRNA can be better preserved, particularly when EDTA is needed for decalcification [33], and this also allows for better morphological preservation and immunostaining in conjunction with the RNAscope labeling.

3.1   Sample Collection and Fixation

3.1.1   Embryonic Samples

To collect embryonic samples, the pregnant dam is euthanized, and the ovarian ducts are removed via abdominal incision. The embryos are quickly dissected out from the ducts and placed in ice cold HBSS supplemented with 1 mM HEPES buffer. The chorionic membrane is removed, and the embryos are immersed in freshly prepared, 4%PFA (made in 1× sterile PBS) in either 15 mL or 50 mL tubes which are then put on an orbital shaker for 3–6 h at room temperature (RT). For embryonic samples older than E16, temporal bones can be better fixed by being dissected separately and then immersed in 4% PFA in 2 mL centrifuge tubes and incubated for 4 h at RT on a paddle rotator. In either case, once the tissues are fixed, wash the tissues once with RNAlater and then store the samples in RNAlater at 4 °C until ready for further processing.

3.1.2 Neonatal/Postnatal Samples

Neonatal/postnatal mice, younger than postnatal day P21 are euthanized by CO_2 asphyxiation, then decapitated and the temporal bones are rapidly removed. The temporal bones are perfused through the round window with 4% PFA and then the samples are immersed in 4%PFA in 2 mL tubes and incubated for 3–6 h in 4% PFA on an orbital shaker at RT. At the end of fixation, the samples are washed with RNAlater and stored in fresh RNAlater solution at 4 °C until ready for further use.

3.1.3 Adult Samples

To collect samples from mice older than P21, mice are euthanized by CO_2 inhalation to effect, confirmed by decapitation, and temporal bones are quickly resected. Using sharp forceps, an opening is made in the bone at the helicotrema, and the cochlear duct is then perfused with 4%PFA through the round window. The samples are then immersed in 4%PFA and incubated for 3–6 h on a paddle rotator at RT. Samples are then washed once in RNAlater and stored in RNAlater at 4 °C. It must be noted for samples coming from older mice (>4–5 months) longer fixation time and cardiac perfusion are recommended and should be optimized accordingly.

3.2 Decalcification

The temporal bone is one of the densest bones in the body. In mice older than P12-P14, decalcification to some extent is generally required for sectioning or the fine dissection of sensory organs. Mouse temporal bones can be decalcified by immersion in EDTA solution prepared in RNAlater (pH adjusted to 5.2 with HCl) and put on an orbital shaker at 23 °C. Decalcification time and concentration of EDTA required for this process depend on three factors: (1) age of the mice, (2) type of sample preparation such as whole-mount dissection, vibratome, or cryosectioning, (3) the organ of interest in the inner ear such as cochlea or vestibular organ or both (Table 2). For whole-mount dissection or cryosectioning of temporal bones from mice aged P14-P30, samples are generally decalcified in 0.125 M EDTA in RNAlater for 12–36 h at room temp. For vibratome sectioning, P14-P30 samples are generally decalcified in 0.2 M to 0.25 M EDTA in RNAlater for 4–5 days at 4 °C. After decalcification, samples are washed thoroughly with 20% RNAlater solution (made in 1× sterile PBS) for three times 10 min followed by three times 24 h-long washes in RNAlater to completely remove EDTA before proceeding with further processing. At this stage, the sample is stored in 100% RNAlater in 4 °C until ready for use and then washed once more with RNAlater before being used for RNAscope. Samples can be stored for up to 1–2 weeks in RNAlater at 4 °C if necessary.

3.3 Sample Preparation

3.3.1 Whole-Mount Tissue

The inner ear organs including cochlear and/or vestibular epithelia are dissected out from the temporal bone in a sylgard-coated petri dish filled with cold RNAlater solution and the tissues from each ear are stored in individual wells of a 96 well plate in RNAlater and stored at 4 °C until ready to proceed with RNAscope. Methods for performing the dissection of inner ear tissues from the temporal bone have been reported elsewhere [34–37]. An example of RNAscope performed on whole-mounted cochlear and vestibular tissues is presented in Fig. 4.

3.3.2 Cryosections

Sufficiently decalcified temporal bones are immersed in 30% sucrose solution (made in 1× PBS and filter sterilized) for 3–4 days in 4 °C, until the samples completely sink to the bottom. If samples remain buoyant, they either have not equilibrated to 30% sucrose internally, or they contain air bubbles. If samples are floating in the sucrose, check for and remove any air bubbles, replace the sucrose with fresh 30% sucrose solution, and incubate for additional days if necessary. Samples that have equilibrated to 30% sucrose content are then embedded in OCT freezing medium in a tissue mold and frozen by placing the mold on dry ice. Tissue sections are cut on cryostat and collected on SuperFrost positively charged slides. Sections can be cut to desired thickness on a cryostat; however, due to the antigen retrieval steps, and repeated washes in the RNAscope protocol, it is recommended that sections not be cut thinner than 15 μm as thin sections can disintegrate and/or de-adhere from the charged slides during processing. Tissue sections can be used immediately for RNAscope or subsequently stored at −80 °C. Optimal results from stored frozen sections can be obtained with up to 1–2 months of storage at −80 °C. By 3–6 months, however, RNAscope signal tends to decrease and may be undetectable beyond 6 months even when stored at ultralow temperatures (*see* Fig. 6).

3.3.3 Vibratome Sections

When thicker sections or greater preservation of cytoarchitecture are desired, Vibratome sectioning of agarose-embedded samples may be desirous. Immediately after decalcification and rinsing with RNAlater, temporal bone samples are embedded in molds with 5% agarose made in sterile 1× PBS and cut using a Vibratome or similar vibrating tissue slicer. The section thickness typically varies from 30 to 100 μm. It is recommended not to exceed 100 μm thickness as the ability of the RNAscope probes to permeate through the full depth of thicker sections diminishes at this point. After sectioning, samples are collected and stored in a 96 or 48 well plate, immersed in RNAlater, and stored in 4 °C until ready to proceed for RNAscope. Again, long term storage beyond 1 month is not recommended. If longer term storage is necessary, it is recommended that the sections be immersed in 30% sucrose or

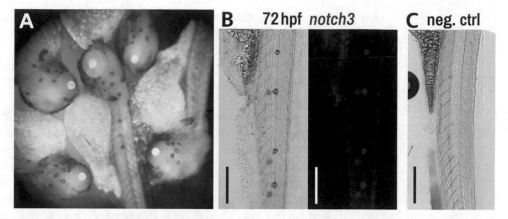

Fig. 7 RNAscope labeling of *notch3* in zebrafish embryos. (**a**) A low power image of RNAscope labeling using the zebrafish-specific *notch3* probe and the Red manual kit demonstrate the enriched expression of *notch3* in the neuromasts and also the utility of RNAscope in different model organisms important for inner ear research. (**b**) Higher power images of *notch3* labeling demonstrate how the red chromogen can be visualized using either brightfield (left panel) or fluorescence (right panel) microscopy. (**c**) The negative control probe, with sequences complementary to the bacterial *DapB* gene, did not hybridize nor generate any detectable signal in zebrafish embryos. Scale bars = 200 μm

other antifreeze solution and stored at −20 °C to −80 °C. Examples of RNAscope performed in vibratome sections are provided in Figs. 5 and 9.

3.3.4 Zebrafish Embryos

We have found that zebrafish embryos can be effectively labeled using RNAscope probes specific for that species (Fig. 7). The methods presented here for whole-mounted mouse inner ear tissues appear to work well with only one small modification which is the pretreatment of the embryos prior to dehydration by immersion in 100% methanol for 20–30 min at −20 °C. All other steps are identical to what is outlined for whole-mount murine inner ear tissues.

3.4 RNAscope

Here we first describe a detailed protocol of a single-channel chromogenic assay to study the expression of a single transcript of interest using RNAscope 2.5 HD Assay-RED kit.

3.4.1 Pretreatment

Before beginning with the RNAscope pretreatment processing, the Isotemp oven is set to 65 °C. Target retrieval (TR) solution/wash buffer is prepared by diluting sodium citrate buffer (low pH antigen unmasking solution, Vector labs #H3300) 1:100 in nuclease free water (NFW) in a 50 ml conical tube. The amount of this solution needed to cover all of the samples to be used (~0.5 mL per ear or ~0.25 mL per slide) is aliquoted into a 15 mL tube and warmed to 65 °C. If working with cryosections they should be taken out from −80 °C and a barrier drawn around the section with hydrophobic pen and washed three times (5 min each) with ultrapure

NFW to remove any traces of OCT. After the final wash, dehydration steps are performed. For whole-mount tissues and vibratome sections, RNAlater is thoroughly removed from the wells, the samples are washed one to two times with NFW, and then immediately moved through dehydration steps, free-floating in their respective wells.

Samples are dehydrated with a series of graded ethanol solutions starting with 50% ethanol for 5 min at RT. At the end of the incubation the solution is carefully removed, and samples are subsequently treated with 70% ethanol, 90% ethanol, and 100% ethanol for 5 min each at RT. Typically, each slide can take approximately 200 μL of each ethanol solution, though it depends on the size of the box drawn with the barrier pen on the slide. Similar volumes of ethanol solution can be used for each well of a 96 well plate, and larger volumes may be required if using 48 well plates. It is important that, regardless of the volume used, the tissue be completely submerged in the ethanol. Upon completion of the dehydration steps, the slides are air-dried for 5 min, though the tissue should not be allowed to dry out completely during this step. Next, samples are immersed in pre-treat 1 (hydrogen peroxide solution) for 30 min at RT. If the H_2O_2 purchased from ACDBio becomes a limiting reagent, H_2O_2 can be purchased from another vendor and diluted to 0.3% in NFW and substituted here. After H_2O_2 treatment, the samples are washed 3×5 min. in NFW, then the warmed TR solution is added, covering the samples, which are then incubated at 65 °C for 5 min in the humidifying chamber in the oven followed by three 5-min washes in ultrapure water. This step, 5 min TR treatment at 65 °C and three washes, is then repeated once. After the last wash, water is removed completely, and the samples are then treated with "protease plus" for 25 min at RT followed by 2×5 min. Washes in NFW. At this point, the samples can be stored in NFW overnight at 4 °C. Alternatively, hybridization and amplification steps can be continued to complete the entire processing in one day. Overnight incubation in NFW is actually recommended here as it generally yields better results, perhaps due to residual protease activity promoting greater tissue accessibility for the RNAscope probes.

3.4.2 Hybridization and Signal Amplification

The oven is preheated to 40 °C. A humidified chamber is placed in the oven and warmed to 40 °C. Probes are taken out from 4 °C and warmed for ~5 min at 40 °C. NFW is carefully aspirated and disposed of and then the RNAscope probe for the transcript of interest is added in sufficient volume to completely cover each sample. The slides or well plates are then placed inside the humidified chamber and incubated at 40 °C for 2 h. During incubation, AMP1-6 reagents are warmed to RT. Upon completion of probe hybridization, the samples are then washed three times (5 min each) in TR solution at room temp and then incubated with AMP1 for 30 min at 40 °C, followed by three more washes in TR solution at RT. AMP2 is then added to the samples and incubated at 40 °C for 15 min, followed by three washes in TR solution at RT. AMP3 is subsequently added to the samples and incubated at 40 °C for 30 min and then washed three times in TR solution at RT. The samples are then incubated in AMP4 for 15 min at 40 °C. For all of the above amplification reactions, the samples are to be placed in the humidifying chamber in the oven. After the AMP4 step, the samples are washed 3×5 min. in the TR solution at RT. During these last three washes, the humidifying chamber can be left to cool to room temperature. The tissues are then incubated in AMP5 for 25–30 min at RT in the cooled humidified chamber and then washed in TR solution three times at RT. AMP6 is subsequently added to the samples and incubated for 10–15 min at RT in the humidified chamber, followed by three washes with TR solution.

3.4.3 Color Reaction for Detection of a Single RNA of Interest by Brightfield or Fluorescence

Amplification probes from the RNAscope Red kits are conjugated with alkaline phosphatase (AP) which allows for the deposition of an AP-sensitive chromogen and subsequent visualization. The color reaction can be accomplished using either Fast RED-A and Fast RED-B solutions provided with the ACDBio Red kit or, if those reagents become limiting, Vector® Red Substrate Kit (or similar). If Fast RED solutions from the RNAscope kit are used, Fast RED-B is spun down, and RED working solution is prepared by combining Fast RED-B into Fast RED-A at a 1:75 ratio. An adequate amount of RED working solution is added to each sample to cover the entire tissue surface and incubated for 3–10 min at RT. Once the RED reaction is completed (*see* below), RED reagent is entirely removed and quickly washed at least 3×5 min with NFW. It must be noted that complete removal of the RED reagent by thorough washing is critical since remnant RED reagent can precipitate onto the tissue causing high background. If ImmPACT Vector Red Substrate is used for the chromogenic reaction, the manufacturer's protocol is followed. Briefly, one drop of Vector Red Reagent 1 and one drop of Vector Red Reagent 2 are added to 2.5 ml of Vector Red Diluent and mixed well by vortexing. The Red solution mixture is added to the samples and incubated for 3–10 min at RT. The color reaction is terminated by thorough washing (again at least 3×5 min) with NFW. One of the advantages to using these Red substrates is that they can be visualized under brightfield conditions, thus samples can be observed under a stereomicroscope to determine the optimal amount of time for the color development reaction. Areas where the RNA of interest is present should turn a deep red color (*see* Fig. 3). When background areas start to turn pink, the reaction should be terminated. The use of positive and negative control probes can also assist with this determination where incubation should persist at least until positive control samples turn red and the reaction should be terminated before the negative control samples begin to turn noticeably pink. The samples can be mounted at this point to visualize RNA expression or co-immunolabeled with antibodies (Subheading 3.4.4). Another advantage to these Red reagents is that they can be visualized not only under brightfield illumination, but also by fluorescence microscopy using TRITC or similar optical filtration (*see* Figs. 3 and 7). However, the Red reagents have broad excitation and emission spectra, emitting at wavelengths >560 nm and so may interfere with imaging other targets or immunostaining in far-red channels (e.g. Alexa Fluor 647). However, if the signal-to-noise ratio is strong for whatever is to be imaged in the 647 channel, this usually is not a problem. At this point, if one is not planning to co-label the samples using ICC, the sections or whole mounts can be coverslipped on slides using Fluorogel or other suitable mounting media. It must be noted that the red chromogen is soluble in alcohols and organic solvents; therefore, alcohol based mounting media must not be used when coverslipping the samples. If one wishes to co-label sections or tissue with antibodies, the tissue should be kept in NFW or buffered saline (PBS or TBS) at 4 °C until ready to proceed with ICC.

3.4.4 Immunostaining and Mounting

RNAscope 2.5 HD Assay-RED allows for the visualization of RNA expression via brightfield microscopy, but deposition of the red reagent can also be visualized via fluorescence using TRITC or similar optical filtration (excitation 365–560 nm/emission >560 nm). This provides the opportunity to co-immunolabel samples with antibodies that can be visualized with fluorescence microscopy in the blue (401/421), green (488/520 nm), and far-red (647/670 nm) channels and possibly an even greater number of channels depending on the spectral separation capabilities of one's microscope. As mentioned earlier, the color reaction is ended by thorough washing with NFW three or more times. After the last wash with NFW, the samples are additionally washed twice more with 1× PBS, followed by incubation with primary antibodies (diluted in PBS containing 0.5% triton X-100) overnight (~12 h) at 4 °C. Note that we have found that a blocking step is not necessary after the RNAscope procedure and may even hinder subsequent immunostaining or compromise tissue integrity, particularly for thin sections (<30 μm). After incubation in primary antibodies, the samples are then washed with 1× PBS three times (10 min each) at RT and incubated with secondary antibodies diluted in PBS containing 0.1% triton X-100 for 2–3 h at RT in the dark, followed by three washes with 1× PBS. The cell nuclei in the samples can be then counterstained with Hoechst/DAPI (diluted in PBS, 1:1500 for 25 min at RT) and then thoroughly

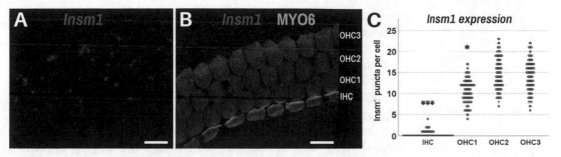

Fig. 8 Quantification of *Insm1* expression in different cochlear hair cell types using RNAscope. (**a**) A representative, single plane confocal image of RNAscope puncta using the probe against murine *Insm1* (magenta) in a P0 mouse cochlea. (**b**) Localization of *Insm1* RNAscope puncta in different cochlear hair cell types can be accomplished by co-labeling using an antibody against myosin VI (MYO6) protein (green). (**c**) A scatter plot of the number of *Insm1* puncta per hair cell counted from 120 hair cells per cochlea, for a total of 480 hair cells from four mice. These data demonstrate that the number of *Insm1* transcripts in any given hair cell at this timepoint can be variable. Additionally, *Insm1* counts using this method conform to previously published results that suggest much higher expression of *Insm1* in outer hair cells (OHC) compared to inner hair cells (IHC). Indeed, IHCs have little to no *Insm1* expression at P0 such that the mean number of puncta per IHC is significantly less than that of OHCs in any of the three rows (***$p \leq 0.001$, Bonferroni adjusted, as compared to OHC1, OHC2, or OHC3). Interestingly, what was not apparent from previously reported RNA-seq data is that OHCs in the most medial row (OHC1) also had significantly less mean *Insm1* puncta per cell than OHCs in the more lateral rows (*$p < 0.05$, Bonferroni adjusted, as compared to OHC2 or OHC3). Scale bars = 10 μm

washed three times (10 min each) with $1 \times$ PBS. Following the final wash, PBS is carefully removed and the samples can be mounted and coverslipped with any aqueous-based mounting media designed for use with fluorescent imaging (e.g. Fluorogel, Prolong gold, or simply 50% glycerol in PBS). Once mounted, the edges of the coverslip can be sealed with clear nail polish. Again, it should be noted that if either of the RED chromogens are used, the mounting media should not contain either ethanol or any other organic solvents as this can cause the red chromogen to dissolve back into solution, dissipating the signal into the mounting media. Once mounted, and the mounting media has set, samples can be imaged using fluorescence microscopy. When puncta are to be counted (*see* Fig. 8), we recommend imaging at higher magnification (e.g. $63\times$ objective) and, if possible, using confocal microscopy, particularly for whole-mounted samples or for thick sections as the enhanced resolution in *x*, *y*, and *z* planes makes it easier to resolve and count individual RNAscope labeled puncta.

3.5 Transcript Detection by Duplex Kit, Multiplex Fluorescent V2 Kit, and HiPlex Kit

RNAscope 2.5 HD duplex assay, RNAscope multiplex fluorescent v2 kit, and RNAscope HiPlex kit allow for the detection of 2, 4, and 12 or more transcripts in the same sample, respectively. These assays therefore grant an opportunity to simultaneously study co-regulation of multiple genes and profile or compare their expression patterns in the same sample. For the duplex kit, the protocol outlined above can be followed from tissue preparation all the way through the red color reaction with one important exception: a C2 probe must be diluted into the C1 probe at 1:50 prior to the 2 h hybridization step as outlined in the manufacturer's protocol. After AMP6 and the red color reaction have been performed, the manufacturer's protocol can be followed for AMP7 through to mounting and coverslipping with the recommendation that the TR solution be used for all washes in place of the commercially available wash buffer. It is important to note that the duplex kit, when performed this way will allow for visualization of two target RNA molecules of interest in red and green using brightfield microscopy. While the red reagent can also be visualized with fluorescence, the green chromogen is not fluorescent. When wishing to visualize multiple transcripts of interest via fluorescence, it is recommended that you use either the multiplex fluorescent V2 kit or the HiPlex kit depending on the number of targets needing to be visualized. However, if so desired, one can replace the green chromogen step with the following steps to utilize a green fluorescent reagent instead. After AMP10, wash the samples 3×5 min in TR solution. While the samples are washing, prepare the fluorescent green reagent solution by adding H_2O_2 to TSA buffer solution (ACDBio cat#322810) to obtain a 0.0015% H_2O_2 solution. Reconstitute Opal Dye 520 (Akoya Biosciences cat#FP1487001KT) in 75 µL of DMSO as recommended by the manufacturer. Dilute the reconstituted Opal dye 1:1000 in the 0.0015% H_2O_2 TSA solution. Remove the wash buffer from the samples and immediately immerse them in the opal dye solution and incubate at RT for 10 min. It is important to note that if following this modification, the Opal Dye will have fluorescence properties similar to green fluorescent protein, but will not be visible under the stereomicroscope or other brightfield microscope. While both the channels have relatively comparable sensitivity, it is recommended to use the probe against the lower expressing gene in the red channel as the FAST Red chromogen has high intensity fluorescence and minimal bleaching. After incubation in the Opal Dye/TSA solution, wash the samples 3×5 min in NFW and proceed to mounting and coverslipping or to subsequent Hoechst or immune-labeling.

If it is desirable to label more than two different transcripts in the same tissue, the multiplex fluorescent V2 kit should be used (for up to four different transcripts) or the HiPlex kit should be used (for more than four different targets and possibly up to 48 targets). For these fluorescent kits, tissue can be prepared as described above (Subheadings 3.1 to 3.4.1) and after this, the manufacturer's protocols can be followed with the exception that the wash buffer provided by the kit should be replaced by the TR solution. In the multiplex fluorescent V2 assay, the labeled probes are conjugated to a peroxidase and Opal 520, Opal 570, Opal 620, and Opal 690 tyramide dyes are deposited at the site of each probe amplification tree.

This V2 assay differs from the RNAscope fluorescent multiplex kit (ACDBio cat# 320850) in two key ways. First, the fluorescent multiplex kit uses labeled probes that have fluorescent molecules directly conjugated to them, whereas the V2 kit uses tyramide signal amplification giving greater signal and brighter fluorescence. Second, the V2 kit can be used to detect up to four different transcripts simultaneously, while the non-V2 fluorescent multiplex kit can only label up to three different targets. Here, we recommend to use the V2 kit, especially for transcripts of low to moderate abundance. Fluorescence microscopy can then be used to visualize the labeling of each transcript by four distinct filters including FITC (green), Cy3 (orange), Texas red (red), and Cy5 (red). For experiments where it is desirous to label more than four different transcripts in the same tissue, the RNAscope HiPlex assay was designed to detect up to 12 transcripts in a single sample (*see* Fig. 9), and with sturdy tissue samples it is possible to run through the entire protocol for up to 4 iterations, labeling as many as 48 different transcripts in a single sample. Unlike the V2 kit, the HiPlex kit uses labeled probes which are directly conjugated to fluorescent dyes. This allows for dyes in four different channels (Alexa 488, ATTO 540, ATTO 647, and Alexa 750) to be added to different

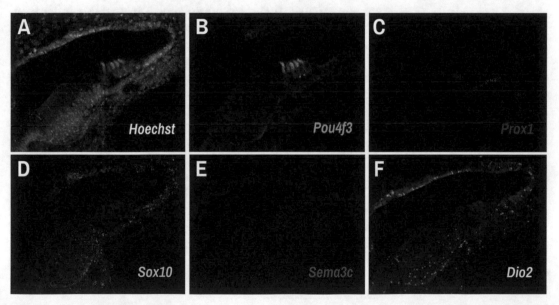

Fig. 9 RNAscope HiPlex assays can be used to label more than four targets in the same tissue. (**a**) A single vibratome section of a P0 mouse cochlea was stained with Hoechst dye (white) and labeled for six different target mRNAs using RNAscope HiPlex. Transcripts of interest were labeled using probes against (**b**) *Pou4f3* (green) which is enriched in hair cells, (**c**) *Prox1* (red) which is enriched in pillar and Deiters cells, (**d**) *Sox10* (blue) which is enriched in many nonsensory cells around the *scala media*, (**e**) *Sema3c* (magenta) which is moderately expressed in the greater epithelial ridge and in the hair cells, (**f**) *Dio2* which is moderately expressed in many different cells around the *scala media*, and *Mbp* (not shown) which is enriched in glial cells medial to the area shown in the images

RNAscope probe-preamplifier trees simultaneously based on sequence differences. After the first four transcripts are so labeled, the fluorescent tags are cleaved and washed out and another round of four labeled probes can be hybridized and imaged and then the process can be repeated. Samples are isolated and subjected to pretreatment steps as described above and processed using the manufacturer's protocols, though, again, the TR solution is recommended as wash buffer. Critically, the samples must be counterstained with Hoechst or DAPI so that image registration software can align the images from the different rounds using the nuclear labeling to properly overlap the tissue images. ACDBio can provide HiPlex Image Registration software for this purpose, though, alignment can be accomplished manually and performed using ImageJ or other similar software packages if so desired. It is important to note when using the HiPlex method that the successive rounds of fluorophore cleaving, and in particular, multiple rounds of hybridization for more than 12 targets, can hasten tissue degradation and may generate increasing levels of background which can make it harder to detect the lower expressing targets. Therefore, it is advisable to use fresh tissue if possible and to assign the order of T1 to T12 probes for detecting the lowest to the highest expressing transcripts to help circumvent this problem. Though, for inner ear tissues in particular, the green fluorescence channel can suffer from higher levels of endogenous autofluorescence [38, 39]; therefore, it is advised to use a probe against a high expressing target in the green channel. While HiPlex assay and the associated numbers of probes needed can be costly, the labeling workflow takes a similar amount of time or even less time than the single RED or multiplex fluorescent kits and reduces the number of animals and samples needed when investigating multiple genes of interest, thus leading to an overall time savings for these types of experiments. The HiPlex assay therefore offers a suitable platform for the study of multiple transcripts in the same sample which may make it amenable to the validation and extension of RNA-seq or scRNA-seq data. Similarly, it can be useful for experiments where animals or samples may be limited, such as with complex genetic models involving multiple knock-in, knock-out, or transgenic manipulations. Also, it may be possible to immunostain the samples after completing the HiPlex assay, though our lab has not yet attempted this.

Important technical notes

1. It is not recommended that tissue samples be dried at any point during the RNAscope assay. While other RNAscope or ISH protocols may call for this, our experience has not yielded robust results whenever sample tissues were allowed to dry out completely during any step of the procedure.

2. The target probes (and AMP probes) must be warmed at 40 °C for 5 min (or possibly longer) to ensure dissolving of any

precipitate. It is recommended that upon first warm-up, the probes should be aliquoted in sterile, nuclease free tubes to avoid multiple cycles of cooling and reheating.

3. Over and under fixing of samples can interfere with the outcome of the technique. Under-fixed samples will likely suffer from accelerated RNA degradation and probe penetrance can be hindered by overfixation. Therefore, tissue fixation strategies may need to be optimized for different sample types.

4. For best results, it is generally advisable to make the TR buffer solution and any other buffer solutions that require dilution (HiPlex buffers, etc.) as fresh as possible when performing RNAscope. Do not use TR solution if it has been one week or longer since it was diluted from the stock.

5. It is also recommended, when working with cryosections, not to use coplin jars for any of the washing steps throughout the protocol, since the tissue sections may lose adherence and fall off of the slides.

6. It is recommended that samples be kept covered as much as possible (lids on well plates or slides placed into humidifying chamber) for all steps of the procedure to minimize the accumulation of dust or other particles which may cause spurious signal.

7. Whenever samples are being stored (e.g. overnight at 4 °C) they should be immersed in sufficient volumes of solution so as not to dry out. In the case of cryosections on charged slides, it is generally best to store these in humidifying chambers.

4 Discussion

Mammalian inner ear tissues, including cochlear and vestibular epithelia, are composed of multiple heterogenous cell types such as HCs, SCs, fibrocytes, mesenchymal cells, neurons, glia, etc. Despite the elegant organization and distinct compartmentalization of the tissues of the inner ear, much is still unknown about the development and functions of many of these different cell types. In the past decade, NGS approaches like scRNA-seq have been increasingly used to gain insight into the transcriptomic profiles of these various cell types at different developmental stages, postnatal ages, and under various experimental and pathological conditions. Subsequently, RNAscope is emerging as a highly useful, complementary technique to validate spatiotemporal expression and regulation of many transcripts of interest in situ. Here we demonstrate that this technique can be applied to detect transcripts in both the cochlea and the vestibular organs from mice of different age groups starting from embryonic stages to adulthood in various types of sample preparations including whole mount, fixed frozen sections and vibratome sections.

Data from our lab along with other publications clearly demonstrate that RNAscope ISH can be accompanied by immunohistochemistry for protein detection and therefore can aid in identifying specific cellular structures and subcellular localization of RNA [12, 27, 40, 41]. For example, here we show that RNAscope samples can be labeled with antibodies against myosin VIIa, HuD, and connexin29, which selectively label and reveal the morphology of HCs, neuronal soma, and myelinating glial cells. Also, though not shown here, our lab has utilized other antibodies such as those raised against SOX2 and neurofilament heavy chain to label cochlear SCs and spiral ganglion neurons (*see* Table 1). However, sometimes, the RNAscope procedure can interfere with the efficacy of certain antibodies for immunocytochemical labeling (Table 1). It is likely in these cases that the protease treatment and/or the target unmasking step likely degrade or alter the structures of certain proteins to a degree where they are no longer optimally recognized by the antibodies. Furthermore, inner ear tissues, particularly when aldehyde fixed, tend to exhibit fairly bright autofluorescence in the green channel, and the AP-sensitive red chromogens described herein have fluorescence emission spectra that extend into the far-red channel [38, 39]. Due to the potential issues with antibody labeling, and with fluorescence imaging in the remaining available channels, we recommend using either pre-validated antibodies (Table 1) or using antibodies that are robust (i.e. work well at low concentrations), that target proteins that are highly expressed, and/or that are known to work following sodium citrate and heat antigen retrieval procedures. If a suitable antibody cannot be readily found to highlight the location or structures desired, RNAscope is also compatible with fluorescent proteins expressed by transgenic animal models which can also allow researchers to identify cell types or subcellular locations that are expressing the RNA of interest [12]. However, it should be noted that in our experience the intensity of endogenous fluorescent reporters does appear to be somewhat diminished after the RNAscope procedure and so again using a reporter with high levels of expression, or intense fluorescence, is beneficial.

When possible, RNAscope probes are generally designed to span intron-exon boarders, which allows probes to be selective for mature mRNA over pre-mRNA and enhances overall specificity for target transcripts. In this chapter we show practical application of RNAscope probes to detect mRNA molecules with robust accuracy but have made no attempts to label splice variants or microRNA molecules. Though certain splice variants for larger gene products can be identified using RNAscope, the detection of single exons or shorter sequences that are specific to splice variants may require custom probe generation via ACDBio's BaseScope technology. Similarly, the detection of even smaller RNAs such as microRNAs can be accomplished with "miRNAscope" probes and reagents. Though these probes and reagents differ in some ways from standard RNAscope, many of the principles and even procedural steps of these approaches are similar. While a nuanced discussion of the key differences or the practical application of BaseScope and miRNAscope in the inner ear is outside the purview of this chapter, general information regarding BaseScope and miRNAscope, as well as manufacturer provided protocols, can be found on ACD's website (https://acdbio.com/basescope%E2%84%A2-assays and https://acdbio.com/mirnascope-assay-red-overview).

In order to maximize success using an RNAscope approach, several factors must be taken into consideration. Positive and negative control probes must be chosen carefully in order to complement your experimental gene of interest. In general, housekeeping genes or genes with known expression patterns that have been pre-validated (e.g. *Insm1* or other probes shown here or published elsewhere) are sufficient for use as positive controls. However, it is ideal to use positive control probes that recognize RNA molecules with similar levels of expression as the target. Failure to match expression levels could lead to misinterpretation of mRNA expression data (i.e. false positives or false negatives) if the gene of interest is not as robustly expressed as the chosen positive control. To alleviate this, ACDBio offers multiple different types of positive control probes in various species. For mice, three readily available positive controls that target RNA with differing levels of expression are probes against: Ubiquitin C for medium to high expression, Cyclophilin B for medium expression, and DNA-directed RNA polymerase II subunit RPB1 for low expressing gene targets. As a negative control, researchers can again choose to utilize probes against an RNA that is known to be absent from the tissue of interest or can pay to have custom sense or randomized probes generated based on the target sequence. However, this latter option can become costly, and as such, ACDBio offers at a much more minimal cost, negative control probes such as probes that target the bacterial DapB gene which should not be expressed in vertebrate animal tissues. Another important consideration for adopting RNAscope can be the costs. While each vial of RNAscope target probes available from the catalog is similar in cost to commercial antibodies, having a custom probe created for a target not already in the catalog can add significantly to the cost. Additionally, the kits containing the amplification and other reagents add to the cost of performing the experiments, though the kit reagents can be stored at 4 °C for many months and used for a number of experiments. Also, similar to approaches that have been used with antibodies, some cost savings can be attained through reuse of probes for multiple rounds of labeling or even dilution. Our lab has found that most probes can be diluted 1:1 in NFW or in probe diluent (cat#300041) and still provide robust labeling. However, we recommend that if the experiment requires absolute mRNA quantification that it may be preferable to neither dilute nor reuse the reagents.

One of the key advantages of RNAscope over traditional ISH is that it offers the possibility of absolute quantification of RNA molecules since there is a one-to-one ratio of visible puncta and labeled transcripts. However, this approach to RNA quantification is optimal for transcripts with low and medium levels of expression. When transcripts of interest are highly expressed, or too densely packed in one area, it can be difficult to resolve individual puncta that are separated by diffraction limited distances of 200 nm or less. While this can be overcome to some extent with super-resolution microscopy and/or deconvolution algorithms, not all labs may have access to or expertise with such technology. That said, even for low to mid-level expressing transcripts, in order to image for RNAscope quantification, researchers should use microscopy methods which better control the depth of field than traditional widefield microscopy techniques. Thus approaches such as confocal microscopy, light sheet fluorescent microscopy, structured illumination microscopy, or other similar approaches are recommended. Visualization is best done with higher power objectives (40× or greater) in order to obtain sufficient distinction of individual puncta [38, 42, 43]. Here we have demonstrated successful quantification of Insm1 transcripts in cochlear hair cells from mice at postnatal day 0 using confocal microscopy and a 63× objective (Fig. 8). Consistent with what has been previously published [26], our quantification confirms that Insm1 expression is highly enriched in outer HCs compared to inner HCs and even refine or extend this finding to demonstrate that the lateral two rows of outer HCs contain significantly more Insm1 puncta than the medial row of outer HCs suggesting a possible mediolateral gradient of expression (Fig. 8). This example highlights the improved resolution of RNA quantification that may be obtained from this method compared to standard RNA-seq approaches.

Given that puncta size can vary based on the number of probes that bind to a cell, researchers should take note that a greater numbers of probes binding to one transcript will result in more intense fluorescence (or a larger diameter chromogen deposit) that is not necessarily indicative of multiple transcripts or higher expression levels. Again, however, researchers should remain aware that high expression levels for a gene of interest can result in overlapping of fluorescent signals and due to the variability of puncta size, this provides another reason why RNAscope quantification can be difficult for abundant or densely packed transcripts. The variability in punctal size could also mean that quantification based on fluorescence intensity or mean gray values could result in the introduction of bias or added variability. However, if the variability can be tested or assumed to be similar across comparison groups, then relative quantification of RNA expression can be achieved using average intensity measures and several reports have sought to optimize such methods for RNAscope [12, 13, 18, 40]. Still, when resolution allows, counting individual puncta will give more accurate and unbiased expression quantification, particularly if investigators are blinded. Such counting can be done manually if desired; however, there are also several options for automated counting. Freeware such as CellProfiler and ImageJ have built-in counting features and support the use of plugins that allow for automation and more specialized computing. For instance, Stapel et al. developed an ImageJ plugin that is specialized for single-molecule FISH quantification [44]. FISH quantification plugins are also available for applications such as HALO through the work of Jolly et al. [18]. Trcek et al. demonstrated a powerful automated quantification software for counting single mRNA molecule by spot detection approach in drosophila, whereas Maynard et al. developed another automated quantification approach called "dotdot" which can quantify mRNA transcript in multiplexed smFISH assay platform though it requires users to have MATLAB [43, 45].

A potential limitation of this technique is that some aspects of the technique are not transparent. The commercial supplier, Advanced Cell Diagnostics (ACD), makes all of the RNAscope probes and the accompanying kits; and the sequences that are used to create the linker and tail regions of the "Z" shaped probes, as well as the sequences for the amplifier and labeled probes, are not disclosed. While other commercial options exist (e.g. https://www.molecularinstruments.com/hcr-v3 and https://www.pixelbiotech.com/) these also generally require probes and reagents with proprietary aspects. Though for all of these options, the sequence to be hybridized is known, pretty much all of the other sequences are not disclosed to maintain the proprietary nature of each company's technology. Another potential caveat to the adoption of RNAscope in a given lab is the cost of kits, reagents, and probes, particularly if custom probes are required or a large number of transcripts are to be investigated. However, this protocol obviates the need to purchase specialized ovens or other equipment, thereby reducing some of the costs, and prices for RNAscope probes have reduced since its inception and the reagents may perhaps continue to become more affordable in the future. Alternatively, labs that are looking for other similar methods may find alternative single-molecule FISH (smFISH) techniques that can be largely carried out "in-house" without the need to purchase proprietary reagents [46, 47]. Using these or other published smFISH techniques can be more cost effective, but may sacrifice multiplexing abilities and may require more optimization or troubleshooting compared to pre-validated reagents and protocols. Still, various methods of smFISH have been published and include single-molecule inexpensive FISH [48], single-molecule hybridized chain reaction [49], and enzymatic single-molecule FISH [50].

5   Conclusions

RNAscope provides a set of protocols that are useful for the in situ detection of 1–12 transcripts of interest, and possibly as many as 48 or more, in either fresh frozen or in fixed tissues. It can allow for quantification and for easy co-labeling with immunostaining or endogenous reporters and has already been shown to be useful in the validation of scRNA-seq data in the mouse inner ear. While RNAscope is generally kit-like and allows for easy adoption into labs without much experience in ISH, this chapter and other previous reports, for example, Kersigo et al. [12], demonstrate that some modification of the manufacturer's protocols may be necessary to optimize results. Potential caveats include the cost and proprietary nature of some of the technology, though alternative methods that may reduce costs or provide greater autonomy are available.

References

1. Crick F (1970) Central dogma of molecular biology. Nature 227(5258):561–563. https://doi.org/10.1038/227561a0

2. Xie F, Timme KA, Wood JR (2018) Using single molecule mRNA fluorescent in situ hybridization (RNA-FISH) to quantify mRNAs in individual murine oocytes and embryos. Sci Rep:1–11. https://doi.org/10.1038/s41598-018-26345-0

3. Soares RJ, Maglieri G, Gutschner T et al (2018) Evaluation of fluorescence in situ hybridization techniques to study long non-coding RNA expression in cultured cells. Nucleic Acids Res 46(1):e4. https://doi.org/10.1093/nar/gkx946

4. Wang F, Flanagan J, Su N et al (2012) RNAscope : a novel in situ. RNA analysis platform for formalin-fixed, paraffin-embedded tissues. J Mol Diagn 4(1):22–29. https://doi.org/10.1016/j.jmoldx.2011.08.002

5. Yao RW, Wang Y, Chen LL (2019) Cellular functions of long noncoding RNAs. Nat Cell Biol 21(5):542–551. https://doi.org/10.1038/s41556-019-0311-8

6. Gebert LFR, MacRae IJ (2019) Regulation of microRNA function in animals. Nat Rev Mol Cell Biol 20(1):21–37. https://doi.org/10.1038/s41580-018-0045-7

7. Wang Z, Gerstein M, Snyder M (2009) RNA-Seq: a revolutionary tool for transcriptomics. Nat Rev Genet 10(1):57–63. https://doi.org/10.1038/nrg2484

8. Lee JH, Daugharthy ER, Scheiman J et al (2015) Fluorescent in situ sequencing (FIS-SEQ) of RNA for gene expression profiling in intact cells and tissues. Nat Protoc 10(3):442–458. https://doi.org/10.1038/nprot.2014.191

9. Maniatis S, Äijö T, Vickovic S, Braine C et al (2019) Spatiotemporal dynamics of molecular pathology in amyotrophic lateral sclerosis. Science 364(6435):89–93. https://doi.org/10.1126/science.aav9776

10. Thrane K, Eriksson H, Maaskola J et al (2018) Spatially resolved transcriptomics enables dissection of genetic heterogeneity in stage III cutaneous malignant melanoma. Cancer Res 78(20):5970–5979. https://doi.org/10.1158/0008-5472.CAN-18-0747

11. Judice TN, Nelson NC, Beisel CL et al (2002) Cochlear whole mount in situ hybridization: identification of longitudinal and radial gradients. Brain Res Brain Res Protoc 9(1):65–76. https://doi.org/10.1016/s1385-299x(01)00138-6

12. Kersigo J, Pan N, Lederman JD et al (2018) A RNAscope whole mount approach that can be combined with immunofluorescence to quantify differential distribution of mRNA. Cell Tissue Res 374(2):251–262. https://doi.org/10.1007/s00441-018-2864-4

13. Salehi P, Nelson CN, Chen Y et al (2018) Detection of single mRNAs in individual cells of the auditory system. Hear Res 367:88–96. https://doi.org/10.1016/j.heares.2018.07.008

14. Gold EM, Vasilevko V, Hasselmann J et al (2018) Repeated mild closed head injuries induce long-term white matter pathology and neuronal loss that are correlated with

behavioral deficits. ASN Neuro 10: 1759091418781921

15. John HA, Birnstiel ML, Jones KW (1969) RNA-DNA hybrids at the cytological level. Nature 223(5206):582–587. https://doi.org/10.1038/223582a0

16. Gall JG, Pardue ML (1969) Formation and detection of RNA-DNA hybrid molecules in cytological preparations. Proc Natl Acad Sci 63(2):378–383. https://doi.org/10.1073/pnas.63.2.378

17. Xia L, Yin S, Wang J (2012) Inner ear gene transfection in neonatal mice using adeno-associated viral vector: a comparison of two approaches. PLoS One 7(8):e43218. https://doi.org/10.1371/journal.pone.0043218

18. Jolly S, Lang V, Koelzer VH et al (2019) Single-cell quantification of mRNA expression in the human brain. Sci Rep 9(1):12353. https://doi.org/10.1038/s41598-019-48787-w

19. Chan S, Filézac de L'Etang A et al (2018) A method for manual and automated multiplex RNAscope in situ hybridization and immuno-cytochemistry on cytospin samples. PLoS One 13(11):e0207619. https://doi.org/10.1371/journal.pone.0207619

20. Yin VP (2018) In Situ detection of MicroRNA expression with RNAscope probes. Methods Mol Biol 1649:197–208. https://doi.org/10.1007/978-1-4939-7213-5_13

21. Grandi FC, De Tomassi L, Mustapha M (2020) Single-cell RNA analysis of type I spiral ganglion neurons reveals a Lmx1a population in the cochlea. Front Mol Neurosci 13:83. https://doi.org/10.3389/fnmol.2020.00083

22. Roccio M, Perny M, Ealy M et al (2018) Molecular characterization and prospective isolation of human fetal cochlear hair cell progenitors. Nat Commun 9(1):1–14. https://doi.org/10.1038/s41467-018-06334-7

23. Lingle CJ, Martinez-Espinosa PL, Yang-Hood A et al (2019) LRRC52 regulates BK channel function and localization in mouse cochlear inner hair cells. Proc Natl Acad Sci U S A 116 (37):18397–18403. https://doi.org/10.1073/pnas.1907065116

24. Yu KS, Frumm SM, Park JS et al (2019) Development of the mouse and human cochlea at single cell resolution. Cell Commun Signal 11:78. https://doi.org/10.1186/1478-811X-11-78

25. Sherrill HE, Jean P, Driver EC et al (2019) Pou4f1 defines a subgroup of Type I spiral ganglion neurons and is necessary for normal inner hair cell presynaptic Ca2+ signaling. J

Neurosci 39(27):5284–5298. https://doi.org/10.1523/JNEUROSCI.2728-18.2019

26. Wiwatpanit T, Lorenzen SM, Cantú JA et al (2018) Trans-differentiation of outer hair cells into inner hair cells in the absence of INSM1. Nature 563(7733):691–695. https://doi.org/10.1038/s41586-018-0570-8

27. Shrestha BR, Chia C, Wu L et al (2018) Sensory neuron diversity in the inner ear is shaped by activity. Cell 174(5):1229–1246. https://doi.org/10.1016/j.cell.2018.07.007

28. Oivanen M, Kuusela S, Lönnberg H (1998) Kinetics and mechanisms for the cleavage and isomerization of the phosphodiester bonds of RNA by bronsted acids and bases. Chem Rev 98(3):961–990. https://doi.org/10.1021/cr960425x

29. Järvinen P, Oivanen M, Lönnberg H (1991) Interconversion and phosphoester hydrolysis of 2′,5′- and 3′,5′-dinucleoside monophosphates: kinetics and mechanisms. J Org Chem 56(18):5396–5401. https://doi.org/10.1021/jo00018a037

30. Taylor CR, Shi SR, Cote RJ (1996) Antigen retrieval for immunohistochemistry status and need for greater standardization. Applied Immunohistochemistry and Molecular Morphology. Appl Immunohistochem 4 (3):144–166

31. Urieli-Shoval S, Meek RL, Hanson RH et al (1992) Preservation of RNA for in situ hybridization: Carnoy's versus formaldehyde fixation. J Histochem Cytochem 40 (12):1879–1885. https://doi.org/10.1177/40.12.1280665

32. Yan F, Wu X, Crawford M et al (1992) The search for an optimal DNA, RNA, and protein detection by in situ hybridization, immunohistochemistry, and solution-based methods. Methods 52(4):281–286. https://doi.org/10.1016/j.ymeth.2010.09.005

33. Ryan AF, Watts AG, Simmons DM (1991) Preservation of mRNA during in situ hybridization in the cochlea. Hear Res 56 (1-2):148–152. https://doi.org/10.1016/0378-5955(91)90164-5

34. Cunningham LL (2006) The adult mouse utricle as an in vitro preparation for studies of ototoxic-drug-induced sensory hair cell death. Brain Res 1091(1):277–281. https://doi.org/10.1016/j.brainres.2006.01.128

35. Montgomery SC, Cox BC (2016) Whole mount dissection and immunofluorescence of the adult mouse cochlea. J Vis Exp 107:53561. https://doi.org/10.3791/53561

36. Driver EC, Kelley MW (2010) Transfection of mouse cochlear explants by electroporation. Curr Protoc Neuroscipr Chapter 4:Unit 4.34.1–Unit 4.34.10. https://doi.org/10.1002/0471142301.ns0434s51

37. Ogier JM, Burt RA, Drury HR et al (2019) Organotypic culture of neonatal murine inner ear explants. Front Cell Neurosci 13:170. https://doi.org/10.3389/fncel.2019.00170

38. Bode J, Kruwel T, Tews B (2017) Light sheet fluorescence microscopy combined with optical cleaning methods as a novel imaging tool in biomedical research. EMJ Innov 1(1):67–74

39. Carraro M, Paroutis P, Woodside M, Harrison R V (2015) Improved imaging of cleared samples with ZEISS Lightsheet Z.1: refractive index on demand. Zeiss Appl Note:1–7

40. Sun S, Babola T, Pregernig G et al (2018) Hair cell mechanotransduction regulates spontaneous activity and spiral ganglion subtype specification in the auditory system. Cell 174 (5):1247–1263.e15. https://doi.org/10.1016/j.cell.2018.07.008

41. Hoa M, Olszewski R, Li X et al (2020) Characterizing adult cochlear supporting cell transcriptional diversity using single-cell RNA-Seq: validation in the adult mouse and translational implications for the adult human cochlea. Front Mol Neurosci 13:13. https://doi.org/10.3389/fnmol.2020.00013

42. Gross-Thebing T, Paksa A, Raz E (2014) Simultaneous high-resolution detection of multiple transcripts combined with localization of proteins in whole-mount embryos. BMC Biol 12:55. https://doi.org/10.1186/s12915-014-0055-7

43. Maynard KR, Tippani M, Takahashi Y et al (2020) dotdotdot: an automated approach to quantify multiplex single molecule fluorescent in situ hybridization (smFISH) images in complex tissues. Nucleic Acid Res 48(11):e66. https://doi.org/10.1093/nar/gkaa312

44. Carine Stapel L, Lombardot B, Broaddus C et al (2016) Automated detection and quantification of single RNAs at cellular resolution in zebrafish embryos. Development 143 (3):540–546. https://doi.org/10.1242/dev.128918

45. Trcek T, Lionnet T, Shroff H, Lehmann R (2017) mRNA quantification using single-molecule FISH in Drosophila embryos. Nat Protoc 12(7):1326–1348. https://doi.org/10.1038/nprot.2017.030

46. Femino AM, Fay FS, Fogarty K, Singer RH (1998) Visualization of single RNA transcripts in situ. Science 280(5363):585–590. https://doi.org/10.1126/science.280.5363.585

47. Raj A, van den Bogaard P, Rifkin SA et al (2008) Imaging individual mRNA molecules using multiple singly labeled probes. Nat Methods 5(10):877–879. https://doi.org/10.1038/nmeth.1253

48. Tsanov N, Samacoits A, Chouaib R et al (2016) SmiFISH and FISH-quant—a flexible single RNA detection approach with super-resolution capability. Nucleic Acids Res 44(22):e165. https://doi.org/10.1093/nar/gkw784

49. Shah S, Lubeck E, Schwarzkopf M et al (2016) Single-molecule RNA detection at depth by hybridization chain reaction and tissue hydrogel embedding and clearing. Development 143 (15):2862–2867. https://doi.org/10.1242/dev.138560

50. Gaspar I, Wippich F, Ephrussi A (2017) Enzymatic production of single-molecule FISH and RNA capture probes. RNA 23 (10):1582–1591. https://doi.org/10.1261/rna.061184.117

Chapter 7

A Manual Technique for Isolation and Single-Cell RNA Sequencing Analysis of Cochlear Hair Cells and Supporting Cells

Cody West, Paul T. Ranum, Ryotaro Omichi, Yoichiro Iwasa, Miles J. Klimara, Daniel Walls, Jin-Young Koh, and Richard J. H. Smith

Abstract

Single-cell RNA sequencing is a powerful tool that can be used to characterize the transcriptional profile of low-abundance cell types in many organ systems; however, its application to the inner ear is especially challenging. Access to the membranous cochlea is hampered by the bony labyrinth, and the membranous labyrinth itself offers its own technical challenges due to tissue sparsity and difficulty in dissociating and isolating individual cells. Herein, we present a manual method to isolate individual inner hair cells, outer hair cells and Deiters' cells from the murine cochlea at any postnatal time point. We describe the technique and downstream bioinformatic analysis and, by way of example, show how both short- and long-read single-cell RNA sequencing can be leveraged to profile transcript abundance and structure. This method has afforded us novel insights into the expression profiles of these specific cell types, revealing an unappreciated complexity in isoform variety in deafness-associated genes. Further application of this methodology may provide a more refined view of transcription in the organ of Corti, thereby improving our understanding of the biology of hearing and deafness.

Key words Single cell RNA-seq, Manual isolation, Outer hair cell, Inner hair cell, Deiters' cell

1 Introduction

Hearing loss is the most common sensory deficit, with a prevalence that increases with age from 0.2% among newborns to >45% in adults over the age of 65 years [1, 2]. The impact of hearing loss is exceptionally broad, with debilitating effects on social, financial, and mental health. Whether caused by environmental insults or

The original version of this chapter was revised. The correction to this chapter is available at https://doi.org/10.1007/978-1-0716-2022-9_18

Supplementary Information The online version of this chapter (https://doi.org/10.1007/978-1-0716-2022-9_7) contains supplementary material, which is available to authorized users.

Andrew K. Groves (ed.), *Developmental, Physiological, and Functional Neurobiology of the Inner Ear*, Neuromethods, vol. 176, https://doi.org/10.1007/978-1-0716-2022-9_7,

genetic mutations, loss in auditory acuity reflects damage to the critical sensory and supporting cells of the cochlea.

Historically, the greatest advancements to our understanding of the auditory system have developed within the field of anatomy, although in the past three to four decades, scientists around the world have focused on the genetics of hearing loss to provide unparalleled subcellular insights into the biology of hearing and deafness. As a result of these advancements, we are now able to diagnose gene-specific forms of hearing loss, a capability that has changed the clinical diagnosis and evaluation of the person with hearing loss. These discoveries will usher in a new generation of treatments for hearing loss that focuses on targeted gene therapies, the first class of therapies for hearing loss with the potential to correct hearing dysfunction at its origin.

Both genetic diagnosis and the design of targeted gene therapies for hearing loss rely on precise knowledge of cell types and genes involved in hearing function. If, for example, annotation is inaccurate for a gene product and specific exons unique to inner hair cells (IHCs) or outer hair cells (OHCs) are not recognized, failing to screen these exons for deafness-causing variants can compromise genetic testing. Similarly, gene therapy may be unsuccessful if isoforms essential for auditory function are not recognized.

Single-cell-based transcriptomic studies can address these challenges and exquisitely refine our understanding of the pathways, genes, and molecular mechanisms that underlie normal and abnormal auditory function in the different cell types within the cochlea. Cell isolation, however, is difficult. Not only is the cochlea relatively inaccessible due to its position in the temporal bone, but the various cell types are rare. A single murine cochlea is comprised of approximately 415,000 cells that fall into potentially 40 different cell types, with IHCs and OHCs representing only ~0.19% and ~0.59%, respectively, of this cell count. The consequence of both inaccessibility and diversity has been the exclusion of cochlear tissue from datasets that make up most gene annotation databases.

Undeterred by these challenges, over the past two decades, researchers have regularly undertaken transcriptomic studies of cochlear tissues [3]. Early work utilized microarray technology, which at the time represented a transformative advance by enabling simultaneous expression profiling of large numbers of genes. One limitation, however, was the reliance on known transcript sequences; discovery of new transcripts and transcript features was not possible. RNA sequencing (RNA-seq) overcomes this shortcoming as global transcriptome profiling is not limited by probe design and prior knowledge. However, microarray and early RNA-seq technologies required large amounts of complementary DNA (cDNA) as input, often requiring researchers to pool many whole cochleae to obtain sufficient starting material. With pooling, the ability to resolve the signal from low abundance cell types like IHCs and OHCs is lost.

Low-input reverse transcription technology has made it possible to generate transcriptome profiles from small amounts of tissue and even single cells, making tissue- and cell-type-specific profiling of the membranous labyrinth possible [4, 5]. Several recent studies have utilized this technology to gain insights into cochlear development, cell fate determination, and for the first time, to define the transcriptomes of IHCs and OHCs [6, 7]. A variety of methods have been adopted for cell-type-specific profiling of cochlear cells, and multiple cell-type-specific transcriptomic datasets are now publicly available on the University of Maryland's umgEAR.org, the University of Iowa's morlscrnaseq.org, and Harvard's Shield interactive databases. The majority of these dataset profile expression at embryonic and early postnatal timepoints in mice. Additionally, most datasets reflect the cumulative expression profiles of many cells with only a limited number providing single-cell resolution.

In this chapter, we describe the methodology we use for the manual isolation of single cells from the cochlea. The method is based on the simple isolation technique described by Liu et al. [6]. Pulled glass micropipettes are used to capture individual cells from the membranous labyrinth, which are then preserved and prepared for single-cell RNA-seq (scRNA-seq). Differential expression analysis can be completed on any cell at any time point, providing unprecedented ability to identify cell-type-defining genes and pathways. Downstream analysis by both short- and long-read sequencing permits expression profiling of specific cochlear cell types and accurate assessment of transcript abundance and structure.

2 Materials

Materials and tools necessary for preparing and executing manual single cell isolation are described here in detail.

2.1 Equipment Required for the Described Surgical Procedure

Tank with compressed nitrogen and CO_2:

- Compressed nitrogen tank with regulator and tubing.
- Compressed CO_2 tank with regulator and tubing.

    Stereotaxic instruments:

- Leica M165FC Dissecting Scope (Leica, Model M165FC) (Fig. 1).
- Leica DMI 3000B Compound Light Microscope (Leica, Model DMI 3000B) (Fig. 2).

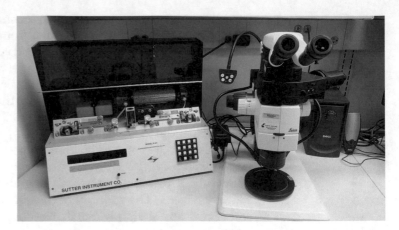

Fig. 1 Dissection and isolation equipment. Glass micropipettes were crafted using a Sutter Instrument Flaming/Brown Micropipette Puller Model P-97 (left). To access the cochlear membranous labyrinth, the apical portion of the bony labyrinth was visualized under a Leica Microsystems Dissecting Microscope Model M165FC (right)

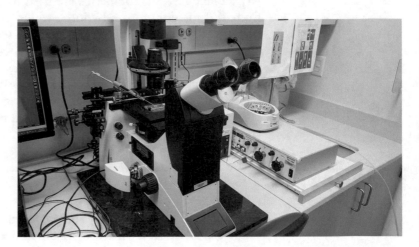

Fig. 2 Single-cell isolation equipment. Cell isolation and harvest were completed using a Leica Microsystems DMI 3000B Compound Light Microscope Model DMI 3000B with 20× and 40× objectives (left). A Warner Instruments Pico-Liter Injector, Model PLI-100A with pressurized nitrogen was used for cell capture (right)

General surgical supplies (Fig. 3):

- Surgical Scissors—Blunt (Fine Science Tools, cat. No. 14000-14).
- Fine Scissors—Sharp (Fine Science Tools, cat. no. 14061-11).
- Student Standard Pattern Forceps (Fine Science Tools, cat. no. 91100-12).
- Dumont #5 Forceps (Fine Science Tools, cat. No. 11252-40).
- Vannas Spring Scissors (Fine Science Tools, cat. no. 15001-8).

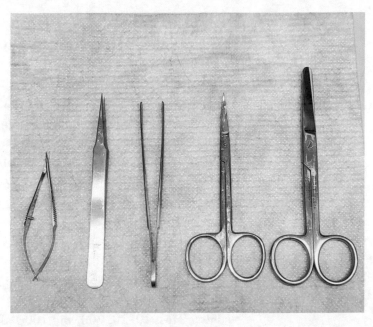

Fig. 3 General surgical supplies. From left to right: Vannas spring scissors, Dumont #5 forceps, student standard pattern forceps, fine scissors, and surgical scissors

Additional instruments and equipment:

- Flaming/Brown Micropipette Puller (Sutter Instrument, Model P-97) (Fig. 1).
- Pico-Liter Injector (Warner Instruments, Model PLI-100A) (Fig. 2).
- Mini Centrifuge (Benchmark Scientific, Model C1012).
- Veriti Thermal cycler (Applied Biosystems, cat. no. 4375786).
- Vortex Genie 2 (Scientific Industries, Model G-560).
- Magnetic stand 96 (Ambion, cat. No. AM10027).
- Borosilicate Glass (Sutter Instruments, cat. no. BF150-86-10).
- P-97 2.0 mm square box filament, 1.5 mm wide (Harvard Apparatus, cat. no. FB215B).
- Super PAP Pen (Polysciences, cat. no. 24230-1).
- Superfrost Plus Microscope Slides (Fisher Scientific, cat. no. 12-550-15).
- TempAssure 0.2 mL PCR 8-Tube Strips (USA scientific, cat. No. 19413).
- DNA LoBind Tube 1.5 mL (Eppendorf, cat. no. 022431021).
- 35-mm TC-treated Easy-Grip Style Cell Culture Dish (Falcon, cat. No. 353001).

Reagents and kits:

- High-sensitivity DNA kit (Agilent Technologies, cat. no. 5067-4626).
- SMART-Seq HT Kit (Takara Bio, cat. No. 634438).
- Accumax (Sigma-Aldrich cat. no. AM-105).
- Dulbecco's phosphate-buffered saline (DPBS) ($1\times$), no calcium, no magnesium (Gibco, cat. no. 14190-144).
- Ethanol 200 PROOF (Decon, cat. no. 64-17-5).
- Nextera XT index kit v2 (Illumina, cat. no. FC-131-2001).
- Nextera XT DNA Library Preparation kit (Illumina, cat. no. FC-131-1096).
- Agencourt Ampure XP beads (Beckman Coulter, cat. No. A 63881).
- NEBNext Ultra II End Repair/dA (New England Biolabs, cat. no. E7546S).
- Filter Tips: 10, 20, 200, and 1000 μL.
- Dry ice.

General laboratory equipment:

- Agilent 2100 Bioanalyzer (Agilent Technologies, cat. no. G2938C).
- Freezers, $-20\,°C$ and $-80\,°C$.
- Refrigerator, $4\,°C$.

Sequencing instrumentation:

- MiSeq, NovaSeq, or equivalent sequencing system (Illumina, Inc).
- MinION nanopore sequencing device (Oxford Nanopore Technologies).

3   Methods

For an overview of the methodology, please *see* the attached video (Video 1).

3.1 Preparing the Micropipettes

We use a Flaming/Brown Micropipette Puller P-97 (Sutter Instrument, Novato, CA) and Borosilicate glass with dimensions of 1.50 mm × 0.86 mm × 10 cm (Sutter Instrument). The pipettes are pulled using the following program (heat = 528; vel = 50; pull = 50; del = 90). We use a 2.0-mm square box filament that is 1.5-mm wide (Harvard Apparatus, Canada). It is important to note that filaments and programs will yield variable results. We cut the

Fig. 4 Micropipette example. Micropipettes should be free of blunt edges to prevent cell lysis. Desired diameter is ~30 μm

edge of the pipettes using fine forceps under microscopic visualization, so that the diameter will be ~30 μm (Fig. 4). It is helpful to visualize the micropipette to ensure that the tip does not have blunt edges, which can lyse cells. Two pipettes are used for each cell. The first pipette is used for picking up the cell and transferring it to a wash buffer; the second pipette is used to recapture the cell and transfer it to the collection tube.

3.2 Preparing Lysis Buffer

Always prepare enough solution of SMART-seq HT (Takara Bio, Kusatsu, Japan) for at least 12 cells to ensure accurate pipetting volumes. Lysis buffer should be dispensed into 0.2-mL SnapStrip PCR tubes (SSIbio, Lodi, CA) and immediately stored at −20 °C freezer until used (note: tubes can be stored at −20 °C for up to 24 h). Do not thaw until immediately prior to use.

3.3 Performing the Dissection

Murine temporal bones are removed after euthanasia with CO_2 and placed into a small petri dish containing 1× DPBS. To access the cochlear membranous labyrinth, the apical portion of the bony labyrinth is visualized under a dissecting stereomicroscope (model M165FC; Leica Microsystems, Wetzlar, Germany). The apical bony covering is chipped off using the Dumont #2 Laminectomy forceps (#11223-20, Fine Science Tools, Foster City, CA) and Dumont #5 forceps (#11254-20, Fine Science Tools). To collect cells from apex, middle, or base turn only, cut the labyrinth into thirds at this time using Vannas spring scissors (#15001-8, Fine Science Tools); transfer the membranous labyrinth into 1.5-mL LoBind tubes using a new tube for each section of tissue (Eppendorf, Hauppauge, NY). Tubes should contain 400 μL of Accumax (Innovative Cell Technologies, San Diego, CA). Complete the dissection and transfer within 5 min.

3.4 Dissociating Cells of the Membranous Labyrinth

Draw one circle on each of two Superfrost Plus Microscope Slides (Thermo Fisher Scientific, Waltham, MA) using a Super PAP Pen (Polysciences, Warrington, PA); the circles create a hydrophobic barrier for specimens and should be 1–2 cm in diameter.

The 1.5-mL tube containing the tissue in 400 µL of Accumax should be incubated at room temperature for 5 min. Thereafter, transfer the tissue in Accumax into the enclosed circle on one of the two prepared slides. Next, dissociate the hair cells by 15–20 triturations using a P200 pipette set to 100 µL. After the triturations, wait for 3–5 min to allow the floating cells to settle. While the cells are settling on the slide, visualize the organ of Corti to determine if dissociation was successful. If further dissociation is required, repeat the process using fewer (3–5) titrations. At this time, prepare the wash buffer by dispensing 400 µL of 1× DPBS into the enclosed circle on the second slide.

3.5 Harvesting Individual Cells

Inspect the cells under an inverted microscope model DMI3000B (Leica Microsystems) using a 20× objective lens. IHCs and OHCs can be distinguished by their morphology if they have not been damaged (Fig. 5). If cell morphology is not obvious, cell type assignment can be determined later bioinformatically [8].

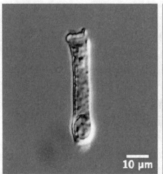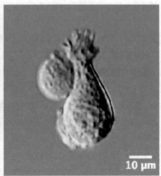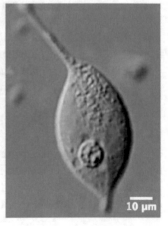

Fig. 5 Cell-type determination and differential expression analysis. Cell types were initially determined by cell morphology. Each cell was imaged during isolation. Images shown are representative of cell morphology of outer hair cells (left) Inner hair cells (middle) and Deiters' cells (right). OHCs are characterized by an elongated tubular shape, relatively short stereocilia, and a distinctively visible nucleus at the very basal end of the cell. IHCs typically have a more flask-like shape with distinct indentations separating the cuticular plate from the cell body, they also have a wider and longer stereocilia bundle relative to outer hair cells. The IHC nucleus is less distinct and usually sits higher in the cell body. Deiters' cells are characterized by their distinctive phalangeal process and foot projections. They typically have a round or lemon shape and are larger than auditory hair cells

Once a candidate cell has been identified, prepare the glass micropipette and change the objective lens to 40×. Using negative and positive pressure control generated by the Pico-liter Microinjector PLI-100A (Warner Instruments, Hamden, CT), gently pull the cell in and out of the pipette a few times to make sure it is not adhering to the walls of the pipette. Next, transfer the cell into wash buffer and discard the first micropipette. Using a new glass micropipette, gently pull the cell in and out of the pipette a few times to ensure it is not adhering to the slide. This step will help to remove excess debris and contaminants. Finally, transfer the cell to a 0.2-mL PCR tube containing the prepared lysis buffer. Avoid air bubbles if possible, even though the bubbles help confirm transfer of the cell. Immediately spin the tube for 10 s in a mini centrifuge and transfer to a box that contains dry ice to flash freeze the cell until isolation is complete and transfer to −80 °C is possible.

3.6 Record Keeping

Record the following data points: date, strain of mouse, age of mouse, time at start of euthanasia, time of death, dissection start and stop times, time of isolation of each cell, as well as hypothesized cell type. Additional data fields should be included based on study parameters and aims. Clear records will enable you to identify potential errors and will facilitate technical improvement.

3.7 Storage of Cells

Transfer all collected cells into a −80 °C freezer until enough cells have been harvested for reverse transcription.

3.8 Reverse Transcription

Stored samples are prepared with SMART-Seq HT Kit (Takara Bio USA) according to the manufacturer's protocol. At One-Step cDNA Synthesis and Amplification step, PCR cycles are set at 18 cycles. After cDNA purification using Ampure XP beads (17022200; Beckman Coulter), cDNA quality should be checked using the Agilent 2100 Bioanalyzer and high-sensitivity DNA chip kit (5067-4626; Agilent Technologies, Santa Clara, CA). Full-length cDNAs show a peak at ~1.5–2 kb.

3.9 Library Preparation and Sequencing

Dilute each cDNA to 100 pg/μl with Nuclease-Free Water for the optimal cDNA input for Illumina sequencing library preparation. Libraries are prepared according to the manufacturer's protocol using the modified Illumina Nextera XT DNA library preparation protocol. After purification of the adaptor-ligated single-cell cDNA libraries, single-cell cDNA libraries are run on the Agilent 2100 Bioanalyzer high-sensitivity DNA chip for quantification and quality control and combined in equimolar concentrations. Pooled single-cell cDNA libraries can be sequenced on a single lane of an Illumina MiSeq, NovaSeq or equivalent sequencing system with 150-bp Paired-End read chemistry (Illumina).

In addition to short-read Illumina sequencing, full-length cDNA should be sequenced from the same cells using the MinION long-read sequencer (Oxford Nanopore Technologies) to identify complex transcript isoforms. Unlike Illumina short-read sequencing, which requires fragments sizes (150–800 bp), Nanopore sequencing generates single reads that span the length of a mRNA transcript. For 1D MinION long-read sequencing, the full-length cDNA product of each cell will have unique index primers to enable multiplexing. Amplify the cDNA with KAPA HiFi Readymix 2× (7958935001; Kapa Biosystems) using the ISPCR primers, as described [9, 10]. Pool the four barcoded single-cell libraries into 10 batches and mix with running buffer provided by Oxford Nanopore Technologies (ONT). The single R9.4 flow cells are run on the nanopore MinION sequencer for 48 h.

4 Bioinformatics

4.1 Computing Environment

Bioinformatic analyses are performed in a Linux environment using the Bash shell. Commodity computing hardware is used for a majority of the analyses and can be used for all parts, but we use an institutional high-performance computing cluster to improve efficiency. The computing cluster comprises many nodes, each with many processor cores, which can be used to perform computations on multiple samples in parallel.

4.2 Expression Quantification

Expression is quantified as a gene expression matrix of read counts by gene and cell, which is constructed as described in the following paragraphs. The gene expression matrix will be used as the primary input to expression analysis software.

Illumina sequencing produces a pair of paired-end sequencing read files per cell in the FASTQ format. For each cell, we map these reads to a reference transcriptome and count the reads per transcript using Kallisto [11]. The following command is used:

```
$ kallisto quant \
--index INDEX_FILE \
--output-dir OUTPUT_DIRECTORY \
--bootstrap-samples 100 \
R1_FASTQ_FILE \
R2_FASTQ_FILE
```

Alternative alignment software such as STAR [12] is commonly used in scRNA-seq. We chose Kallisto because its pseudoalignment algorithm runs several times faster than STAR with a fraction of the memory requirement yet yields similar counts [13]. To shorten turnaround time, Kallisto commands are run in parallel on a high-performance computing cluster.

	Cell 1	Cell 2	Cell 3	...	Cell j
Gene 1	#	#	#	#	#
Gene 2	#	#	#	#	#
Gene 3	#	#	#	#	#
...	#	#	#	#	#
Gene i	#	#	#	#	#

Fig. 6 Format of gene expression matrix. Rows are genes, columns are cells, and values are expression counts. This matrix may be referred to by other names such as "unique molecular identifier (UMI) count matrix"

Next, we use a custom Python script (unpublished, available on request) to create a gene expression matrix (Fig. 6) from the Kallisto output. The operations performed by the script are straightforward: (1) the per-cell transcript expression counts are converted to gene expression counts by summing the counts of all transcripts for the gene; (2) the gene expression counts are arranged in a matrix where columns are the cells and rows are genes; and (3) the matrix is output as a comma-separated values file.

4.3 Expression Analysis

The quantified gene expression data are processed and analyzed using Seurat, an R package for scRNA-seq analysis [14, 15]. Our procedure for using Seurat is developed from the Seurat Guided Clustering Tutorial (https://satijalab.org/seurat/pbmc3k_tutorial.html). The goal is to identify clusters of cells with shared patterns of expression. The clusters represent distinct cell types or conditions whose expression profiles can then be further explored. For example, IHCs, OHCs, and Deiters' cells have different patterns of expression and therefore fall into separate clusters [10].

4.4 Quality Control

Irregularities in isolation or library preparation can result in cells with irregular gene expression data that confound analyses. Therefore, quality-control (QC) filtering is performed to remove poorly performing cells. The standard approach is to plot various QC metrics and develop filtering criteria based upon those metrics. The metrics we have found most useful are number of features (i.e., genes) and percentage mitochondrial reads. We use Seurat `VlnPlot()` to plot these metrics, then use `subset()` to apply criteria that remove outliers on the violin plots (Fig. 7). Cells with an extremely low or high numbers of features are outliers. The usual recommendation is to exclude cells with a mitochondrial percentage > 5%, but this threshold may be lowered for increased stringency.

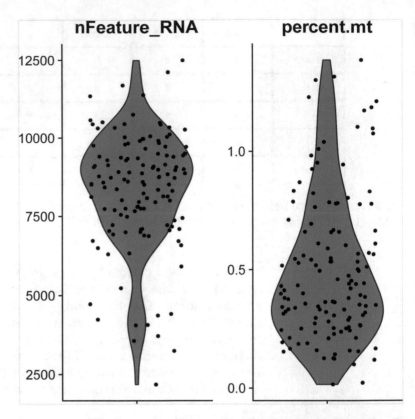

Fig. 7 Seurat violin plots of quality control metrics after filtering. Dots are cells; nFeature_RNA is the number of distinct genes expressed; percent.mt is the percentage of mitochondrial reads. The filtering criteria used here were 2000 $<=$ nFeature_RNA $<=$ 13,000 and percent.mt $<=$ 1.5

4.5 Normalization, Feature Selection, Scaling

Normalization, feature selection, and scaling are all performed per Seurat recommendations and defaults. The number of features in feature selection may be adjusted to improve downstream analysis but we begin by using the default recommendations (2000).

4.6 Linear Dimensional Reduction

Principal component analysis is performed using Seurat RunPCA() as a means of dimension reduction. The principal components are visualized in various ways such as heatmaps and jackstraw plots to determine the dimensionality of the dataset (Fig. 8). These visualizations suggest roughly how many principal components should be used for clustering.

4.7 Clustering

Cells are clustered using Seurat FindNeighbors() and FindClusters(), then visualized using RunUMAP()/RunTSNE() and DimPlot(). If there are no visually distinct clusters, we adjust parameters until clusters emerge (Fig. 9). These parameters are, in the order of decreasing importance per our experience: the

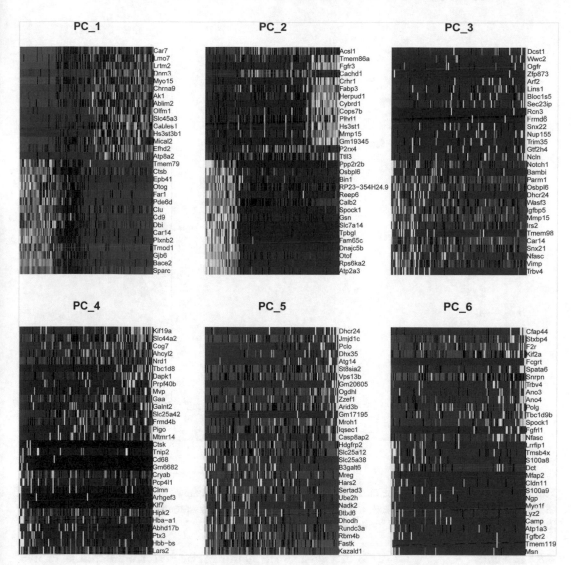

Fig. 8 Heatmap of gene expression. Principal component analysis identify that first six principal components contain major sources of differential expression between cell types. Heatmaps show the top 30 differentially expressed genes between cochlear hair and supporting cells

number of principal components, resolution, the aforementioned QC filtering parameters, the number of UMAP neighbors or the t-SNE perplexity, and the number of features selected. If visually distinct clusters are apparent but not labeled as such, we adjust the resolution until they are labeled appropriately.

As previously mentioned, IHCs, OHCs, and Deiters' cells (and other supporting cells) should be assigned to different clusters. The clustering is refined by removing cells with discordant morphology and cluster assignment.

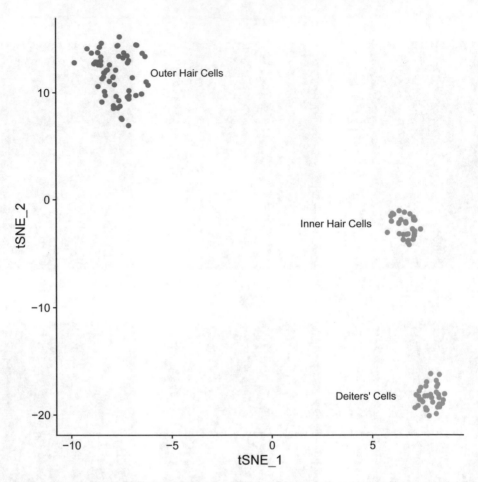

Fig. 9 t-SNE plot generation. t-SNE plots of visually distinct cell type clusters found after repeated adjustment of filtering and clustering parameters. The top two principal components were used. Clusters were biologically identified by cell morphology and further expression analysis

4.8 Exploring Clusters

Once distinct clusters are identified, we perform differential expression analysis between clusters to ascertain their biological identities and explore their unique expression profiles. Markers for each cluster are computed using Seurat `FindAllMarkers()`. Plotting violin plots of the top cluster defining genes for each cluster, with the expression of the genes in each cluster juxtaposed, is particularly effective for developing understanding of the clustered data (Fig. 10). Seurat provides a large number of data visualization and interrogation functions, well documented in the package, which can be used to develop a refined understanding of the clusters and their expression profiles.

4.9 Quantification of Transcript Structure

To identify alternative splicing events, alternative transcription start sites (TSSs), and transcription end sites (TESs), read alignment and transcript assembly are done using a customized Bash script. The customized Bash script contains three parts. First, the Illumina raw

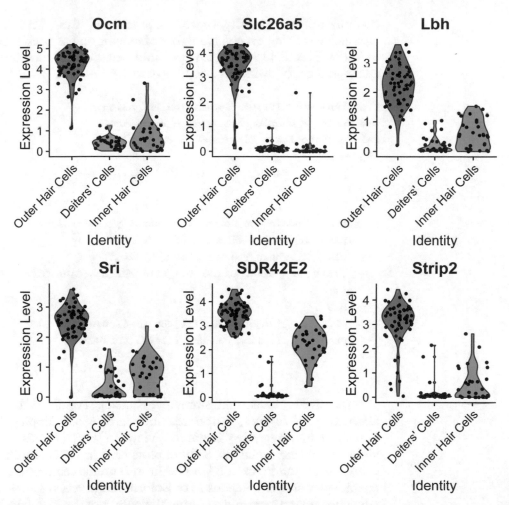

Fig. 10 Violin plots of expression of top six outer hair cell marker genes. Genes are ranked from left to right and top to bottom by descending ROC AUC (receiver operating characteristic area under curve)

sequences are aligned to the mm10 genome (GRCm38) using the STAR (Spliced Transcripts Alignment to a Reference) [12] aligner. Then, SAMtools (Sequence Alignment/Map) [16] is used to create and sort BAM files, a binary format for storing sequence data. And lastly, mapped reads are assembled into transcripts using Cufflinks [17].

As mentioned above, in addition to the short-read Illumina sequencing, full-length cDNAs from the same cells are sequenced using the MinION long-read sequencer (ONT) to complement the limitations of the short-read Illumina sequencing. After the Min-ION long-read sequencing, we used the Mandalorion pipeline (an updated package exists) [9] to align the sequencing reads from FASTQ files to the mm10 genome (GRCm38) and to quantify isoform diversity, which is a product of TSSs, TESs, and alternative splicing events. Note that Mandalorion has been superseded by FLAIR [18], so FLAIR should be considered for future analyses. The following command is used:

```
$ python3 Mandalorion_demultiplex_and_align.py \
 -s /path_to_sample_sheet/sample_sheet.txt \
 -f /path_to_fastq_file/pass.fastq \
 -g mm10 \
 -a /path_to_adapter_fasta/sample_index.fasta \
 -q 7 \
 -t 8
$ python3 Mandalorion_define_and_quantify_isoforms.py \
 -c /path_to_content_file_directory/content_file \
 -p /path_to_ where_you_want_your_output_files/ \
 -a  /path_to_genome_annotation_file/Mus_musculus.GRCm38.88.
chr.gtf \
 -g mm10 \
 -l /path_to_ list_of_genes_for_consensi/gene_list \
 -i /path_to_ illumina_content_file/IlluminaReads.txt \
 -r g
```

The TSSs, TESs, and alternative splicing events, including alternative splice sites, intron retention and exon skipping are visualized in Integrative Genomics Viewer (IGV) [19, 20]. To visualize splicing junctions, Sashimi plots [21] in IGV are utilized using sorted and indexed BAM files from Illumina sequencing. To enable quantitative comparison of isoform usage across samples, a customized R script is utilized using the output of the Mandalorion pipeline as input.

5   Cares and Concerns

While this technique for isolating auditory hair cells and supporting cells for single-cell RNA-Seq has proved to be robust and reproducible by many individuals in our group, there are several areas in which we recommend care be taken to prevent pitfalls.

1. We recommend that reagents be stored in aliquots to prevent unnecessary freeze–thaw cycles. This point is of particular importance for the template switching oligonucleotide (TSO), which contains RNA bases that are susceptible to degradation, reducing the efficiency of the full-length cDNA amplification step.

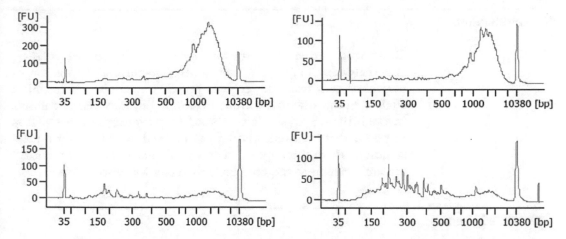

Fig. 11 Bioanalyzer trace examples. The two bioanalyzer traces on the top row reveal desirable results with a strong high-molecular weight peak between 850 and 3000 bp and minimal noise floor. The bottom row shows a lack of successful cell transfer (bottom left) and RNA degradation or contamination resulting in spiky lower molecular weight peaks (bottom right)

2. Close attention should be paid to the bioanalyzer traces of amplified full-length cDNA libraries. Successful libraries will have a strong high-molecular weight peak between 850 and 3000 bp. It is important to confirm this size range. They will also be free of spiky lower molecular weight peaks, which indicate RNA degradation. *See* example of bioanalyzer traces in Fig. 11. Bioanalyzer traces at this post-amplification stage may also be used to screen for contaminating RNA in the wash buffer from which the cells are isolated. A clean isolation will not yield any amplification from wash-buffer-only negative control samples.

3. This technique has proven valuable for the identification of novel exons in hearing loss genes and for the definition and quantification of isoform structure. However, it is our experience that isoform structure assessment needs to be approached with caution to prevent misinterpretation of TSO mispriming PCR products as full-length isoforms. The template switching oligonucleotide used to achieve full-length reverse transcription ends in a 5' - AATrGrG+G - 3'. The three irregular bases at the 3' end—riboguanosine, riboguanisine, and locked nucleic acid modified guanine—are designed to bind to a short CCC overhang with high affinity. We have observed that mispriming events sometimes occur on RNAs with a TACC motif. It is important to keep this fact in mind so as to not misinterpret a TSO mispriming event as a novel isoform that is inexplicably prematurely truncated in the middle of an annotated exon.

6   Conclusions

This chapter presents an effective technique for manual isolation and analysis of single cells from murine cochlea. The isolation of individual cells from the cochlea presents a number of technical challenges that once overcome can yield highly informative datasets through the use of single-cell-based transcriptomic analyses. It is our hope that this methodology can be used to further our understanding of the biology of hearing and deafness and to inform research leading to targeted gene therapies for hearing loss.

Acknowledgments

This work was supported in part by NIDCDs R01s DC002842, DC012049, and DC017955 to R.J.H.S.

References

1. Nash SD et al (2011) The prevalence of hearing impairment and associated risk factors: the Beaver Dam Offspring Study. Arch Otolaryngol Head Neck Surg 137(5):432–439

2. Sloan-Heggen CM et al (2016) Comprehensive genetic testing in the clinical evaluation of 1119 patients with hearing loss. Hum Genet 135(4):441–450

3. Cho Y et al (2002) Gene expression profiles of the rat cochlea, cochlear nucleus, and inferior colliculus. J Assoc Res Otolaryngol 3(1):54–67

4. Tang F et al (2009) mRNA-Seq whole-transcriptome analysis of a single cell. Nat Methods 6(5):377–382

5. Picelli S et al (2014) Full-length RNA-seq from single cells using Smart-seq2. Nat Protoc 9(1):171–181

6. Liu H et al (2014) Characterization of transcriptomes of cochlear inner and outer hair cells. J Neurosci 34(33):11085–11095

7. Burns JC et al (2015) Single-cell RNA-Seq resolves cellular complexity in sensory organs from the neonatal inner ear. Nat Commun 6:8557

8. He DZ et al (2000) Isolation of cochlear inner hair cells. Hear Res 145(1–2):156–160

9. Byrne A et al (2017) Nanopore long-read RNAseq reveals widespread transcriptional variation among the surface receptors of individual B cells. Nat Commun 8:16027

10. Ranum PT et al (2019) Insights into the biology of hearing and deafness revealed by single-cell RNA sequencing. Cell Rep 26(11):3160–3171.e3

11. Bray NL et al (2016) Near-optimal probabilistic RNA-seq quantification. Nat Biotechnol 34(5):525–527

12. Dobin A et al (2013) STAR: ultrafast universal RNA-seq aligner. Bioinformatics 29(1):15–21

13. Du Y et al (2020) Evaluation of STAR and Kallisto on single cell RNA-Seq data alignment. G3 10(5):1775–1783

14. Butler A et al (2018) Integrating single-cell transcriptomic data across different conditions, technologies, and species. Nat Biotechnol 36(5):411–420

15. Stuart T et al (2019) Comprehensive integration of single-cell data. Cell 177(7):1888–1902.e21

16. Li H et al (2009) The sequence alignment/map format and SAMtools. Bioinformatics 25(16):2078–2079

17. Trapnell C et al (2012) Differential gene and transcript expression analysis of RNA-seq experiments with TopHat and Cufflinks. Nat Protoc 7(3):562–578

18. Tang AD, Soulette CM, van Baren MJ et al (2020) Full-length transcript characterization of SF3B1 mutation in chronic lymphocytic leukemia reveals downregulation of retained introns. Nat Commun 11:1438

19. Robinson JT et al (2011) Integrative genomics viewer. Nat Biotechnol 29(1):24–26

20. Thorvaldsdottir H, Robinson JT, Mesirov JP (2013) Integrative Genomics Viewer (IGV): high-performance genomics data visualization and exploration. Brief Bioinform 14(2): 178–192

21. Katz Y, Wang ET, Silterra J, Schwartz S, Wong B, Thorvaldsdóttir II, Robinson JT, Mesirov JP, Airoldi EM, Burge CB. Bioinformatics. 2015 Jul 15;31(14): 2400-2. https://doi.org/10.1093/bioinformatics/btv034. Epub 2015 Jan 22. PMID: 25617416

Additional Resources

https://morlscrnaseq.org/ - interactively browse datasets generated by this technique

Ribosomal Pulldown Assays and Their Use to Analyze Gene Expression in Multiple Inner Ear Cell Types

Maggie S. Matern, Beatrice Milon, Ran Elkon, and Ronna Hertzano

Abstract

Research of inner ear cell development and function has greatly benefited from the advent of next-generation sequencing. Cell type-specific analysis of gene expression by RNA sequencing (RNA-seq) has mediated the discovery of novel transcriptional regulators of cell fate during development, cell type-specific marker genes for the design of transgenic models, and illuminated the inner ear molecular heterogeneity, both within and across cell types. Here we compare different methods of tissue- and cell-specific gene expression analyses, followed by a methodological overview of the use of the RiboTag mouse model for studying the inner ear.

Key words RiboTag, Immunoprecipitation, Cre recombinase, Hair cells, Supporting cells

1 Introduction

Next-Gen sequencing of the cellular transcriptome, also known as RNA sequencing (RNA-seq), allows for detection and characterization of the gene expression profiles of cell types or tissues of interest. Since its introduction to the scientific community in the mid 2000s [1–5], the application of this methodology, along with newly developed analysis pipelines, has broadened our understanding of the transcriptional dynamics that delineate the development, function, and interdependencies of specific cell types within complex tissue environments. The inner ear is one such complex tissue that has been extensively studied using RNA-seq, with over 100 individual published datasets accessible via the gene Expression Analysis Resource (umgear.org) [6]. Traditionally, RNA-seq is divided into "bulk" or "single-cell" sequencing. Bulk sequencing consists of merging signal from groups of cells. Thus, bulk RNA-seq can use whole organs or sorted cell types as starting material. Bulk sequencing has the advantages of larger amounts of genetic material for analysis, and the drawback of signal averaging.

Andrew K. Groves (ed.), *Developmental, Physiological, and Functional Neurobiology of the Inner Ear*, Neuromethods, vol. 176, https://doi.org/10.1007/978-1-0716-2022-9_8,

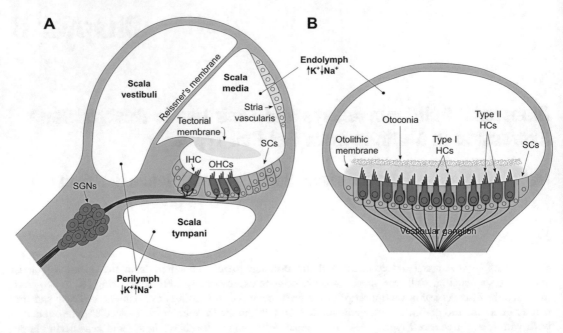

Fig. 1 Schematic representation of cell heterogeneity in the inner ear cochlear and vestibular systems. Nonexhaustive depiction of the various cell types present within the cochlear (**a**) and vestibular (**b**) systems. *HCs* hair cells, *IHCs* inner hair cells, *OHCs* outer hair cells, *SCs* supporting cells, *SGNs* spiral ganglion neurons

Single-cell RNA-seq (scRNA-seq), on the other hand, is the most specific approach as the genetic material from each cell is analyzed independently. However, it suffers from very small amounts of starting material, and at present, higher cost per reaction [7].

Hearing and balance depend on the proper function of the highly complex and heterogeneous epithelia of the auditory and vestibular systems of the inner ear (Fig. 1a, b). This cellular heterogeneity limits the utility of bulk RNA-seq approaches, as the signal from any one population is averaged with the expression of the same gene a multitude of other cell types. It is for this reason that cell type-specific analyses of gene expression, or methods to assess gene expression in single-cell types, have played a pivotal role in increasing our understanding of the development and function of the inner ear. Here we will discuss the advantages and disadvantages of several methods of cell type-specific gene expression analysis, followed by an in-depth overview of the RiboTag mouse and its use for assessing gene expression in the inner ear.

The first and perhaps the most popular method for isolating a specific cell type from a heterogenous tissue is fluorescence-activated cell sorting (FACS) of dissociated cells. For this technique, a tissue is dissociated into a single-cell suspension using proteolytic enzymes that digest the connections between the cells. Live cells of interest can then be isolated using flow cytometry based on the expression of fluorescent transgenes (such as *Atoh1*-nGFP to mark hair cells (HCs)) or by the presence of specific cell

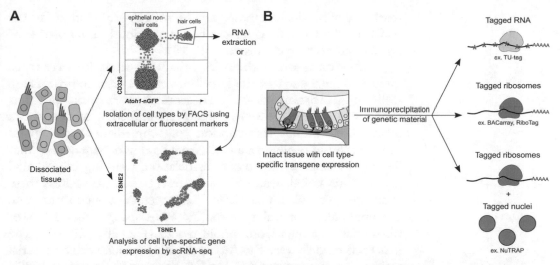

Fig. 2 Tissue dissociation and pulldown from whole-tissue techniques for studying cell type-specific gene expression. (**a**) Cells of interest can be isolated for RNA extraction using fluorescence-activated cell sorting (FACS) on dissociated tissues. Isolation of specific cells by this technique requires the expression of fluorescent markers (such as *Atoh1*-nGFP to detect hair cells) and/or detection of specific cell surface markers using fluorescently conjugated antibodies (such as CD326 conjugated to the fluorescent molecule APC to detect epithelial cells). As an alternative to bulk RNA-seq using RNA extracted from sorted cells, whole tissues or FACS-sorted subpopulations can be analyzed at a single-cell resolution using single-cell RNA-seq (scRNA-seq). (**b**) To avoid secondary changes in gene expression that may result from tissue dissociation, several techniques have been developed to immunoprecipitate cell type-specific RNA from whole tissues. These techniques rely on transgenic animals driving cell type-specific expression of either directly tagged RNA or tagged ribosomes that can be immunoprecipitated. Additionally, more recent animal models such as the NuTRAP mouse allow for both RNA enrichment from a cell type of interest through tagged ribosome pulldown and epigenetic analyses by nuclei pulldown. Asterisks represent encoded tags that are used for immunoprecipitation

surface antigens which are detected by fluorescently conjugated antibodies (such as CD326 for epithelial cells, Fig. 2a) [7, 8]. RNA is then extracted from the isolated populations and used for bulk RNA-seq. FACS is advantageous for studying cell type-specific gene expression as it allows a user to define strict sorting criteria (gates), which can result in highly enriched or pure cell populations—depending on the specificity of the marker. However, there are several drawbacks. First, if the cell type of interest is rare within the tissue, isolating enough cells for an RNA-seq analysis may require a large number of animals for tissue harvest. Additionally, FACS relies on fluorescent markers, which may not be available for the cell type of interest (either as transgenic mice or as cell surface proteins with antibodies to detect their extracellular domain). As an alternative, gene expression from all cells within a tissue or from FACS or hand sorted subpopulations of cells can be analyzed by scRNA-seq (Fig. 2a). This method is discussed more in Chapter 4 and is particularly advantageous for assessing the cellular heterogeneity of a tissue and identifying novel marker genes. For example, recent scRNA-seq studies of the cochlear spiral ganglion have revealed previously unknown subtypes of afferent type 1 neurons [9, 10].

A disadvantage of both FACS and scRNA-seq is the necessity to dissociate the tissue to a single-cell suspension prior to processing. Cellular dissociation results in an acute stress response and thus changes in gene expression, namely the induction of heat shock proteins and the transcription of immediate early genes [11, 12]. Additionally, not all cells from mature or damaged epithelia may be amenable to tissue dissociation, resulting in potential lower yields and skewed cellular representation. To address these challenges, several animal models have been developed to mediate the enrichment of cell type-specific RNA using either RNA or ribosomal pulldown from whole tissues (Fig. 2b) [13–16]. The first such models were the BACarray mice, which utilize bacterial artificial chromosomes (BAC) to drive the expression of an eGFP-tagged ribosome in a tissue-specific manner [13]. Tagged ribosomes and their associated RNAs are immunoprecipitated from homogenized whole tissue samples using an antibody to eGFP, avoiding the need for tissue dissociation. Similarly, the RiboTag mouse model was developed to allow for greater spaciotemporal control of tagged ribosomes via Cre-inducible expression of hemagglutinin (HA) tagged ribosomes [14]. Finally, the NuTRAP mouse [16] allows for cell type-specific ribosomal immunoprecipitation and nuclear isolation from the same tissue sample, mediating concomitant gene expression and epigenetic analyses (Fig. 2b). When NuTRAP mice are crossed with a Cre driver mouse, Cre-positive cells express eGFP-tagged ribosomes in addition to mCherry and biotin co-labeled nuclei, which can be isolated via immunoprecipitation or FACS for downstream analyses [16].

Immunoprecipitation of ribosomes from the BACarray, RiboTag, and NuTRAP models results in enrichment for a population of RNA that is being actively translated in the cell type of interest (i.e., the translatome). Importantly, the immunoprecipitated RNA is by no means a pure population, and as discussed later in the chapter, has to be analyzed with this caveat in mind. Another model, called the TU-tagging model, allows for immunoprecipitation of the total RNA population (i.e., the transcriptome) through thiouracil (TU) incorporation [15, 17]. Cre inducible expression of uracil phosphoribosyltransferase (UPRT) mediates the incorporation of injected 4TU into all newly synthesized RNA (Fig. 2b). 4TU-RNA is then labeled with biotin for purification from the homogenized tissue. In Erickson et al. a TU-tag model was successfully utilized to study gene expression specifically in zebrafish HCs [18]. However, while immunoprecipitation of TU-tagged RNA did show enrichment for known and novel HC-specific transcripts, only few transcripts overall were identified as expressed in HCs compared to RNA extracted from the whole zebrafish [18]. This may be due to incorporation of 4TU into the RNA of off target cells not expressing UPRT and may be improved by varying the time and concentration of 4TU exposure.

Ribosomal immunoprecipitation analyses utilizing the Ribo-Tag model have been successfully utilized to study cell type-specific inner ear gene expression [19–21]. In this chapter, we will discuss the method of ribosomal immunoprecipitation from pooled Ribo-Tag inner ear tissues. We will also review techniques for assessing RNA quality and immunoprecipitation efficiency, as well as provide guidance on processing and analysis of RiboTag RNA samples using RNA-seq.

For this model, a transgene encoding uracil phosphoribosyl-transferase (UPRT) under control of the ubiquitous chicken β-actin/CMV (CAG) promoter is expressed in a Cre recombinase-dependent manner. Therefore, upon injection of the animal with the uracil analog 4-thiouracil (4TU), UPRT-expressing cells incorporate 4TU.

2 Materials

2.1 The RiboTag Mouse Model

The RiboTag transgenic mice have a modified *Rpl22* gene that, in the presence of Cre recombinase, is recombined to encode a hem-agglutinin (HA) tag at the C-terminus of the RPL22 protein (Fig. 3) [14]. RPL22 is a component of the 60S subunit of the ribosome. Therefore, crossing the RiboTag mouse with a Cre recombinase-encoding mouse model results in expression of HA-tagged ribosomes in the recombined cells. These, in turn, can be used to immunoprecipitate cell type-specific ribosome-associated mRNA (the translatome). RiboTag mice are available from the Jackson Laboratory (B6N.129-*Rpl22*$^{tm1.1Psam}$/J, Stock No. 011029) [14]. This model is on a mixed C57BL/6J and C57BL/6N background and is therefore susceptible to the *Cdh23* age-related hearing loss phenotype [22].

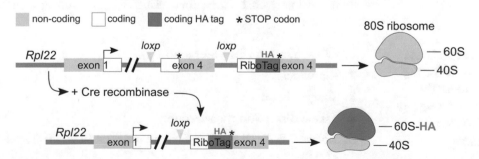

Fig. 3 Cre-mediated recombination of the RiboTag *Rpl22* gene results in HA tagged ribosomes. Exon 4 of the RiboTag *Rpl22* gene is flanked by loxP Cre recombinase recognition sites. When Cre recombinase is expressed within a cell type of interest, the native exon 4 is spliced out and replaced by an exon 4 that also encodes a hemagglutinin (HA) tag. RPL22-HA is incorporated into the 60S subunit of the 80S ribosome, resulting in HA tagged ribosomes that can be used to immunoprecipitate actively translated RNA. RiboTag Schematic adapted from Sanz et al. (2009)

2.2 Selecting a Cre Recombinase Model

Use of the RiboTag model requires crossing with a Cre recombinase model to induce the expression of HA-tagged ribosomes. Cre recombinase is a DNA modifying enzyme that, when present within a cell, is capable of recognizing a 34 bp consensus sequence within DNA called *loxP* [23]. Segments of DNA that are flanked by *loxP* can be excised, inverted, or translocated by Cre recombinase, depending on the directionality of the *loxP* sequence [23]. Therefore, cell type-specific promoter-driven Cre expression has been an immensely valuable tool to study the effects of conditional gene deletion and overexpression in many systems.

Over the past two decades, many models have been developed to induce Cre-mediated recombination specifically in inner ear cells. Information on these models, including the spatial and temporal expression of Cre recombinase in the inner ear, is reviewed in Cox et al. [23], and new models continue to be developed. When planning to utilize a Cre recombinase model for RiboTag studies, we suggest first performing an in-depth characterization of the Cre recombinase-mediated recombination pattern and efficiency at the time point of interest—even if already available in the published literature. This can be done by crossing the desired Cre recombinase model to a robust reporter mouse such as Ai14 (Jackson Laboratory Stock No. 007908) and analyzing the expression pattern of tdTomato at the time point that tissue samples will be collected for RiboTag immunoprecipitation. Importantly, for inducible CreERT2 lines, the timing and dose of Tamoxifen injection can change the pattern of recombination. This analysis may reveal unexpected patterns of recombination which will be important later on in data analysis. For example, by crossing the Gfi1-Cre mouse to Ai14, we have previously shown that Gfi1-Cre-mediated recombination in the inner ear is not restricted to HCs, but is also present in inner ear resident immune cells [24]. This information has since guided our tissue collection and data analysis methods when using Gfi1-Cre for RiboTag immunoprecipitation experiments.

2.3 Materials and Tools for Collecting Mouse Inner Ear Tissues Followed by RiboTag RNA Extraction

2.3.1 Mouse Inner Ear Dissection

1. CO_2 chamber (for older animals).

2. Kimwipes.

3. Sharp scissors.

4. Tissue dissection microscope.

5. RNase AWAY™ Surface Decontaminant (*Thermo Fisher Cat# 7002*).

6. MilliQ or other ultrapure filtered water.

7. Sterile 1× Phosphate-Buffered Saline (PBS), stored at 4 °C.

8. Sterile 35–60 mm plastic dishes.

9. Fine forceps.

10. Ice.

11. Sylgard dish (for recipe, *see* Driver and Kelley [25]).

12. Nuclease-free 1.5-mL microcentrifuge tubes.

13. Dry ice for flash freezing tissues.

2.3.2 RiboTag RNA Extraction

1. 2 mL Dounce homogenizer with a loose and a tight pestle (*such as Millipore Sigma Cat# D8938*).

2. MilliQ or other ultrapure filtered water.

3. RNase AWAY™ Surface Decontaminant.

4. Ice.

5. Dry ice.

6. Homogenization buffer (HB).

   (a) 1 M Tris–HCl, pH 7.4 (*such as Teknova Cat# T5074*, final concentration 50 mM).

   (b) 1 M KCl (*such as Fisher Scientific Cat# 50-842-959*, final concentration 100 mM).

   (c) 1 M MgCl$_2$ (*such as Fisher Scientific Cat# AAJ61014EQE*, final concentration 12 mM).

   (d) Nonidet P40 Substitute (*such as Millipore-Sigma Cat# 11332473001*, final concentration 1%).

   (e) DNAse/RNAse free H$_2$O (*such as Thermo Fisher Cat# 10977015*).

7. Supplemented homogenization buffer (*S-HB*).

   (a) Homogenization buffer (*HB*).

   (b) 1 M DTT (*such as Thermo Fisher Cat# P2325*, final concentration 1 mM).

   (c) Halt™ Protease Inhibitor cocktail (*Thermo Fisher Cat# 87786*, final concentration 1×).

   (d) RNaseOUT™ Recombinant Ribonuclease Inhibitor (*Thermo Fisher Cat# 10777019*, final concentration 200 U/mL).

   (e) 5 mg/mL Cycloheximide dissolved in DNAse/RNAse-free H$_2$O (*such as Millipore Sigma Cat# 01810*, final concentration 100 μg/mL).

   (f) 100 mg/mL Heparin dissolved in DNAse/RNAse free H$_2$O (*such as Millipore Sigma Cat# H3393*, final concentration 1 mg/mL).

8. High salt buffer (*HSB*).

   (a) 1 M Tris–HCl, pH 7.4 (final concentration 50 mM).

   (b) 1 M KCl (final concentration 300 mM).

   (c) 1 M MgCl$_2$ (final concentration 12 mM).

(d) Nonidet P40 Substitute (final concentration 1%).

(e) 1 M DTT (final concentration 1 mM).

(f) 5 mg/mL Cycloheximide (final concentration 100 μg/mL).

(g) DNAse/RNAse-free H₂O.

9. Nuclease-free 50-mL conical tubes.

10. Nuclease-free 1.5-mL and 2-mL microcentrifuge tubes.

11. Mouse anti-HA epitope tag antibody (*such as BioLegend Cat# 901502*).

12. Bemis™ Parafilm™ M Laboratory Wrapping Film (*Fisher Scientific Cat# 13-374-12*).

13. Invitrogen™ Dynabeads™ Protein G for Immunoprecipitation (*Thermo Fisher Cat# 10004D*).

(a) *Note*: the protocol described below is optimized for the mouse IgG1 anti-HA antibody Cat# 901502 from BioLegend, which binds strongly to Protein G. Other anti-HA antibodies, such as those produced in guinea pig or rabbit, may require the use of Protein A Dynabeads.

14. Magnetic separation rack for 1.5-mL and 2-mL microcentrifuge tubes (*such as Thermo Fisher Cat# CS15000*).

15. Qiagen RNeasy Plus Micro Kit (*Qiagen Cat# 74034*).

16. *β-Mercaptoethanol (such as Millipore Sigma Cat# M6250).*

17. *100% ethanol (such as Millipore Sigma Cat# ME7023) diluted to 70% in* DNAse/RNAse-free H₂O.

Table 1
Example RT-qPCR primers that can be used to test enrichment and depletion of inner ear cell type-specific genes

		Forward	Reverse
Housekeeping	*Gapdh*	GGAGAAACCTGCCAAGTATGA	TCCTCAGTGTAGCCCAAGA
Hair cells	*Gfi1*	AATGCAGCAAGGTGTTCTC	CTTACAGTCAAAGCTGCGT
Outer hair cells	*Slc26a5*	GAAAGGCCCATCTTCAGTCATC	GCCACTTAGTGATAGGCAGGAAC
Supporting cells/Glial cells	*Sox2*	CCCACCTACAGCATGTCCTA	GTGGGAGGAAGAGGTAACCA
Mesenchymal cells	*Pou3f4*	CTGCCTCGAATCCCTACAGC	CTGCAAGTAGTCACTTTGGAGAA

2.4 Other Materials	1. Maxima First Strand cDNA Synthesis Kit for RT-qPCR (*Thermo Fisher Cat# K1672*) or similar.
	2. Maxima SYBR Green/ROX qPCR Master Mix (*Thermo Fisher Cat# K0221*) or similar.
	3. RT-qPCR primers for cell type-specific genes (*see* Table 1).

3   Methods

The following ribosomal immunoprecipitation protocol is optimized for inner ear tissues collected from RiboTag mice [14] crossed with an inner ear Cre driver mouse such as Gfi1-Cre [20, 26], Sox2-CreERT2 [27], or Prestin-CreERT2 [19, 28]. This protocol can also be directly utilized for ribosome immunoprecipitation using the Tg(*myo6b:RiboTag*) zebrafish model [12]. Mouse inner ear tissues can be collected at any age following efficient Cre activation and either immediately utilized for immunoprecipitation, or flash frozen on dry ice for storage and later use. We generally recommend pooling inner ear tissues from two to five mice for sufficient RNA yields for downstream analyses such as qPCR and RNA-seq.

3.1   Inner Ear Tissue Collection

Prepare the dissection workspace by first cleaning all surfaces and dissection tools with RNase AWAY Surface Decontaminant. Place a bucket of ice and a container of dry ice next to the dissection microscope for easy access. Label an appropriate number of nuclease-free 1.5-mL microcentrifuge tubes for collecting the tissues and place them on dry ice. Place a Sylgard dish that has been sprayed with RNase AWAY and rinsed with MilliQ water on the ice and fill with cold, sterile 1× PBS.

3.1.1   Neonatal Mice

A detailed inner ear dissection method for neonatal mice can be found in Chapter 2. Briefly, mice up to 6 days old are euthanized by decapitation with sharp scissors. Using forceps, bisect along the sagittal suture and peel away the skull from the skull base. Scoop the brain out by sliding forceps underneath and pushing upwards, exposing the temporal bones. Remove the temporal bones by carefully pressing on the skull base surrounding it and separating it from the surrounding tissues. Transfer the temporal bones onto the Sylgard dish containing cold PBS. Dissect out the tissue of interest by peeling or chipping away the otic capsule and removing the tissues with fine forceps. Transfer the tissues immediately to the prepared nuclease-free tubes on dry ice. The tissues will flash freeze to the side of the tube, ensuring that no further RNA degradation occurs. Proceed to RiboTag immunoprecipitation or store tissues at −80 °C until further processing.

3.1.2 Adult Mice

Mice older than 6 days are euthanized by CO_2 asphyxiation followed by decapitation. Cut the skull longitudinally in half using sharp scissors and remove the brain tissue to expose the temporal bones. Remove the temporal bone by carefully cracking the skull surrounding it and separating the temporal bone from its surrounding tissues. Transfer the temporal bone onto the Sylgard dish containing cold PBS. After both temporal bones have been transferred to cold PBS, dissect out the tissue of interest by chipping away the otic capsule and removing the tissues with fine forceps. Immediately transfer the tissues to the prepared nuclease free tubes on dry ice. The tissues will flash freeze to the side of the tube, ensuring that no further RNA degradation occurs. Cochlear tissues of adult mice will break into pieces during dissection; we therefore recommend flash freezing each piece as it is removed from the rest of the tissue. Proceed to RiboTag immunoprecipitation or store tissues at $-80\,°C$ until further processing.

3.2 Method for Ribosomal Immunoprecipitation and RNA Extraction from RiboTag Mouse Inner Ear Tissues (Fig. 4)

Day 1

1. Prepare 20 mL of homogenization buffer (*HB*) by combining 667 µL 1.5 M Tris–HCl (pH 7.4), 2 mL 1 M KCl, 240 µL 1 M $MgCl_2$, and 200 µL NP40 in 16.9 mL DNAse/RNAse-free H_2O. This solution can be stored at $4\,°C$ for up to 2 months.

2. Each immunoprecipitation reaction requires 1 mL of supplemented homogenization buffer (*S-HB*). To make 2.5 mL *S-HB* (enough for two reactions), combine 2.39 mL *HB* with 2.5 µL 1 M DTT, 25 µL 100× Halt™ Protease Inhibitor cocktail, 12.5 µL 40 U/µL RNaseOUT™ Recombinant Ribonuclease Inhibitor, 50 µL 5 mg/mL Cyclohexamide, and 25 µL 100 mg/mL Heparin. Prepare S-HB fresh, no more than 24 h before starting the immunoprecipitation protocol.

3. Prepare 20 mL of high salt buffer (*HSB*) by combining 667 µL 1.5 M Tris–HCl (pH 7.4), 6 mL 1 M KCl, 240 µL 1 M $MgCl_2$, 200 µL NP-40, 20 µL 1 M DTT, and 400 µL 5 mg/mL Cyclohexamide in 12.5 mL RNAse-free H_2O. This solution can also be stored at $4\,°C$ for up to 2 months.

4. Clean the Dounce homogenizer, as well as the A and B pestles, by first rinsing 3× with MilliQ water. Thoroughly spray the dounce and pestles with RNase AWAY, followed by an additional 5× rinses with MilliQ water. Shake the dounce and pestles to remove excess water and place the Dounce homogenizer on ice. Place the pestles in a clean 50-mL conical tube, also on ice.

    (a) *Note*: it is important that all solutions and equipment are kept cold on ice, and all centrifugation and incubation steps are performed at $4\,°C$.

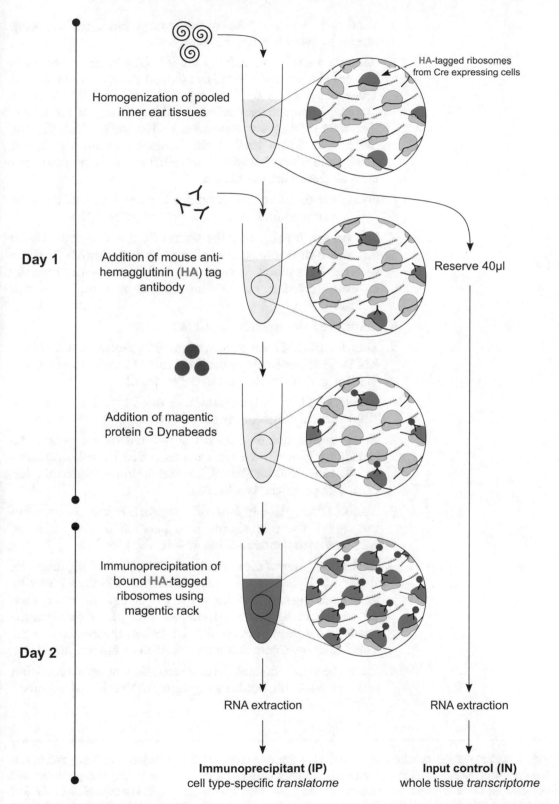

Fig. 4 Stepwise methodology for RiboTag immunoprecipitation from inner ear tissues. Flash frozen inner ear tissues are pooled and homogenized to get HA-tagged ribosomes and their associated cell type-specific RNAs

5. Slowly add 1 mL of *S-HB* to the Dounce homogenizer, being careful to avoid forming bubbles.

6. Remove the tubes containing the RiboTag inner ear tissues to be pooled from the −80 °C freezer and place them on dry ice. Thaw the samples by removing ~200 μL of *S-HB* from the Dounce homogenizer and repeatedly applying to the frozen tissues until they are suspended in the buffer. Transfer the buffer and tissues back to the homogenizer on ice. Repeat rinsing the tube with additional *S-HB* from the homogenizer until all tissues are transferred.

   (a) *Note*: tissue volume should not exceed 10% of the total reaction volume (i.e., tissue volume < 100 μL).

7. Still on ice, homogenize the tissues by first douncing 10–20 times with the loose pestle, followed by douncing 10–20 times with the tight pestle. Douncing should be performed carefully to avoid creating excess foam; however, some foam will still form.

8. Allow the foam to settle for 10 min, on ice.

9. Transfer the solution and any remaining foam to a DNAse/RNAse-free 2-mL microcentrifuge tube. Centrifuge samples at $8000 \times g$, 4 °C for 10 min to pellet debris.

10. Collect 40 μL of the lysate and store at −80 °C. This is the total RNA control sample, or input (*IN*) sample.

11. Being careful not to disturb the pelleted debris, collect the remaining lysate into a clean nuclease-free 1.5-mL microcentrifuge tube. This sample will be used for immunoprecipitation (*IP*). The pellet can be discarded.

12. To the *IP* sample, add 5 μL of 1 mg/mL HA antibody. Wrap the cap of the tube tightly with parafilm to prevent leaks. Incubate with gentle rotation at 4 °C for 4–6 h.

13. In the final hour of the 4–6 h incubation, prepare the magnetic Protein G Dynabeads. Resuspend the beads in their buffer by gently swirling or triturating with a p1000 pipette tip. For each *IP* sample being processed, aliquot 300 μL of resuspended beads into a clean nuclease-free 1.5-mL microcentrifuge tube. Label each bead containing tube for its corresponding *IP*.

14. Place the bead containing tube into the magnetic separation rack and allow the beads to aggregate to the side of the tube.

Fig. 4 (continued) into solution. Mouse anti-HA antibodies recognize the HA-tagged ribosomes and form a complex with magnetic protein G Dynabeads. Washing the beads and associated tagged ribosomes removes non-specific ribosomes from the solution. RNA extraction from this solution results in a population of RNA that is enriched for actively translated RNAs from the cell type of interest (called the translatome)

(a) *Note*: beads can sometimes stick to the side of the tube, especially near the surface of the buffer. Gently tapping the tube can help dislodge the beads.

15. Remove the storage buffer from the tube while still on the magnetic rack, being careful not to disturb the pelleted beads.

16. Wash the beads once with 500 µL of homogenization buffer (*HB*), rotating at 4 °C for 10 min.

17. After 10 min, gently centrifuge the tube to remove the buffer and beads from the cap. Place the tube in the magnetic rack and allow the beads to aggregate to the side of the tube. Remove the buffer and place the tube on ice.

18. At the end of the 4–6 h incubation, transfer the *IP* sample to the bead containing tube. Again, wrap the cap of the tube with parafilm to prevent leaks. Incubate the *IP* sample with the beads overnight at 4 °C, rotating.

Day 2

1. Prepare the lysis buffer (*LB*) by adding 3.5 µL of *β-mercaptoethanol to 350 µL* Buffer RLT from the Qiagen RNeasy® Plus Micro Kit for each *IP* sample to be processed.

2. Place the high salt buffer (*HSB*) on ice.

3. Retrieve the *IP* samples from 4 °C. Moving quickly to ensure the sample remains cold, briefly spin down the *IP* sample to remove liquid from the cap and place the sample in the magnetic rack. After the beads have aggregated to the side of the tube, either transfer the supernatant to a new tube or discard. Immediately wash the beads by gently applying 800 µL of ice cold *HSB* and incubating at 4 °C for 10 min, rotating. *Do not vortex.*

(a) *Note*: at this point in the protocol, the ribosome-associated mRNAs from the cell type(s) of interest are bound to the Protein G Dynabeads. Keeping the samples cold and using gentle rotation are important to prevent the ribosomes from dissociating from the beads.

4. Perform two more 10-min washes with 800 µL of ice cold *HSB*, for a total of three washes.

5. At the end of the final 10-min wash, briefly spin down the *IP* sample to remove liquid from the cap, place the sample in the magnetic rack and allow the beads to aggregate to the side of the tube. Once aggregated, remove the *HSB* and immediately add 350 µL of *LB* to the beads.

6. Vortex the *IP* sample for 30 s to dissociate the ribosome bound mRNAs from the beads.

7. At this point in the protocol, the *IP* sample can be frozen and stored with the *IN* sample at −80 °C or can be processed immediately for RNA extraction (next step).

RNA Extraction

1. Prepare an aliquot of *LB* for the *IN* sample by adding 3.5 µL of *β-mercaptoethanol to 350 µL* Buffer RLT from the Qiagen RNeasy® Plus Micro Kit.

2. Remove the 40 µL *IN* sample from −80 °C and place on ice. Thaw the *IN* sample by applying 350 µL of *LB* and vortexing for 30 s.

3. Centrifuge the *IN* sample at 14,000 × g for 3 min at room temperature. Transfer the supernatant to a labeled gDNA Eliminator spin column included in the Qiagen RNeasy® Plus Micro Kit.

4. If needed, remove the *IP* samples from −80 °C and thaw on ice.

5. Briefly spin down the *IP* sample and place in the magnetic rack. When the beads have aggregated, transfer the sample into a clean 1.5-mL microcentrifuge tube. Place this new tube on the magnetic rack and allow any remaining beads to aggregate. Transfer the sample to a labeled gDNA Eliminator spin column included in the Qiagen RNeasy® Plus Micro Kit.

6. Proceed with steps 2–9 of the Qiagen RNeasy® Plus Micro Kit protocol to extract RNA from the *IP* and *IN* samples. Store RNA samples at −80 °C.

    (a) *Note*: elute RNA in 16 µL RNAse-free H_2O.

3.3 Assessing the Quality and Concentration of RiboTag IP and IN Samples

Based on our experience, the above ribosome immunoprecipitation protocol will yield 16 µL of RNA at a concentration of between 500 pg/µL and 10 ng/µL, depending on the Cre driver used and the number of tissues pooled for the extraction. The quality and quantity of RNA in such low concentration samples is difficult to assess using standard RNA quantification methods such as Nanodrop. We therefore recommend assessing RNA quality of the IP and IN samples using the Agilent Bioanalyzer RNA 6000 pico assay. This assay can detect RNA concentrations as low as 50 pg/µL and is usually standard equipment in genomics core facilities. Up to ten samples (five sets of IP and IN) and one control can be run per "pico chip" using a minimal amount of RNA (1 µL) from each sample. The assay will output the concentration of the loaded sample as well as an RNA Integrity Number (RIN), a score indicative of RNA quality that is based on the 28S to 18S rRNA ratio of the sample (Fig. 5a, b). For best results, we recommend using IP and IN samples with RIN scores between eight and ten for downstream analyses.

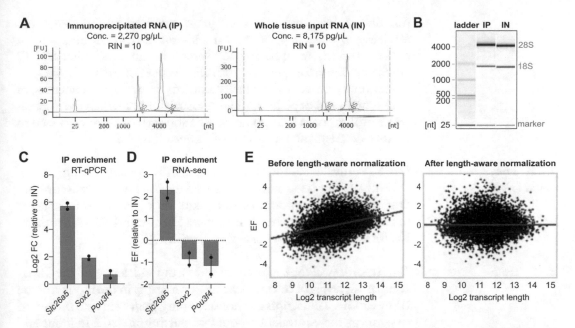

Fig. 5 RiboTag immunoprecipitation from inner ear tissues results in high-quality RNA suitable for downstream gene expression analyses. (**a**) Immunoprecipitation of hair cell and inner ear macrophage RNA from two pooled P0 *Gfi1^cre/+;RiboTag^HA/HA* cochlea results in high-quality immunoprecipitated (IP) and whole-tissue (IN) RNA for downstream analyses (measured by Agilent Bioanalyzer RNA 6000 pico assay). (**b**) RNA gel view of (**a**) showing strong 18S and 28S bands for both IP and IN samples with minimal RNA degradation. (**c**) RT-qPCR results from replicate IPs of 10-week-old *Prestin^creERT2/+;RiboTag^HA/+* cochlea showing enrichment of the outer hair cell gene *Slc26a5*, but also slight enrichment of the supporting cell gene *Sox2* and mesenchymal gene *Pou4f3*. (**d**) RNA-seq results of the same samples in (**c**) showing less-pronounced enrichment of *Slc26a5* and depletion of *Sox2* and *Pou4f3*. (**c**, **d**) demonstrate that RT-qPCR tends to overestimate enrichment and underestimate depletion compared to RNA-seq, and therefore, we use RT-qPCR as a qualitative rather than quantitative methods to test for successful immunoprecipitation. (**e**) Graphical representation of the technical transcript length bias on IP enrichment, with longer transcripts having systematically higher enrichment factors. Length-aware normalization removes this bias. *FU* fluorescent units, *nt* nucleotide length, *FC* fold change, *EF* enrichment factor

3.4 Testing the Efficiency of IP RNA Enrichment Using qPCR

After ensuring that the RNA content of the RiboTag IP and IN samples are of high quality and sufficient concentration, it is important to next assess the efficiency of cell type-specific RNA immunoprecipitation in the IP compared to the corresponding IN samples. By this, we mean testing the relative enrichment of transcripts from the cell type of interest and depletion of transcripts from other cell types in the IP compared to whole tissue IN RNA. The efficiency of immunoprecipitating the HA-tagged ribosomes and their associated RNAs from the cells of interest will vary from one immunoprecipitation experiment to the next. These variations in IP efficiency can be attributed to both tissue-specific as well as technical factors. For example, the prevalence of a specific cell type in a tissue and the expression of the same gene in other cell types within the tissue affect the measured gene enrichment and depletion. Similarly, overheating of the samples during processing, RNAse contamination, or antibody lot variation can also affect the IP efficiency. Therefore, testing the effectiveness of each immunoprecipitation is a necessary quality control step. These concepts of enrichment and depletion, the understanding of which are imperative to appropriate interpretation of RiboTag data, are discussed in further detail at the beginning of Subheading 3.6.

3.4.1 Reverse Transcription

Use an equivalent amount of RNA from the corresponding pairs of IN and IP samples as starting material for the reverse transcription. Due to the small amount of RNA, kits with a ready-to-use master mix are preferable, such as the Maxima First Strand cDNA Synthesis Kit for RT-qPCR (*Thermo Fisher Cat# K1672*).

3.4.2 qPCR

The resulting cDNA is used to assess the efficiency of the IP by comparing levels of cell type-specific transcripts in the input and the IP by qPCR. Transcripts for housekeeping genes are also measured to confirm the equivalent amount of starting material in input and IP samples.

qPCR master mixes such as the Maxima SYBR Green/ROX qPCR Master Mix (*Thermo Fisher Cat# K0221*) can be used to quickly assess the enrichment or depletion of cell type-specific genes, such as *Gfi1* for all HCs, *Slc26a5* for outer HCs, *Sox2* for supporting cells and *Pou4f3* for mesenchymal cells. Examples of qPCR primers for these cell types are listed in Table 1. The enrichment/depletion factor is calculated using the standard $\Delta\Delta$Ct method, with the IN serving as the normalizing sample (IN $\Delta\Delta$Ct values will be normalized to 1).

When validating enrichment/depletion by RT-qPCR prior to sequencing, it is important to keep in mind that the results will be qualitative but not quantitative. As shown in Fig. 5c, d, RT-qPCR results overestimate enrichment while underestimating depletion when compared to the final RNA-seq results.

3.5 Library Kit Selection and RNA-Seq

The amount of RNA obtained following immunoprecipitation varies depending on the prevalence of the cell type of interest in the tissue as well as the number of tissues that were pooled on Day 1, **step 6**. Often, the quantity of RNA that can be submitted for sequencing is limited. Additionally, the IP enriches for immature RNA, therefore increasing the intronic reads compared to the IN sample. For these reasons, the choice of an adequate library preparation kit is critical. Song et al. [29] compared the use of five different kits for RiboTag RNA-seq, varying the RNA starting amounts from 70 ng to as low as 250 pg. The results indicated that two of the kits tested produced a much higher percentage of reads mapping to intronic regions, reaching up to 41% for the IP

samples, which decreased the overall value of the data. Consequently, it is important to choose a library preparation kit carefully
to ensure the obtention of meaningful data. While Song et al.
suggested several kits appropriate for RiboTag sequencing, new
improved products are released regularly.

3.6 Data Analysis and Enrichment Factors

It is important to understand that, as RiboTag is based on immunoprecipitation of ribosomes, it is only a method of enrichment for,
rather than isolation of, cell type-specific ribosome-bound RNA. In
fact, for rare cell populations (e.g., HCs) or even for a population of
cells that is fairly predominant within the tissue, the majority of
mRNA transcripts within the IP samples will originate from the
surrounding cells. Here we will further explain the terms "enrichment" and "depletion" that are critical for the full understanding
and analysis of RiboTag samples.

Cell type-expressed mRNA are expected to show an enrichment (expression level in the IP/IN >1), whereas mRNA that are
not expressed by the IP'd cell type are expected to show a depletion
(expression level in the IP/IN <1). The exact cutoff for enrichment and depletion is somewhat arbitrary and a topic for debate,
and candidate cell-type-specific transcripts with high levels of
enrichment often require validation by another technique such as
in situ hybridization. This is because, in addition to the expression
level of a transcript in the cell type of interest, the prevalence of that
cell type within the tissue and the overall level of expression of the
transcript in all other cell types within the tissue will also affect the
measured enrichment and depletion. By convention, our team uses
a twofold enrichment or depletion cutoff, along with measures to
ensure consistent enrichment or depletion between samples, to
determine expression in a cell type of interest. This is further outlined below.

After RNA sequencing and read alignmsent, normalized read
count data from replicate samples are analyzed for differential gene
expression using bioinformatics software such as DESeq2 [30]. The
IN sample should be treated as the control. This analysis will result
in a fold change value and associated false discovery rate (FDR) of
gene expression between the IP and IN samples. Log_2 transforma-
tion of the fold change value provides what we call an "enrichment
factor" (EF) of the IP compared to the IN. While a standard cutoff
of $EF > 1$ or $EF < -1$ (corresponding to a twofold difference in
expression between IN and IP) along with an $FDR < 0.05$ is
indicative of significant enrichment or depletion of transcripts in
the IP sample compared to the IN, we also recommend assessing
the normalized read data from each individual sample for overlap
between the IP and IN replicates. For example, a gene that meets
EF and FDR cutoffs would only be considered "enriched" in the IP
sample if the normalized read counts from each replicate of the IP

were higher than the highest read count measured in the IN replicates. Conversely, a significant gene would be considered "depleted" in the IP only if the read counts from each replicate of the IP were lower than the lowest read count in the IN. This added step of filtering is advised due to the inherent variability in immunoprecipitation of cell type-specific transcripts between samples, therefore causing read counts among IP replicates to also be variable. Introducing this additional overlap control helps to increase the confidence that a significantly enriched or depleted gene will validate in the tissue.

Importantly, one caution that should be taken into account when calculating EF values between IP and IN samples is the possible impact of technical biases that affect the comparison between expression levels. For example, we have observed, in multiple datasets, a systematic correlation between EF and transcript length, with longer transcripts characterized by higher EF levels and vice versa. The scope and magnitude of this relationship points to a technical rather than biological source (i.e., longer transcripts are not more highly expressed, but rather are more efficiently immunoprecipitated due to the existence of polysomes). Such global coupling between EF and transcript length can lead to misinterpretation of the results and to false calls by popular bioinformatics analysis tools that seek enrichment of functional gene sets, such as GO enrichment or GSEA analyses [31]. We therefore recommended checking RiboTag RNA-seq datasets for such a correlation, and if found, to remove it out of the data by advanced normalization methods that account for sample-specific length effects (e.g., the conditional quantile normalization method, Fig. 5e) [32].

In summary, genes with no overlapping IP and IN read count values, an EF >1 and an associated FDR <0.05 would be considered enriched in the cell type of interest, whereas genes with no overlapping IP and IN read count values, an EF < -1 and an associated FDR <0.05 would be considered depleted from the cell type of interest.

3.7 Validation

As with all RNA-seq experiments, findings from RiboTag enrichment analyses should be validated in the tissue of interest. We have utilized several techniques to validate candidate cell type-specific genes and proteins, including qPCR using independent RiboTag IP and IN samples, in situ hybridization, RNA-scope (methodology explained in Chapter 5), and immunohistochemistry in sectioned or whole mounted inner ear tissues. Additionally, databases of published inner ear transcriptomic datasets such as the SHIELD [33] and umgear.org [6] can help to corroborate gene expression findings and prioritize genes of interest for validation.

4   Conclusion

Cell type-specific analyses of gene expression such as bulk RNA-seq of sorted inner ear cells, and more recently, scRNA-seq have mediated an increase in our understanding of the development, function and heterogeneity of inner ear cell types. The RiboTag method of studying inner ear tissue or cell type-specific transla-tomes has the added advantage of avoiding secondary gene expression changes that result from tissue dissociation and cell sorting. This method is also highly useful for analyzing adult, mutant or previously damaged inner ear tissues which may not be amenable to the tissue dissociation process. RiboTag immunoprecipitation can not only be used to assess gene transcript enrichment and deple-tion, as was the focus of this chapter but also be utilized to compare a cell type-specific translatomes between two conditions. For exam-ple, we have recently utilized RiboTag to assess differences between the HC translatomes of *Gfi1* mutant and control mice [20]. Addi-tionally, the method discussed here can also be adapted for con-comitant analyses of cell type-specific gene expression and epigenetics using the novel NuTRAP mouse. Overall, ribosomal immunoprecipitation methods represent a useful addition to an inner ear researcher's armamentarium of techniques to studying cell type-specific gene expression.

References

1. Emrich SJ, Barbazuk WB, Li L, Schnable PS (2007) Gene discovery and annotation using LCM-454 transcriptome sequencing. Genome Res 17(1):69–73. https://doi.org/10.1101/gr.5145806

2. Weber APM, Weber KL, Carr K, Wilkerson C, Ohlrogge JB (2007) Sampling the arabidopsis transcriptome with massively parallel pyrose-quencing. Plant Physiol 144(1):32–42. https://doi.org/10.1104/pp.107.096677

3. Cheung F, Haas BJ, Goldberg SMD, May GD, Xiao Y, Town CD (2006) Sequencing Medi-cago truncatula expressed sequenced tags using 454 Life Sciences technology. BMC Genomics 7:272. https://doi.org/10.1186/1471-2164-7-272

4. Bainbridge MN, Warren RL, Hirst M et al (2006) Analysis of the prostate cancer cell line LNCaP transcriptome using a sequencing-by-synthesis approach. BMC Genomics 7:246. https://doi.org/10.1186/1471-2164-7-246

5. Nagalakshmi U, Wang Z, Waern K et al (2008) The transcriptional landscape of the yeast genome defined by RNA sequencing. Science 320(5881):1344–1349. https://doi.org/10.1126/science.1158441

6. Orvis J, Gottfried B, Kancherla J et al (2020) gEAR: gene Expression Analysis Resource por-tal for community-driven, multi-omic data exploration. *bioRxiv*:2020.08.28.272039. https://doi.org/10.1101/2020.08.28.272039

7. Hertzano R, Gwilliam K, Rose K, Milon B, Matern MS (2020) Cell type–specific expres-sion analysis of the inner ear: a technical report. Laryngoscope:lary.28765. https://doi.org/10.1002/lary.28765

8. Elkon R, Milon B, Morrison L et al (2015) RFX transcription factors are essential for hearing in mice. Nat Commun 6:8549. https://doi.org/10.1038/ncomms9549

9. Sun S, Babola T, Pregernig G et al (2018) Hair cell mechanotransduction regulates spontane-ous activity and spiral ganglion subtype specifi-cation in the auditory system. Cell. https://doi.org/10.1016/j.cell.2018.07.008

10. Petitpré C, Wu H, Sharma A et al (2018) Neu-ronal heterogeneity and stereotyped

connectivity in the auditory afferent system. Nat Commun 9(1). https://doi.org/10.1038/s41467-018-06033-3

11. van den Brink SC, Sage F, Vértesy A et al (2017) Single-cell sequencing reveals dissociation-induced gene expression in tissue subpopulations. Nat Methods 14 (10):395–396

12. Matern MS, Beirl A, Ogawa Y et al (2018) Transcriptomic profiling of zebrafish hair cells using RiboTag. Front Cell Dev Biol 6:47. https://doi.org/10.3389/fcell.2018.00047

13. Heiman M, Schaefer A, Gong S et al (2009) Development of a BACarray translational profiling approach for the molecular characterization of CNS cell types. Cell Cell 14 (1354):738–748. https://doi.org/10.1016/j.cell.2008.10.028

14. Sanz E, Yang L, Su T, Morris DR, Mcknight GS, Amieux PS (2009) Cell-type-specific isolation of ribosome-associated mRNA from complex tissues. PNAS 106(33):13939–13944

15. Gay L, Karfilis KV, Miller MR, Doe CQ, Stankunas K (2014) Applying thiouracil tagging to mouse transcriptome analysis. Nat Protoc 9 (2):410–420. https://doi.org/10.1038/nprot.2014.023

16. Roh HC, Tsai LT-Y, Lyubetskaya A, Tenen D, Kumari M, Rosen ED (2017) Simultaneous transcriptional and epigenomic profiling from specific cell types within heterogeneous tissues in vivo. Cell Rep 18:1048–1061. https://doi.org/10.1016/j.celrep.2016.12.087

17. Gay L, Miller MR, Ventura PB et al (2013) Mouse TU tagging: a chemical/genetic intersectional method for purifying cell type-specific nascent RNA. Genes Dev 27(1):98–115. https://doi.org/10.1101/gad.205278.112

18. Erickson T, Nicolson T (2015) Identification of sensory hair-cell transcripts by thiouracil-tagging in zebrafish. BMC Genomics 16 (1):842. https://doi.org/10.1186/s12864-015-2072-5

19. Chessum L, Matern MS, Kelly MC et al (2018) Helios is a key transcriptional regulator of outer hair cell maturation. Nature 563 (7733):696–700. https://doi.org/10.1038/s41586-018-0728-4

20. Matern MS, Milon B, Lipford EL et al (2020) GFI1 functions to repress neuronal gene expression in the developing inner ear hair cells. Development 147(17):dev186015. https://doi.org/10.1242/dev.186015

21. Sadler E, Ryals MM, May LA et al (2020) Cell-specific transcriptional responses to heat shock in the mouse utricle epithelium. Front Cell Neurosci 14. https://doi.org/10.3389/fncel.2020.00123

22. Kane KL, Longo-Guess CM, Gagnon LH, Ding D, Salvi RJ, Johnson KR (2012) Genetic background effects on age-related hearing loss associated with Cdh23 variants in mice. Hear Res 283(1–2):80–88. https://doi.org/10.1016/j.heares.2011.11.007

23. Cox B, Liu Z, Mellado Lagarde MM, Zou J (2012) Conditional gene expression in the mouse inner ear using Cre-loxP. J Assoc Res Otolaryngol 13:295–322. https://doi.org/10.1007/s10162-012-0324-5

24. Matern MS, Vijayakumar S, Margulies Z et al (2017) Gfi1Cre mice have early onset progressive hearing loss and induce recombination in numerous inner ear non-hair cells. Sci Rep 7:42079. https://doi.org/10.1038/srep42079

25. Driver EC, Kelley MW (2010) Transfection of mouse cochlear explants by electroporation. Curr Protoc Neurosci Chapter(suppl. 51): Unit. https://doi.org/10.1002/0471142301.ns0434s51

26. Yang H, Gan J, Xie X et al (2011) Gfi1-Cre knock-in mouse line: a tool for inner ear hair cell-specific gene deletion. Genesis 48 (6):400–406. https://doi.org/10.1002/dvg.20632.Gfi1-Cre

27. Walters BJ, Yamashita T, Zuo J (2015) Sox2-CreER mice are useful for fate mapping of mature, but not neonatal, cochlear supporting cells in hair cell regeneration studies. Sci Rep 5 (March):11621. https://doi.org/10.1038/srep11621

28. Fang J, Zhang W-C, Yamashita T, Gao J, Zhu M-S, Zuo J (2012) Outer hair cell-specific prestin-CreER T2 knockin mouse lines. Genesis 50(2):124–131. https://doi.org/10.1002/dvg.20810

29. Song Y, Milon B, Ott S et al (2018) A comparative analysis of library prep approaches for sequencing low input translatome samples. BMC Genomics 19(1):696. https://doi.org/10.1186/s12864-018-5066-2

30. Love MI, Huber W, Anders S (2014) Moderated estimation of fold change and dispersion for RNA-seq data with DESeq2. Genome Biol 15(12):550. https://doi.org/10.1186/s13059-014-0550-8

31. Mandelboum S, Manber Z, Elroy-Stein O, Elkon R (2019) Recurrent functional misinterpretation of RNA-seq data caused by sample-specific gene length bias. PLoS Biol 17(11). https://doi.org/10.1371/journal.pbio.3000481

32. Hansen KD, Irizarry RA, Wu Z (2012) Removing technical variability in RNA-seq data using conditional quantile normalization. Biostatistics 13(2). https://doi.org/10.1093/biostatistics/kxr054

33. Shen J, Scheffer DI, Kwan KY, Corey DP (2015) SHIELD: an integrative gene expression database for inner ear research. Database:1–9. https://doi.org/10.1093/database/bav071

Part III

Hair Cell Function and Physiology

In Vivo Analysis of Hair Cell Sensory Organs in Zebrafish: From Morphology to Function

Saman Hussain, Roberto Aponte-Rivera, Rana M. Barghout, Josef G. Trapani, and Katie S. Kindt

Abstract

Hair cells are the sensory receptors of the vertebrate auditory and vestibular systems. In aquatic vertebrates, hair cells are present in the inner ear, where they are required for hearing and balance, and in the lateral line, where they detect fluid flow. In mammals, hair cell epithelia are embedded within a bony labyrinth in the skull, which makes access challenging for in vivo studies. In larval zebrafish, however, both inner ear and lateral line hair cells can be easily studied in vivo as they are present in accessible locations and in transparent tissue. In addition, zebrafish hair cells have remarkable genetic conservation with humans, and recent advances in reverse genetics have streamlined the process to generate novel zebrafish mutants. Thus, zebrafish are suitable models for studying the genetics underlying hearing and balance and contributing a mechanistic understanding of human sensorineural hearing loss. This methods chapter serves as a guide to creating and studying novel zebrafish mutants, covering a range of topics from zebrafish breeding and propagation to embryo microinjections for targeted mutagenesis (morpholino, CRISPR-Cas9, and tol2-based transgenesis). Once these mutants are created, we discuss ways to test auditory-vestibular phenotypes. We also outline several powerful in vivo methods to study inner ear and lateral line morphology, assess hair cell function, and examine subcellular structures within hair cells. We showcase several previously characterized zebrafish mutants that can be used as controls and exemplars (*emx2*, *cdh23*, *pcdh15a*, *myo7aa*, *ca$_V$1.3a*) for these approaches. Finally, we discuss recent in vivo advances being applied to study and manipulate hair cells with optogenetics techniques such as genetically-encoded indicators and actuators for functional studies.

Key words Hair cells, Zebrafish, Lateral line, In vivo imaging, Morpholino, CRISPR-Cas9, Optophysiology, Electrophysiology

1 Introduction

In vertebrates, specialized mechanoreceptors called hair cells, located within the inner ear, are required for hearing and balance. In aquatic vertebrates such as zebrafish (*Danio rerio*), hair cells are also utilized by lateral line organs to detect local water movements [1]. In mammals, the inner ear is embedded within a small, bony

Andrew K. Groves (ed.), *Developmental, Physiological, and Functional Neurobiology of the Inner Ear*, Neuromethods, vol. 176, https://doi.org/10.1007/978-1-0716-2022-9_9,

labyrinth in the skull, which makes it challenging to access the hair cells for in vivo functional and developmental studies. Typically, acute or cultured explants have been developed and used for live imaging of the inner ear sensory epithelia. Although valuable, these techniques require challenging dissections, and to study inner ear development, both maternal and embryonic dissections are required [2, 3]. Furthermore, these explants may not mimic the exact native environment contained within the inner ear. These limitations to studying hair cells in mammals have made it challenging to investigate the morphology, development, and function of hair cell sensory epithelia in vivo.

In contrast to mammals, the zebrafish is a model system that is amenable to the study of inner ear and lateral line sensory epithelia in vivo. Zebrafish eggs are externally fertilized and require no dissection. After fertilization, the developing embryos (0–72 h post fertilization (hpf)) are transparent through late larval stages (20 days post fertilization (dpf)) [4, 5]. Zebrafish develop rapidly—with regard to hair cell systems, the first hair cells form at 24 hpf, and both the inner ear and lateral line are functional when the larvae are 5 dpf [6–9]. Transparency along with rapid embryonic and larval development has made zebrafish a valuable model system to visualize and study hair cell epithelia in vivo. These benefits have been further enhanced by the fact that zebrafish are genetically tractable, with a large number of molecular and genetic tools available to create transgenic lines or disrupt the function of a gene of interest [10–17]. For example, transgenic lines expressing genetically encoded fluorescent proteins (such as GFP) are invaluable tools to visualize cellular structures in the context of development, disease, and function.

Transgenic approaches have also been used to understand the in vivo localization of proteins within cells, including proteins associated with human hearing loss [18–21]. These transgenic approaches are also used to apply optogenetics to study and manipulate hair cell function in zebrafish [22–24]. In addition to transgenic approaches, initial forward genetics studies in zebrafish revealed that the same core molecules are required for hearing and balance in humans, mice, and zebrafish [25–32]. Furthermore, advances in reverse genetics, through the use of morpholinos, TALENs, and more recently, the CRISPR-Cas9 system, have provided a powerful way to create mutations in a gene of interest [11, 13]. These reverse genetic methods have provided researchers a more targeted way to create zebrafish mutants. For example, reverse genetics to create zebrafish mutants has proved a valuable way to verify and study genomic loci associated with human deafness and vestibular dysfunction [33, 34]. These zebrafish mutants provide excellent in vivo models to study these loci and can complement mouse models when studying the role of genes in hair cell epithelia [35].

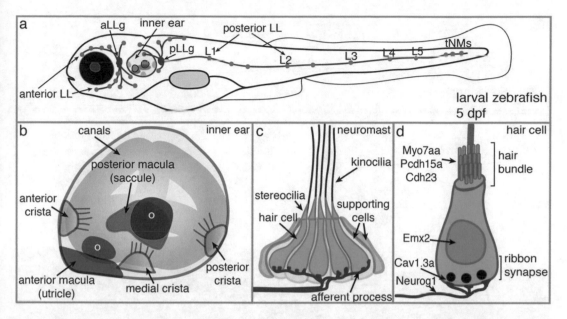

Fig. 1 Location of hair cell sensory epithelia in larval zebrafish. (**a**) A diagram showing the locations of hair cells in larval zebrafish at 5 dpf. The anterior lateral line (LL) and posterior LL neuromasts are marked in green. The locations of the anterior lateral line ganglion (aLLg) and posterior lateral line ganglion (pLLg) are shown in blue. The anterior and posterior lateral lines are responsible for the detection of fluid flow. The location of the zebrafish inner ear, which is essential for hearing and balance, is also shown. (**b**) A zoomed in representation of the zebrafish inner ear showing the locations of the sensory epithelia. The posterior macula (magenta) is primarily involved in hearing, and the anterior macula (blue) is primarily involved in balance. The otoliths are marked with "o's". The locations of the three cristae (anterior, medial, and posterior), which detect angular acceleration, are shown in yellow. (**c**) Hair cells in neuromasts are arranged in a cluster surrounded by supporting cells, which play important roles in the protection, proper development, and regeneration of hair cells. At the apex of hair cells, kinocilia and stereocilia form the mechanosensory hair bundle. Afferent neuronal processes (blue) innervate the basal end of the hair cells. (**d**) Shown is a schematic of an individual hair cell. The hair cell has several important subcellular compartments. At the apex lies the mechanosensitive organelle, called the hair bundle, which consists of actin-rich stereocilia and the tubulin-filled kinocilium (gray). Hair bundles also contain myosins (e.g., Myo7aa) and other components essential for mechanotransduction (e.g., the tip link components Pcdh15a and Cdh23). The base of the hair cell contains the ribbon synapse and presynaptic Cav1.3a clusters. For proper development, the transcription factor Emx2 is present in the nucleus. At the postsynapse, transcription factors such as Neurog1 are required for development of afferent neurons

In zebrafish, hair cell epithelia are found in two main locations: the inner ear and lateral line (Fig. 1a). Similar to mammals, the zebrafish inner ear has two maculae, a saccule and utricle, as well as three cristae (Fig. 1b; [36]). Zebrafish do not have a cochlea, the specialized auditory organ used by mammals. Instead, zebrafish primarily use their saccule to detect sound [37] along with the lagena that develops at later stages (*see* below). The zebrafish utricle is used for balance and to detect gravity [38, 39]. The three cristae, with associated semicircular canals, are used to detect angular acceleration [40]. While the saccule and utricle are functional in larvae

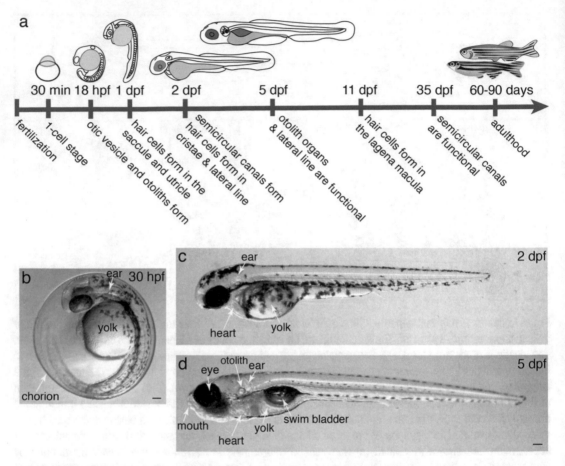

Fig. 2 Developmental trajectory of hair cell epithelia in zebrafish. (**a**) A timeline outlining the key stages of hair cell sensory epithelia development. The otoliths and otic vesicle begin to form as early as 18 hpf. Hair cells first form in the inner ear anterior and posterior macula at 1 dpf and in the cristae and lateral line at 2 dpf. By 5 dpf, the hair cells in the otolith organs and lateral line are functional. Hair cells in the lagena macula start to form later at 11 dpf. The cristae of the semicircular canals become functional at 35 dpf. (**b**) A zebrafish embryo is shown at 30 hpf. This embryo is unhatched and encompassed in the chorion. At 30 hpf, the yolk sac is large and prominent, and the head is curved around the yolk at an angle to the trunk. Even at this stage, the zebrafish inner ear is already present. (**c**) Larvae at 2 dpf begin to hatch, still retaining a prominent but smaller yolk sac. The inner ear continues to develop. Melanophores are beginning to organize at the lateral stripe. (**d**) At 5 dpf, the swim bladder is inflated, and the yolk has almost disappeared. The inner ear and heart are easily visible. More melanophores have been deposited along the lateral stripes. Images were acquired under a dissecting microscope, scale bars = 0.1 mm in **b–d**

by 5 dpf, the cristae are not functional until the zebrafish are ~30 dpf, when the canals are large enough to sustain the inertial forces to drive endolymph flow within the canals (Fig. 2a; [40]). Along with these five inner ear epithelia, the zebrafish forms an additional macula, the lagena, around 11 dpf (Fig. 2a). Similar to other teleost and elasmobranch fishes, the lagena is proposed to contribute to sound detection [41, 42]. In addition to the inner ear, zebrafish also rely on hair cells present in the lateral line. The lateral line system is specialized for detecting fluid flow and is important for many behaviors including shoaling, feeding, and evading predators [8, 43–46]. The lateral line is composed of patches of neuroepithelium just beneath the skin called neuromasts (Fig. 1c). Similar to the inner ear sensory epithelia, neuromasts are composed of hair cells, glia-like supporting cells, and afferent and efferent neurons (Fig. 1c). The development of the lateral line system occurs rapidly, and is functional by 5 dpf [8]. While the zebrafish inner ear has proved advantageous for studying hair cell epithelia in vivo, generally the lateral line is the more commonly studied hair cell system. This popularity is largely due to the superficial location of lateral line neuromasts with the hair bundles projecting directly into the aqueous environment. This location makes the lateral line system ideal for straightforward in vivo imaging, stimulation, physiology, pharmacology, and toxicology, including ototoxicity analyses [47–51].

Regardless of the hair cell epithelium, mammalian and zebrafish hair cells function using two conserved and specialized structures: a mechanosensory hair bundle and a ribbon synapse (Fig. 1d). At the apex of all hair cells is a specialized structure called the mechanosensory hair bundle. The hair bundle is composed of microvilli-like structures called stereocilia that are tightly packed with actin and arranged asymmetrically in rows of increasing height [52]. Extracellular tip links (made of Cadherin 23 (Cdh23) and Protocadherin 15 (Pcdh15)) link each stereocilium in the taller row to an adjacent stereocilium in the neighboring shorter row. In zebrafish hair cells and mammalian vestibular hair cells, a single tubulin-based kinocilium is located adjacent to the tallest row of stereocilia (Fig. 1c; [47]). Mechanical deflection of hair bundles toward the tallest row of stereocilia excites the hair cell [53].

When hair bundle deflection occurs, there is an increase in tension on the tip links, which leads to the opening of the mechanoelectrical transduction (MET) channels located at the lower end of each tip link [29, 32]. When open and due to the large pore size of the channel, these nonselective MET channels permit the passage of cations, particularly an influx of K^+ and Ca^{2+}, which leads to graded changes of the hair cell receptor potential [54]. The hair cell receptor potential determines the open probability of voltage-gated L-type calcium channels (VGCCs, $Ca_V 1.3$) located at the presynapse. Therefore, graded mechanotransduction leads to graded changes in Ca^{2+} at the basally located presynapses, which triggers varying amounts of synaptic vesicle fusion and neurotransmitter release. Diffusion of neurotransmitter (glutamate) leads to activation of postsynaptic receptors on afferent neurons ultimately evoking action potentials (spikes) at rates that are correlated with the frequency and magnitude of hair bundle deflections. These sensory-encoded spikes are then propagated to the central nervous system for perception and reflexes [28, 55–57].

In order to accomplish the transduction of precise and sustained hair cell stimuli, zebrafish and mammalian hair cells both rely on a specialized ribbon synapse [58, 59]. This synapse is defined by a dense body known as a "ribbon" at the presynapse. The ribbon presumably acts to tether vesicles at the active zone, ensuring the presence of a pool of vesicles for sustained neurotransmitter release. In zebrafish, the ribbon synapse is defined both presynaptically by the ribbon body and postsynaptically by afferents with bouton endings (Fig. 1d; [58]). Each single afferent can innervate multiple hair cells similar to the innervation pattern of type 2 vestibular hair cells in mammals [60].

Currently, next-generation sequencing (whole-genome and exome sequencing, RNA sequencing, single-cell RNA sequencing) has advanced the genetics of hearing and balance at both the genomic and the transcriptomic levels. Cumulatively, these analyses have dramatically increased the number of genes associated with hair cell sensory systems [61–69]. The zebrafish is a compelling model system to verify this new genetic information in a relatively high throughput, yet inexpensive and rapid way. Below, we outline several ways to generate a zebrafish mutant model by disrupting a gene of interest (morpholino, CRISPR-Cas9, and transgenic approaches). We then highlight in vivo methods that can be used to characterize the novel mutant, such as examining auditory and vestibular phenotypes. In addition, we outline how to mount, image, and visualize hair cell epithelia in vivo. We also discuss how to use the vital dye FM 1–43 to label and assess hair cell mechanotransduction. Finally, we conclude by describing more advanced in vivo methodologies used to explore how a gene of interest impacts hair cell systems in zebrafish. Comprehensively, these methods provide a useful primer to initiate in vivo functional and morphological analyses of hair cell epithelia in zebrafish.

2   Materials and Methods

2.1 Propagation and Maintenance of Embryonic and Larval Zebrafish

Current NIH policy states that zebrafish 3 dpf or older (after hatching from their chorion or surrounding eggshell, Fig. 2b) are vertebrate animals and experiments be approved by an Institutional Animal Care and Use Committee (IACUC) (https://oacu.oir.nih.gov/sites/default/files/uploads/arac-guidelines/zebrafish.pdf). Our work here was approved by NIH ASP #1362-19 and #3925-1 (Amherst College).

Prior to generating or examining a zebrafish mutant for hair cell-related phenotypes, it is important to understand how to maintain and propagate zebrafish. The majority of in vivo work on hair cell sensory systems in zebrafish is performed during embryonic (0–72 hpf) and larval (3–7 dpf) stages. To produce embryos and larvae for the methods outlined in this chapter, a breeding colony of adults is maintained to produce zebrafish of the appropriate age. The colony of adult male and female breeders (Fig. 2a; 3 months to 1.5 years of age) are housed in an aquatics facility at 28.5 °C with a 14 h/10 h light/dark cycle [70]. Adult pairs are set up in the evening in breeding tanks to produce embryos for analyses (Aqua-neering, #ZHCT100; Aquatic Habitats/Pentair, BTANK2D, SBTANK2). For timed spawning (required for microinjections for transgenesis, morpholino, and CRISPR-Cas9-related work, *see* Subheading 2.2), a divider is placed in the breeding tanks to sepa-rate the male and female zebrafish. In the morning, when the lights turn on, the breeders are cued to spawn. For divided pairs, spawn-ing occurs after the lights turn on, but only after the divider separating the pair is removed. Each adult pair can be set up weekly, and a spawning between a single pair can yield hundreds of eggs [70]. Fertilized eggs are collected in E3 media from the mating tank. To make 1 L of a 60× E3 stock, combine: 17.2 g NaCl, 0.76 g KCl, 19.8 mL 1 M $CaCl_2$, 19.8 mL 1 M $MgSO_4$, 6 mL 1 M HEPES (MilliporeSigma) along with Milli-Q water to a final vol-ume of 1 L. Dilute to 1× E3. The final concentration of the 1× E3 working solution is: 5 mM NaCl, 0.17 mM KCl, 0.33 mM $CaCl_2$, 0.33 mM $MgSO_4$, and 10 mM HEPES, pH 7.2. E3 media is used for many experiments including injections, mounting, and dye labeling (Subheadings 2.2, 2.3, 2.4, 2.5, and 2.6).

After collection in E3 media, fertilized embryos are placed in 100 mm petri dishes. Immediately after fertilization, embryos can be used for microinjection (Subheading 2.1) or simply propagated in an incubator (Benchmark Scientific, H2505-130) at 28.5 °C in E3 media from 0–7 dpf. Incubators can be placed on a 14 h/10 h light/dark cycle matching that of the aquatics facility using LEDs (Westek LED Ultra-Thin White Strip Light, USL30HBCC) set on a timer (Intermatic, TN111K) placed inside the incubator.

After hatching, disposable plastic transfer pipettes (Fisher Sci-entific, 13-711-7M) are used to place embryos in plates with new E3 media (*see* **Note 1**). At 5–7 dpf, larvae can be placed on the nursery and grown in the aquatics facility. At 7 dpf, larvae no longer have yolk to sustain them and must be propagated or euthanized. For euthanasia in our animal protocol, we place embryos and larvae (0–7 dpf) in an ice bath and then transfer them directly into a −20 °C freezer.

This methods chapter describes the following zebrafish trans-genic lines: $Tg(myo6b:actb1-EGFP)^{vo8}$, $Tg(myo6b:emx2-P2A-NLS-mCherry)^{idc4}$, $Tg(myo6b:GCaMP6s-CAAX)^{idc1}$, $TgBAC(neurod1:EGFP)^{nl1}$, $Tg(myo6b:Tmc2a^{1-117}-GFP-CAAX)^{vo11}$, $Tg(myo6b:RGECO1)^{vo10}$, $Tg(myo6b:ChR2-YFP)^{ahc1}$, $Tg(myo6b:KDEL-mkate2)$ (unpublished) [23, 28, 35, 47, 50, 71], and zebrafish mutants: $emx2^{idc5}$, $cdh23^{aj64a}$, $pcdh15a^{th263b}$, $myo7aa^{ty220d}$, $ca_V1.3a^{tn004}$ [25, 29, 31, 32, 35, 50].

Wild-type zebrafish (we use Tubingen (Tu), Tubingen longfin (TL), and Tubingen AB (TAB) strains) are available from the Zebrafish International Resource Center (ZIRC), University of Oregon: http://zebrafish.org/zirc/home/guide.php. In addition to wild-type zebrafish, ZIRC also has many zebrafish mutants that have been created using chemical ENU or retroviral mutagenesis [10, 15]. A search for your gene of interest on the ZIRC website or on the Zebrafish Information Network (ZFIN), https://zfin.org/ may reveal an existing mutant. It is important to point out that there was a whole-genome duplication event in the teleost lineage after it diverged from mammals [72]. This duplication means that in some instances, your gene of interest may have two paralogous copies (e.g., two paralogs for *pcdh15a* and *pcdh15b*, versus one ortholog for *cdh23*). If both paralogs are not coexpressed in your cell types of interest (e.g., hair cells), it is possible that creating a single mutant is sufficient to assess for a potential phenotype (*pcdh15a* is expressed in lateral line hair cells and *pcdh15b* is not). If both paralogs are expressed in your cell type of interest, it is possible that both genes copies may function redundantly, and a double mutant may be required to assess for a potential phenotype. The expression pattern of your gene(s) of interest, if known, may be documented on ZFIN. RNA sequencing can also help determine expression patterns [73]. Many zebrafish mutants are available from ZIRC.

2.2 Microinjection of Zebrafish Embryos to Investigate Gene Function

Microinjections of zebrafish embryos are required for many important applications, including dye labeling, transgenesis, random and targeted mutagenesis, and mRNA knockdown [10, 11, 13, 14, 74]. Microinjection is an important starting point in targeted approaches to generate zebrafish mutants. We highlight three ways in which embryo microinjections can be used to manipulate gene function in zebrafish: (1) morpholino-induced mRNA disruption, (2) CRISPR-Cas9 genome mutagenesis, and (3) DNA-based transgenesis to create gain-of-function models.

For all three applications, the equipment and overall procedure for embryo microinjection is quite similar [75, 76]. A microinjection setup consists of a dissecting microscope (Subheading 2.3), equipped with a mechanical micromanipulator (Narishige, M-152) mounted on a magnetic stand (Tritech research, MINJ-HBMB). In addition, a pressure injector (WPI, PV 820) is used, which is connected to pressurized air and equipped with a needle holder (WPI, S420-ALL) that is then mounted on the micromanipulator. To create microinjection needles, glass capillaries (WPI, TW100-4) are fabricated with a micropipette puller (e.g., Sutter P97) to obtain long and fine-tipped needles. The needle tips are broken with fine forceps to create a ~2–3 μm opening. A P10 pipette equipped with a microloader pipette tip (Eppendorf, 930001007) is used to fill the tip of the injection needle with 2–3 μL injection solution.

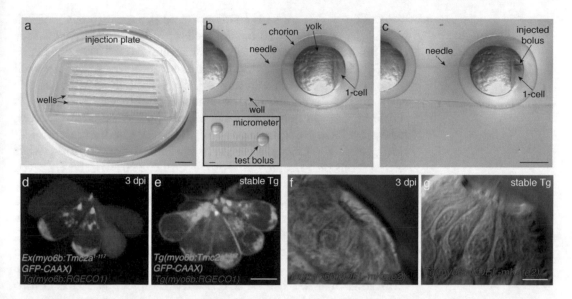

Fig. 3 Microinjection of zebrafish embryos to create transgenic, morphants, or CRISPR-Cas9 mutants. (**a**) An injection plate made of 1.5% agarose solution in E3 media is shown here inside a petri dish. The plate contains wells in which the embryos are arranged for injections. (**b**) A typical one-cell stage embryo that is ready for injection inside a well. The chorion, yolk, and embryo are visible with a dissecting microscope. The injection needle can also be seen here out of focus, containing the required solution for transgenics, morphants, or CRISPR-Cas9 mutations. A typical injection bolus is 100–200 μm in diameter (inset). (**c**) An injected bolus can be visualized due to the presence of phenol red in the injection solution, making it easy to determine whether a successful injection has occurred. The injection is typically carried out at this stage, and the solution is injected directly into the single cell. Images in **b**–**c** were acquired by Katherine Pinter. (**d**) Example of transient expression in neuromast hair cells expressing a gain of function Tmc2a clone (*myo6b:tmc2a$^{1-117}$ GFP-CAAX*) at 3 days post injection (dpi). Embryos were injected at the one-cell stage with 30 ng/μL *myo6b:tmc2a$^{1-117}$ GFP-CAAX* DNA and 25 ng/μL tol2 RNA. In this image, another transgene *Tg(myo6b:RGECO1)* is used to show all hair cells present. Here 11 hair cells express *Tg(myo6b:RGECO1)*, but only three hair cells express *myo6b: tmc2a$^{1-117}$ GFP-CAAX*. (**e**) After larvae transiently expressing the *myo6b:tmc2a$^{1-117}$ GFP-CAAX* transgene were grown to adulthood, a stable transgenic (Tg) line was identified. Here all hair cells now express both *Tg (myo6b:RGECO1)* and *Tg(myo6b:tmc2a$^{1-117}$ GFP-CAAX)* transgenes. (**f**) Example of transient expression of an ER-localized red fluorescent protein mKate2 transgene in neuromast hair cells 3 days post injection (dpi). Embryos were injected at the one-cell stage with 50 ng/μL *myo6b:KDEL-mkate2* DNA and 25 ng/μL tol2 RNA. In this image, only two hair cells express *myo6b:KDEL-mkate2*. (**g**) After larvae transiently expressing the *myo6b:KDEL-mkate2* transgene were grown to adulthood, a stable transgenic (Tg) line was identified. Here all hair cells now express the *myo6b:KDEL-mkate2* transgene. Images in **a**–**c** were taken under a dissecting microscope. For images in **d**–**g**, larvae were mounted in LMP agarose in a glass depression slide and imaged using a confocal microscope at 63×. Scale bar in **a** = 10 mm, inset in **b** = 100 μm, **b**–**c** = 250 μm, **d**–**g** = 5 μm

Embryos are held and injected in the wells formed within a microinjection plate using a plastic mold (Fig. 3a). The plates are made using a molten 1.5 % agarose solution prepared with E3. After pouring 20–50 mL of the agarose solution into a 100 × 15 mm petri dish, a plastic mold (Adaptive Science Tools, TU-1 or I34) is placed on top of the agarose to create the embryo wells. The plastic mold is removed after the agarose is solidified. Microinjection plates can be wrapped with Parafilm (Fischer Scientific, S37440) and stored at 4 °C for several days.

For microinjection, newly fertilized embryos at the one-cell stage are injected (Figs. 2a and 3b, c). To obtain embryos at this early stage, timed mating of adult zebrafish using dividers is required (*see* Subheading 2.1). Using plastic transfer pipettes, newly fertilized embryos are deposited into the wells of microinjection plates. The embryos can be oriented and secured in the wells with a dissection needle (ThermoScientific, 19010). The embryos are oriented to align the single cell of the embryo with the trajectory of the microinjection needle (Fig. 3b, c). The manipulator containing the injection solution is then advanced through the protective chorion surrounding the embryo and into the fertilized, single cell (Fig. 3b, c). The pressure injector is then triggered to deliver a 500 pL to 4 nL bolus (100–200 μm diameter) into the embryo. Bolus size can be determined by injecting solution into mineral oil on a slide micrometer and calculating volume from the diameter of the droplet formed within the oil (Fig. 3b inset; Fine Science Tools, MA285). After injections, the embryos are washed with E3 and stored in an incubator at 28.5 °C (Subheading 2.1). Embryos and larvae can be screened for expression and phenotypes days after injection (Figs. 3d, f and 6h, i). In addition, after microinjection to create CRISPR-Cas9 mutants or after DNA-based transgenesis, larvae can be grown to adulthood and progeny can be tested for germline mutagenesis or transgenesis, respectively (Fig. 3e, g).

Morpholino-based gene knockdown is used to target and disrupt mRNA in developing embryos and larvae. Morpholinos are synthetic oligos that bind and disrupt mRNA [11]. Morpholino can be designed to bind to the ATG of mRNA and block translation. Morpholinos can also be designed to bind to splice site junctions and disrupt splicing. Morpholinos that target your gene of interest can be designed and ordered with assistance from Gene tools (https://www.gene-tools.com/). Morpholino-injected embryos, called "morphants," can be examined hours and days after injection (e.g., Fig. 6h, i). Because morpholinos dilute with each cell division, some phenotypes are no longer apparent past 3–4 dpf as the amount of morpholino present is insufficient to disrupt mRNA transcripts. Furthermore, morpholinos target the transcriptome, are not stable germline mutants and therefore cannot be propagated in subsequent generations. It is important to point out that morpholinos should be used with care—recent studies indicate that morpholino phenotypes are not always recapitulated in stable germline mutants [77]. Currently the zebrafish community highly recommends that morpholino analyses be validated with a stable germline mutant line whenever possible [78]. In addition, there are several guidelines to follow: (1) validate phenotypes using multiple morpholinos, along with control morpholinos, (2) whenever possible validate gene knockdown using RT-PCR, qPCR, or immunohistochemistry, (3) rescue the morpholino-associated phenotype with an mRNA not targeted by the morpholino, and lastly (4) be a robust experimentalist: perform a dose–response curve, use a sufficient number of animals for statistical power, and employ blinding strategies [78].

Sample morpholino injection solution

10–1000 µM Morpholino (Gene tools).

0.5 µL Phenol red (MilliporeSigma P0290).

Dilute to 10 µL with sterile water.

Store at room temperature.

CRISPR-Cas9 is a powerful way to create stable, germline zebrafish mutants [13, 17, 79]. This method relies on single-guide RNAs (sgRNAs) that work with Cas9 protein to cut the genome in a targeted manner. SgRNAs can be used to create small insertions and deletions (INDELs) in the genome that ultimately disrupt gene function [17]. In addition, multiple sgRNAs can be combined to create larger genomic deletions [80] or to create overlapping knockdown of multiple genes simultaneously [81, 82]. Several methods highlight how to design and either generate or purchase sgRNAs [12, 81]. If effective sgRNAs are used and optimized, phenotypes in "crispants" (CRISPR-Cas9-injected embryos) can be assessed hours and days after microinjection, similar to morphants [12]. Unlike morphants, however, mutations created using CRISPR-Cas9 target the genome and can be propagated in subsequent generations. It is important to point out that there is evidence for genetic compensation in CRISPR-Cas9 stable mutant lines. In these stable mutants, there is evidence that gene homologs can be upregulated and can mask phenotypes or phenotype severity [83–85]. This genetic compensation is not present when using morpholinos. Therefore, in some instances, complementary use of CRISPR-Cas9 and morpholinos can an effective way to study gene function.

Sample CRISPR-Cas9 injection solution

1000 ng/µL Cas9 protein (IDT, 1081058).

500–850 ng/µL sgRNA (for synthesis options *see*: Hoshijima et al. [12]; Varshney et al. [17]).

Dilute to 5 µL with KCl, phenol red and sterile water for a final concentration of 300 mM KCl and 5% phenol red (MilliporeSigma P0290).

Store at room temperature and use same day.

Morpholinos and CRISPR-Cas9-based mutagenesis are predominantly used to provide targeted gene disruption to create loss of function mutants. Alternatively, gene product

Table 1
List of existing promoters for expression in hair cell epithelia

Promoter/enhancer	Cell type	References
myo6b, ctbp2a, pvalb9, atoh1a, pou4f3, myo7aa	Hair cell	[28, 88–92]
she, etv5b	Supporting cells	[93, 94]
neurod, SILL1	Primary afferents	[95, 96]

overexpression or expression of a dominant negative gene product can also be a powerful way to investigate gene function [86]. One way to overexpress a gene product is through injection of wild-type or dominant negative mRNA at the one-cell stage [75, 87]. While fast and powerful, mRNAs are not long lasting (past 2–3 dpf) and are best suited to study early developmental events. Alternatively, DNA expression constructs can be used in zebrafish to drive expression of your gene of interest using an appropriate promoter (e.g., *myo6b*: hair cells; *see* Table 1) (*see* **Note 2**).

In the zebrafish community, the Gateway cloning system is often used to share and generate DNA expression clones [14]. For maintained expression of DNA constructs, tol2-based transgenesis methods are used [14]. In this scenario, the DNA expression construct (*promoter:gene product*) is flanked by tol2 sites. The DNA clone is coinjected with tol2 mRNA to facilitate genome integration and stable gene expression. After microinjection, embryos and larvae with transient expression can be examined for phenotypes hours and days after injection (Fig. 3d, f). In addition, injected larvae can be propagated, and germline transgenesis can be assayed in the subsequent generation (Fig. 3e, g). For example, in a preliminary study, we transiently expressed the N-terminus of the mechanotransduction channel Tmc2a in zebrafish hair cells (*Ex(myo6b:Tmc2a$^{1-117}$-GFP-CAAX)*; expression is only in a subset of hair cells; Fig. 3d). Our preliminary analysis revealed that this fragment acted as dominant negative and could suppress mechanotransduction. For our final published study, we grew up these fish and confirmed the experiment in a stable transgenic line (*Tg(myo6b:Tmc2a$^{1-117}$-GFP-CAAX)^{vo11}*; expression is in all hair cells; Fig. 3e [71]).

Sample tol2 Transgenesis Injection Solution

10–20 ng/µL DNA expression construct flanked by tol2 sites.

10–20 ng/µL Capped tol2 mRNA (synthesized from Addgene clone # 133032; [14]).

0.5 µL Phenol red (MilliporeSigma P0290).

Dilute to 10 µL with sterile water.

Keep frozen or on ice.

2.3 Viewing Gross Inner Ear Morphology and Auditory-Vestibular Behavior Under a Dissecting Scope

After creating, identifying, or obtaining a novel zebrafish mutant, initial screening of embryos and larvae can be achieved with a dissecting microscope or stereo microscope (Figs. 2b–d and 4a, b). Dissecting microscopes are used daily when working with zebrafish including for: developmental staging, examining gross morphology, mounting fish in agarose for higher magnification viewing (Subheading 2.4), and microinjection (Subheading 2.2). Dissecting microscopes with a transmitted light base and an adjustable mirror are preferred (Zeiss, Stemi2000). While stereoscopes with just transmitted light are useful, fluorescent dissecting microscopes (Zeiss, Discovery.V20; Nikon, SMZ25 or SMZ18) are also powerful options for advanced techniques such as sorting and selecting transgenic zebrafish to be used for further experimentation and analyses. Fluorescent dissecting microscopes are also used to select for zebrafish with transient expression of a fluorescent DNA construct after microinjection (Subheading 2.2) and to assess the results of hair cell labeling with vital dyes (Subheadings 2.6 and 3.4).

2.4 Mounting and Immobilizing Embryonic and Larval Zebrafish

To assess whether there are more subtle morphological defects in a zebrafish mutant, it is necessary to view embryos and larvae at a higher magnification. Higher magnification is also useful to assess the expression of DNA constructs after microinjection (Subheading 2.2, Fig. 3d–g). To stably view and image zebrafish embryos and larvae at a higher magnification, they must be anesthetized and immobilized. To anesthetize and paralyze zebrafish, tricaine is added to the E3 or mounting media. A common media to mount and immobilize zebrafish for live imaging is low melting point (LMP) agarose.

To make a stock of 0.4% w/v buffered tricaine solution, dissolve 400 mg tricaine powder (Tricaine methanesulfonate: FDA-approved Tricaine-S, aNaDa #200-226) and 800 mg Na_2PO_4 (MilliporeSigma, S7907) in Milli-Q or double-distilled water (ddH$_2$O). Adjust the pH to 7 and store at 4 °C. Aliquot the 0.4% w/v tricaine stock solution into amber glass dropper bottles (Kimble, 15040P-30) for daily use. Use tricaine at a working concentration of 0.02–0.04% w/v (1:10 or 1:20; 100 μL or 50 μL tricaine in 1 mL E3 media).

To create a 1% working concentration of LMP agarose to embed zebrafish, melt LMP agarose powder (Promega, V2111) in E3 media (e.g., 500 mg LMP agarose in 50 mL E3). This solution can be aliquoted in 2-mL Eppendorf tubes and stored at room temperature for up to 3 months. Prior to use, melt an aliquot in a beaker of boiling water. Place the melted aliquot at 42 °C on a heat block and do not use the aliquot until after the LMP agarose has cooled to 42 °C. Add tricaine to the aliquot to ensure zebrafish are paralyzed when mounted in LMP agarose. Use aliquots stored at 42 °C within 2–3 days.

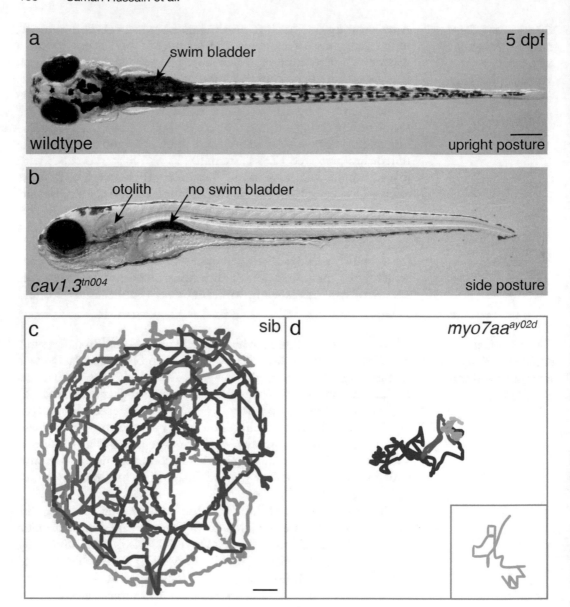

Fig. 4 Overall posture and behavior of wild-type and zebrafish mutants with hearing and balance defects. (**a**) When viewed under a dissecting microscope, wild-type larvae at 5 dpf maintain an upright posture with their dorsal side up. This posture is aided by an inflated swim bladder located above the yolk (yolk is on the ventral side). (**b**) $Ca_V1.3$ is required for hair cell neurotransmission, hearing, and balance in zebrafish. When viewed under a dissecting microscope, $ca_V1.3a$ zebrafish mutants rest on their sides and do not inflate their swim bladder. (**c, d**) Tracks showing swim trajectories of four wild-type and four *myo7aa* zebrafish mutants. *Myo7aa* zebrafish mutants lack hair cell mechanotransduction and have strong auditory and vestibular defects (5 dpf). Trajectories were monitored using a Zantiks MWP (Zantiks, Cambridge, UK) behavior system in 9.5 cm well plates over 2 min. While wild-type larvae make long swim bouts that cover the entire well, *myo7aa* mutants are less motile, make shorter swim bouts, and often swim in circles. Tracking data was kindly provided by Candy Wong. Inset in **d** highlights the trajectory of a single *myo7aa* larvae magnified 5×. Scale bar in **a–b** = 250 μm and in **c–d** = 1 cm

There are numerous chambers and dishes that can be used to mount embryonic or larval zebrafish in LMP agarose for in vivo imaging. Here we describe two approaches. In the first approach, embryos or larvae are mounted in the concave well of a depression slide or in a small plastic petri dish. This type of mounting is useful for viewing fish under a dissecting microscope or an upright microscope (Subheading 2.5). In the second approach, larvae are mounted in a petri dish with a coverslip at the bottom, which is useful when imaging from below on an inverted microscope (Subheading 2.5).

To mount larvae in depression slides (Globe Scientific, 1344), use a glass Pasteur pipette (Fisherbrand, 13-678-20B) to pick up one to two larvae in a minimal amount of E3 media and pipette them into an Eppendorf tube containing 1% LMP agarose (at 42 °C) with added tricaine (at 0.02–0.04% w/v). Make sure the temperature of the agarose has cooled down to 42 °C before adding the larvae. Immediately draw the larvae back into the pipette, along with 200–300 μL of agarose and deposit the entire volume onto the glass depression slide (Fig. 5a). The angle (e.g., top down or side view) of the larvae can be adjusted by gently repositioning with a dissection needle (ThermoScientific, 19010) or forceps for optimal imaging. Once the agarose is set (approximately 5 min), the slide is ready for imaging. Add E3 media with tricaine (at 0.02–0.04% w/v) on top of the agarose prior to imaging to prevent the agarose from drying and also serve as immersion media when using a water immersion or dipping objective. For longer imaging sessions using an upright microscope, larvae can be mounted in LMP agarose in a similar fashion but in plastic petri dishes (Corning, 430558). Once the agarose has set, the petri dish can be partially filled using E3 media with tricaine at 0.02–0.04% w/v (*see* **Note 3**).

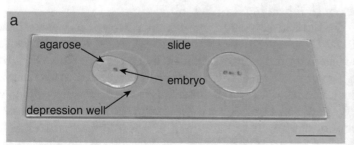
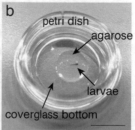

Fig. 5 Mounting and immobilization of larval zebrafish in LMP agarose. (**a**) A depression slide with mounted embryos at 30 hpf is shown. Each depression well contains 1% low melting point (LMP) agarose with added tricaine (at 0.02–0.04% w/v). The larvae were immobilized and positioned in the agarose to acquire the images in Fig. 2b. (**b**) A larva mounted in a glass-bottom petri dish (35 mm Iwaki, 9.380190) is shown. The larva is immobilized in 1% low melting point (LMP) agarose with added tricaine (at 0.02–0.04% w/v), and its position is adjusted to be as close to the cover glass as possible for optimal viewing on an inverted microscope. Scale bars = 10 mm in **a–b**

To mount larvae in a glass-bottom petri dish (35 mm Iwaki, 9.380190; 27 mm ThermoScientific, 1141351), use a glass pipette to pick up one to two larvae and place them in 1% LMP agarose (at 42 °C) with added tricaine (at 0.02–0.04% w/v). Draw the larvae back into the pipette, along with 100–200 μL of agarose and place them in a glass-bottom petri dish (Fig. 5b). Using a dissection needle, press down on the larva gently so it touches the coverslip and hold it down until the agarose begins to solidify. The closer the larvae are to the coverslip, the better the quality of imaging. Once the agarose has set, the chamber must be saturated with media to prevent the agarose from drying out, which is achieved by filling the dish using E3 media with tricaine (at 0.02–0.04% w/v). Another method is to saturate a Kimwipe (Kimtech, 34155) with E3 media containing tricaine (at 0.02–0.04% w/v) and place it around the edges of the dish in a circle. Using the saturated Kimwipe rather than filling the dish with E3 media helps prevent the agarose from slipping off the glass. For time-lapse imaging, seal the dish with parafilm (*see* **Note 3**).

2.5 Viewing and Imaging Zebrafish Hair Cell Epithelia

To capture high-resolution images of hair cell epithelia in your zebrafish mutant in vivo, a compound microscope with a high magnification objective is required. For stability during imaging, compound microscopes are mounted on an anti-vibration table. To visualize larvae on a compound microscope, larvae can be mounted in LMP agarose (Subheading 2.4). Our preference is to use an upright microscope with a water-immersion (dipping) objective that reaches the sample from above. Here the objective is directly immersed in the solution surrounding the larvae (for example, E3 media on the LMP agarose in a depression slide) (Fig. 5a). Alternatively, an inverted microscope can be used along with oil-immersion objectives. On an inverted system, larvae are imaged from below, through a coverslip, for example in a petri dish with a coverslip bottom (Subheading 2.4; Fig. 5b).

On both inverted and upright microscopes, low-magnification (10–20×) air objectives are used to view and image larger structures such as the entire zebrafish inner ear or brain (Fig. 6a–c). High-magnification (40–60×) objectives are used to visualize individual hair cell epithelia or subcellular structures (Figs. 3d–g, 6e–i, 7, 8, and 9). For high-magnification objectives (on either an inverted or upright microscope), it is recommended to use those specialized to acquire fluorescence. In addition, condensers with a polarizing filter, objectives equipped with prisms, and analyzers mounted above the objective (either within filter cubes or as independent sliders) combine to provide a valuable way to visualize cells and structures using differential interference contrast (DIC) optics. On an upright microscope, water-immersion objectives with a longer working distance are preferred in order to image deeper into the larvae (e.g., Nikon Apochromat CFI 60× water-immersion

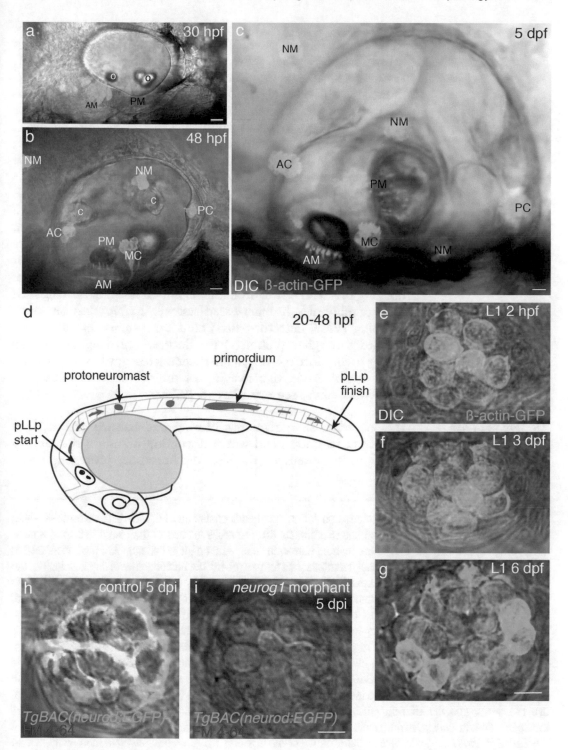

Fig. 6 In vivo images depicting inner ear and lateral line development. (**a–c**) The sensory epithelia of the inner ear develop over several days. The onset of hair cell formation can easily be seen using an established transgenic line such as *Tg(myo6b:actb1-EGFP)*. Using this line, hair cells in the anterior and posterior macula (AM, PM) can be observed at 30 hpf (**a**). The anterior and posterior otoliths are marked with "o's." At 48 hpf,

objective, N.A. 1.0, W.D. 2.0 mm; Nikon Apochromat CFI 40× water-immersion objective, N.A. 0.8 W.D. 3.5 mm; Nikon Apochromat CFI75 25× water-immersion objective, N.A. 1.1, W.D. 1.43–2.04 mm; Zeiss Plan-Apochromat 63× water-immersion objective, N.A. 1.0, W.D. 2.1 mm). For inverted microscopes, resolution (numerical aperture, N.A.) and working distance must be weighed to achieve optimal imaging quality and depth (e.g., Zeiss Plan-Apochromat 63× oil-immersion objective, N.A. 1.4, W.D. 0.19 mm; Zeiss Plan-Apochromat 40× oil-immersion objective, N.A. 1.3, W.D. 0.21 mm).

In vivo imaging of embryonic and larval zebrafish is amenable to many types of compound microscopes. The simplest system is a wide-field microscope. A wide-field epifluorescence microscope can be equipped with a camera (*see* **Note 4**), computer, and fluorescence illumination source. Wide-field microscopy is suitable for many applications, such as DIC-based imaging and fluorescence imaging of lateral line neuromasts. Confocal imaging systems such as a spinning-disk (camera-based systems, *see* **Note 4**) or point-scanning (PMT detector-based) confocal systems that use lasers focused through a pinhole(s) for fluorescence imaging are also widely used. More recently, light-sheet microscopy has been particularly useful for in vivo imaging of the zebrafish inner ear and projections of the sensory afferents [97–99].

2.6 FM 1–43 Dye Labeling of Lateral Line Hair Cells

If your zebrafish mutant has hearing and balance defects, it is important to understand where disruption occurs in the sensory pathways. Hair cells are the first step in sensory processing and a

---

Fig. 6 (continued) hair cells in the anterior, posterior, and medial cristae (AC, PC, MC) can also be seen, along with hair cells of two anterior lateral line neuromasts (**b**). The cross section of the pillars that form around the developing semicircular canals are marked here with "c's." At 5 dpf, hair cell numbers have increased in the sensory epithelia, and an additional neuromast has formed (**c**). (**d**) The posterior lateral line develops as the posterior lateral line primordium (pLLp) migrates toward the tail (shown by orange arrows) from 20–48 hpf. As the pLLp migrates, it deposits protoneuromasts along the lateral line, which develop into mature neuromasts. (**e–g**) The development of a single L1 neuromast at 2, 3, and 6 dpf (*see* Fig. 1a for the position of L1 in the lateral line). Images in **e–g** were acquired by Natalie Mosqueda. During this time window, the number of hair cells have increased from 8 (2 dpf) to 16 (6 dpf). A mature lateral line neuromast contains 16–20 hair cells. (**h, i**) Additional transgenic lines can be used to assess the development and morphology of hair cell epithelia in zebrafish. For example, the *TgBAC(neurod1:GFP)* line labels the afferents that innervate hair cells in zebrafish. Shown in **h** is a wild-type lateral line neuromast at 5 dpf. The hair cells are labeled with the vital dye FM 4–64, and the afferent process beneath the hair cells is labeled with the *TgBAC(neurod1:GFP)* transgenic line. In *neurog1a* morphants (5 days post morpholino injection, dpi), hair cells are still present but the *TgBAC(neurod1:GFP)* line reveals the absence of the afferents (**i**). To create *neurog1a* morphants, a start codon MO was used: 5′ATCGGAGTATACGATCTCCATTGTT3′ (Gene Tools). 1 nL of a 900 μM MO solution was injected. Injections were done by Alisha Beirl. To acquire the images in **a–i**, larvae were mounted in LMP agarose in a glass depression slide and imaged on an upright confocal microscope. Images **a–c** were taken at 10× while **e–i** were taken at 60×. Scale bars in **a–c** = 10 μm and in **e–i** = 5 μm

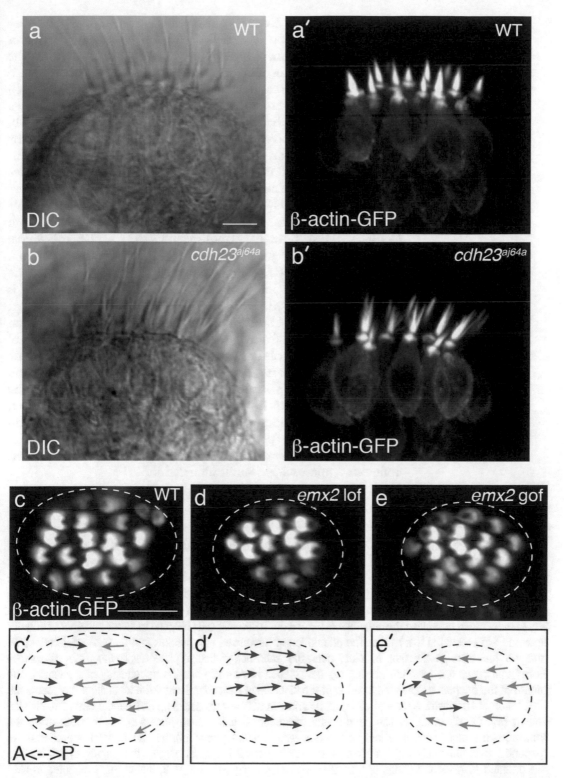

Fig. 7 Visualizing the morphology of zebrafish inner ear and lateral line hair bundles. (**a**) Wild-type hair bundles have a staircase morphology, containing rows of stereocilia with descending heights (Fig. 1d). The stereocilia are actin-rich and hair bundle morphology can be easily observed using DIC optics (**a**) or using the *Tg(myo6b:*

good place to start. Even if your mutant has intact hair cells, they may not function properly at sites of either mechanotransduction or synaptic transmission. A powerful, yet simple way to assess the mechanotransduction function of hair cells in your zebrafish mutant is to use the cationic dye FM 1–43 (*see* **Note 5**). FM 1–43 is a vital dye that enters and fluorescently labels hair cells by passage through intact mechanotransduction (MET) channels (Fig. 8b, [100, 101]). In wild-type larvae, FM 1–43 rapidly and robustly labels lateral line hair cells in whole larvae (Fig. 8a, [102]). This approach has been used to characterize the mechanosensitive defects in numerous zebrafish mutants that lack mechanotransduction [19, 26, 71, 103, 104].

Prepare a 1 mM stock solution of FM 1–4: dissolve 1 mg lyophilized FM 1–43 (Thermofisher, T3163) in 1.5 mL sterile water. Make 10–50 μL aliquots of the stock solution and store parafilmed at −20 °C.

To label lateral line hair cells, a working 1 μM solution is created by diluting the stock solution 1:1000 in E3 media (e.g., 10 μL of the 1 mM stock in 10 mL E3). This working solution should be protected from light and used the same day. Put 1 mL of the 1 μM FM 1–43 solution into an Eppendorf tube. Using a plastic or glass transfer pipette, place the larva into the dye solution and incubate for 30–45 s at room temperature (*see* **Note 6**). Wash the larva three times with E3 media to remove the dye. After labeling, the larvae can be mounted in LMP agarose in a glass depression slide or glass-bottom petri dish (Subheading 2.4) and viewed under a GFP dissecting microscope (Subheading 2.3), or a fluorescent, compound microscope (Subheading 2.5).

Fig. 7 (continued) actb1-EGFP) transgenic line (**a**′) when viewed in a side view. In wild-type (WT) hair cells, hair bundles are held together with cadherin and protocadherin-based tip links. (**b**) In a *cadherin 23* mutant (*cdh23^aj64a^*), the hair bundles are splayed and have lost their typical staircase morphology. These hair cells also lack mechanotransduction (MET) currents. Hair bundles of the medial cristae are shown in **a**–**b**′. (**c**, **c**′) To detect bidirectional flow in zebrafish, lateral line neuromasts contain hair cells that are organized bidirectionally to detect fluid flow along the anterior-posterior axis. These hair cells respond to either posterior or anterior flow and are present in a 1:1 ratio. The orientation of neuromast hair bundles can be easily determined by imaging the base of the hair bundles, near the actin-rich cuticular plate using the *Tg(myo6b:actb1-EGFP)* transgenic line (**c**) when viewing top down. Each hair bundle has the morphology of a pacman. The mouth of the pacman denotes the location of the kinocilium. Deflection toward the kinocilium represents the orientation of sensitivity. A diagrammatic representation with arrows depicting hair bundle orientation is also shown (**c**′). (**d**, **d**′) In lateral line neuromasts, bidirectional hair bundle orientation is determined by the transcription factor Emx2. In a loss of function *emx2* mutant (*emx2* lof), all hair bundles are oriented to respond to posterior fluid flow. (**e**, **e**′) In a gain of function *emx2* mutant (*emx2* gof), the majority of hair bundles are oriented to respond to anterior fluid flow. For image acquisition in **a**–**e**, larvae were mounted in LMP agarose in a glass depression slide and imaged on an upright confocal microscope at 63× or 60×. Scale bar in **a**–**e** = 5 μm

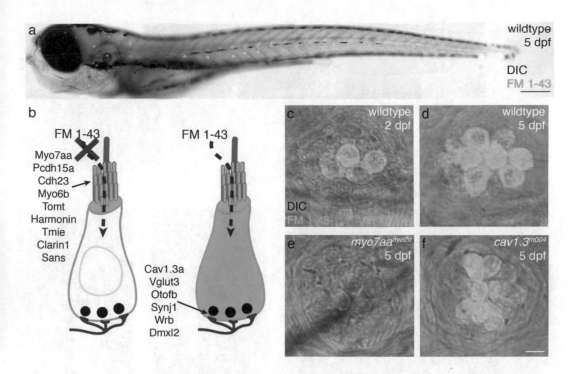

Fig. 8 FM 1–43 labeling of lateral line hair cells. (**a**) Wild-type zebrafish larva at 5 dpf treated with 1 µM FM 1–43 for 30 s. The neuromasts are seen as bright puncta present around the head (anterior lateral line) and along the body (posterior lateral line) of the larva. (**b**) FM 1–43 vital dye rapidly enters hair cells from the aquatic media through mechanotransduction channels. In hair cells without intact mechanotransduction (*myo7aa*, *pcdh15a*, *cdh23*, *myo6b*, *tomt*, *harmonin*, *tmie*, *clarin1*, *sans* mutants), hair cells do not label with FM 1–43. FM 1–43 label still occurs in mutants that impair hair cell neurotransmission (*ca$_V$1.3*, *vglut3*, *otofb*, *synj1*, *wrb*, *dmxl2* mutants). (**c**, **d**) In wild-type larvae, neuromast hair cells at 2 dpf start to label with FM 1–43. Labeling at 2 dpf is reduced compared to neuromasts at 5 dpf due to fewer mature hair cells with intact mechanotransduction at this early stage. FM 1–43 robustly labels the majority of neuromast hair cells in wild-type larvae at 5 dpf when the lateral line system is functional. (**e**, **f**) No FM 1–43 labeling is seen in *myo7aa* zebrafish mutants which lack mechanotransduction (**e**). However, FM 1–43 labeling and mechanotransduction are intact in *cav1.3a^{tn004}* mutants that have impaired hair cell neurotransmission (**f**). For image acquisition in **a–f**, larvae were mounted in LMP agarose in a glass depression slide and imaged on an upright confocal microscope. The image in (**a**) was taken at 10× while the images in (**c–f**) were taken at 60×. Scale bar in **a** = 250 µm and in **b–e** = 5 µm

3 Results

3.1 Examination of Gross Development and Behavior

After obtaining or generating a zebrafish mutant (obtained from ZIRC or created from injected embryos: morphants, CRISPR-Cas9 mutants, or transgenic dominant-negative mutants (Subheading 2.2)), comparison of the gross development of your mutant zebrafish with wild-type is recommended. Prior to 2 dpf, zebrafish live in their chorion or eggshell (Fig. 2b). At these early stages, embryos can easily be observed using a dissecting microscope as they are contained within this environment. While

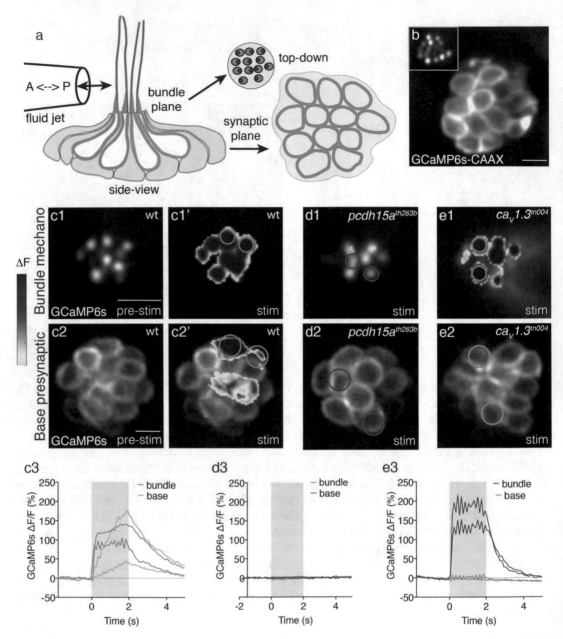

Fig. 9 Calcium imaging to detect mechanotransduction and presynaptic activity in lateral line hair cells. (**a**) Schematic of the setup used to stimulate lateral line hair cells while imaging mechanotransduction or presynaptic calcium signals, using the membrane-localized calcium sensor GCaMP6s-CAAX (green). For posterior lateral line hair cells, a fluid jet is used to stimulate hair cells along the A-P axis. Calcium signals are imaged in one of two planes, either an apical plane containing hair bundles or a synaptic plane at the base of hair cells. (**b**) Example apical (inset in upper left corner shows 16 hair bundles) and synaptic plane from the same neuromast. (**c1–c3**) Example calcium signals from a wild-type neuromast during stimulation. In wild-type, there is an increase in GCaMP6s signal in all hair bundles during stimulation (**c1′**). In contrast, there are only presynaptic calcium signals in a subset of these same hair cells (**c2′**). Apical and basal calcium signals (circles in **c1′** and **c2′**) from two hair cells are plotted in **c3**. (**d1–d3**) Example of calcium signals from a single neuromast in a *pcdh15a* mutant that lacks mechanotransduction. During stimulation, no increase in GCaMP6s signal is observed in the hair bundles (**d1′**) or at the presynapse (**d2′**). Apical and basal calcium signals (circles

zebrafish are still in their chorion, the developing otic vesicle (Figs. 2a, b, and 6a, first detectable ~18 hpf) can be observed. Within the otic vesicle, two otoliths (stone like deposits of calcium carbonate and protein) associated with the anterior and posterior macula rapidly form (Fig. 2a, 18–24 hpf) and can be easily detected under a dissecting microscope by 30 hpf (Fig. 2b). Some mutants including *rock solo/otog* and *hmx3a* have missing or fused otoliths, respectively, that disrupt hearing and balance [7, 105]. In many zebrafish mutants such as those with disrupted hair cell mechanotransduction or neurotransmission, the otic vesicle and otoliths form normally (e.g., $ca_V1.3a$ (Fig. 4b)).

Around 2–3 dpf, zebrafish embryos hatch and are free swimming (Fig. 2c). At this stage, the continued growth of the inner ear can easily be observed as wild-type larvae predominantly rest on their sides (Fig. 2c). Around 4 dpf, zebrafish larvae inflate their swim bladder, an air-filled pocket just above the yolk that helps zebrafish swim upright (Fig. 4a, top-down view) rather than on their side (Fig. 2d, side view). At 4–5 dpf, the zebrafish auditory, vestibular, and lateral-line sensory systems become functional [6–8]. Many zebrafish mutants with hearing or balance defects do not swim upright; when viewed under a dissecting microscope, these mutants rest on their sides in the petri dish (e.g., $ca_V1.3a$ (Fig. 4b); see **Note 7**). In addition, many of these mutants are unable to reach the surface to sip air and inflate their swim bladder (Fig. 4b).

Upright posture and the presence or absence of a swim bladder can easily be observed under a dissecting microscope. When viewing wild-type larvae (after 4 dpf), they are observed to be quite active and make long swim bouts (Fig. 4c). In contrast, mutants with balance defects are less motile, make smaller swim bouts (e.g., *myo7aa* (Fig. 4d)), and often swim in circles (see **Note 8**). To determine if your mutant has strong auditory defects, you can test the integrity of the acoustic startle reflex and ensuing escape response [6]. A cursory test is made by tapping a pen on the rim of the petri dish. Wild-type larvae will exhibit an escape response to this acoustic stimulus, while auditory mutants will fail to respond. To quickly test whether somatosensation is intact, larvae can be lightly touched with a dissecting needle. If your mutant escapes from a touch, this indicates that the escape response is intact and that the loss of acoustic startle reflex is likely not due to a locomotor or other downstream escape circuit defect.

Fig. 9 (continued) in **d1′** and **d2′**) from two hair cells are plotted in **d3**. (**e1–e3**) Example calcium signals from a single neuromast in a *$ca_V1.3a$* mutant that lacks presynaptic calcium channels. During stimulation there are robust increases in GCaMP6s signal in the hair bundles (**e1′**) but not at the presynapse (**e2′**). Apical and basal calcium signals (circles in **e1′** and **e2′**) from two hair cells are plotted in **e3**. For calcium imaging in **c1–e3** a 2 s-5 Hz stimulus alternating in the A-P direction was applied. The yellow to red heat maps reflect increases in GCaMP6s fluorescence (ΔF) that occurred in each cell during the stimulus. All images were taken at 60×. Scale bar in **b**, **c1**, **c2** = 5 μm

3.2 Closer Examination of the Zebrafish Inner Ear and Lateral Line

After determining whether your mutant has any gross inner ear or behavioral defects, a closer inspection of the hair cell epithelia is warranted. To view the inner ear in more detail and to view the lateral line, embryos and larvae must be mounted in LMP agarose (Subheading 2.4) and viewed at higher magnification (10–60×) using a compound microscope (Subheading 2.5). Due to the transparency of embryonic and larval zebrafish, the cells and structures of the inner ear and lateral line can easily be observed using DIC optics (Figs. 6 and 7a, b′). DIC can be augmented by using established transgenic lines that express fluorescent proteins in hair cells (Fig. 6) or other associated cells or structures (e.g., afferents, efferents, and supporting cells, *see* Table 2). In embryonic and larval zebrafish, these transgenic lines can be used to assess the developmental trajectory of hair cell epithelia by counting cell numbers at different developmental stages.

For example, using a transgenic line with GFP-labeled hair cells along with DIC optics on a wide-field microscope, the onset of hair cell formation within the five inner ear hair cell epithelia (anterior and posterior macula, and anterior, medial and posterior crista; Fig. 6a–c) can be observed. At all ages, in vivo imaging of hair cells of the posterior macula is challenging because the epithelium is located deeper, beneath a large otolith (Fig. 4a–c). Within the ear, hair cells in each of the two maculae can first be detected around 27 hpf (Fig. 6a; [9]). By 2 dpf, hair cells begin to form in each of the three cristae (Fig. 6b). Around 11 dpf, hair cells in another macula, the lagena, start to form (Fig. 2). These hair cells cannot be visualized as easily in vivo due to their location and reduced transparency

Table 2
Established transgenic lines that label cells in zebrafish hair cell epithelia

Transgenic line	Cell type labeled	References
$Tg(myo6b:EGFP)^{w186}$ $Tg(myo6b:mkate2)^{w232}$ $Tg(myo6b:Tomato)^{vo13}$ $Tg(pvalb3b:GFP)^{ru1001}$ $Et(krt4:EGFP)^{sqet4ET}$ $Tg(atoh1a:dTomato)^{nns8}$	Inner ear and lateral line hair cells	[73, 90, 106–109]
$Et(tnks1bp1:EGFP)^{y229Gt}$ $Et(krt4:EGFP)^{SqGw57a}$	All supporting cells	[73, 110]
$TgBAC(neurod1:EGFP)$ $Et(T2KHG)^{nkhgn39d}$	Inner ear and lateral line afferents Lateral line afferents	[28] [95]
$Tg(slc6a3:EGFP)^{ot80}$	Dopaminergic efferents	[108]
$Tg(phox2bb:EGFP)^{w37}$ $Tg(isl1:GFP)^{rw0}$	Cholinergic efferents	[111, 112]
$Tg(cldnb:LY-EGFP)^{zf106}$	Lateral line primordium	[113]

of larvae at this age. To date, several zebrafish mutants have been identified that impair inner ear formation. Many of these impairments can be readily observed in vivo: malformed semicircular canals (*gpr126, sox10, eya1, and ugdh1*; [114, 115]), excess of hair cells (*delta(dx2)* and *pax2.1*; [116]), and loss of cristae (*eya1*; [115, 117]).

The lateral line system forms along a similar timeline as the zebrafish inner ear. The lateral line is formed from a collection of migrating cells called a primordium. The primary posterior lateral line primordium (pLLp) is the most commonly studied primordium. The pLLp originates from a location just posterior to the zebrafish inner ear and migrates posteriorly along the trunk until it reaches the tail (Fig. 6d; 20–48 hpf). As the primordium migrates, it deposits clusters of cells called protoneuromasts that ultimately form neuromasts. Just hours after being deposited, hair cells form within each protoneuromast (Fig. 6e, 2 dpf). Additional hair cells are added each day to neuromasts (Fig. 6e–g show the same neuromast at 2, 3, and 6 dpf). Transgenic lines with fluorescent labeling of structures such as the pLLp, hair cells, and afferent neurons (Table 2) can be used to evaluate whether the pLLp has migrated and the number and location of hair cells or afferents neurons in the lateral line. For example, embryos injected with a morpholino targeted against *neurog1a* lack an acoustic startle response, although hair cells are present and functional in *neurog1a* morphants (Fig. 6h, i). Neurog1a is required for the formation of hair cell afferents and efferents in zebrafish [118]. The *neurod:GFP* transgenic line reveals that in *neurog1a* morphants, neuromasts are not innervated by afferent neurons (Fig. 6h, i).

3.3 Subcellular Examination of Hair Cell Morphology

In addition to transgenic lines that label hair cells and associated structures, there are also lines that label subcellular structures within hair cells (Table 3). These lines can be used to ascertain whether there are more subtle hair cell phenotypes in a zebrafish mutant. For example, in a subset of zebrafish mutants with hearing and balance defects, mechanotransduction is disrupted. Hair cell mechanotransduction is the first step in the sensory processing pathway and in humans, its dysfunction commonly leads to hearing and balance defects [123]. In many instances, mechanotransduction functional defects are accompanied by structural defects such as altered hair bundle integrity including splayed hair bundles. Splayed hair bundles can be visualized in vivo using DIC and also by using the *myo6b:actb1-EGFP* transgenic line that labels the actin-rich hair bundles and the cuticular plate (Fig. 7a, b'). In zebrafish, splayed hair bundles are most easily visualized in the cristae where hair bundles are the tallest. When fish are mounted in LMP agarose on their sides (Fig. 2c, d), hair bundles in the anterior, posterior, and medial cristae can easily be visualized in a side view at high magnification (Fig. 7a, b'). In wild-type cristae expressing the

Table 3
Established transgenic lines that label subcellular structures in zebrafish hair cells

Transgenic line	Substructure labeled	References
Tg(myo6b:actb1-EGFP)^{vo8}	Actin, including hair bundles	[47]
Tg(pvalb9:EGFP-fscn2b)^{cwr3}	Hair bundle	[119]
Tg(pou4f3:GAP-GFP)^{s356t}	Membrane, including hair bundles	[92]
Tg(myo6b:ctbp2a-mcherry)^{idc2} *Tg(ctbp2a:mCherry-ctbp2a)*^{lmb7}	Presynaptic ribbons	[50, 120]
Tg(myo6b:YFP-tubulin)	Tubulin, including the kinocilium	[121]
Gt(macf1a–citrine)^{ct68a/+}	Cuticular plate	[18]
Tg(myo6b:KDEL-Crimson)^{b1319}	Endoplasmic reticulum	[122]
Tg(myo6b:mgat1a-mKate2)^{vo24}	Golgi	[19]

myo6b:actb1-EGFP transgenic line, hair bundles have a carrot-like appearance (Fig. 7a, a'). In contrast, in *cdh23* zebrafish mutants that lack the tip links that hold the hair bundle together, hair bundles in the cristae are splayed and take on a rabbit ear-like appearance (e.g., *cdh23* (Fig. 7b, b')). Numerous zebrafish mechanotransduction mutants have splayed hair bundles (e.g., *myo7aa* (Fig. 8e), *cdh23* (Fig. 7b, b'); also: *pcdh15a, myo6a, tmie, clarin1, ush1c, ush1ga*, [25, 29, 30, 32, 103, 104, 122, 124]).

The *myo6b:actb1-EGFP* transgenic line can also be used to assess whether a zebrafish mutant has defects in hair bundle orientation. Hair bundle orientation determines the directional sensitivity of hair cells. Hair cells are optimally mechanosensitive when their hair bundles are deflected toward the tallest part. In mammalian vestibular hair cells and all zebrafish hair cells, the tallest part of the hair bundle is the kinocilium (Fig. 1c). In zebrafish, the lateral line is a popular system to study hair bundle orientation [95, 125]. Each neuromast has two populations of hair cells, each with opposing hair bundle orientations. For example, in neuromasts of the primary posterior lateral line (Fig. 1a, L1–L5 neuromasts), one population of hair bundles is oriented to respond to posterior and the other to respond to anterior fluid flow. In the *myo6b:actb1-EGFP* transgenic line, when fish are mounted in LMP agarose on their sides (Fig. 2c, d), the orientation of hair bundles in a neuromast can be visualized in a top-down view (Fig. 7c, c'). Several proteins have been identified that are required for proper hair bundle orientation (Emx2, Notch1a, Vangl2; [35, 126–129]). For example, the transcription factor Emx2, is required in the macular organs of mice and zebrafish, as well as the zebrafish lateral line to establish hair cell populations with opposing orientations [35]. In zebrafish mutants with loss of Emx2 function, the primary posterior lateral line contains one population of hair cells with all the hair bundles oriented in a single direction (Fig. 7d, d', all oriented to respond to posterior fluid flow). Similarly, in a gain of function approach, when Emx2 is expressed in all neuromast hair cells (*myo6b:emx2-p2a-nls-mcherry*), the majority of hair cells were all oriented in the opposite direction (Fig. 7e, e', all oriented to respond to anterior fluid flow).

These examples demonstrate how a single transgenic line, *myo6b:actb1-EGFP*, can be used to examine two features of hair bundle morphology (hair bundle integrity and orientation). Additional established transgenic lines can and have been used to examine other morphological features in wild-type and mutant zebrafish hair cells (e.g., presynaptic ribbons, cuticular plate, ER (Fig. 3g), mitochondria, Golgi, tubulin; *see* Table 3). Together, these transgenic lines represent a powerful toolkit to study the function and morphology of hair cells in vivo.

3.4 FM 1–43 Labeling of Hair Cells in the Lateral Line

To understand how a mutation disrupts hearing, balance, or lateral line function, it is important to determine what part of the sensory pathway is functionally disrupted. As mentioned above, hair cell mechanotransduction is the first step in the sensory processing pathway. The vital dye FM 1–43 can be used to assess hair cell mechanotransduction in many species including zebrafish (Subheading 2.6; [101, 102]). FM 1–43 enters and labels hair cells with intact mechanotransduction channels (Fig. 8b). FM 1–43 has been used to examine and refine stages of hair cell development in the zebrafish lateral line [47]. In wild-type larvae, FM 1–43 starts to label lateral line hair cells around 2 dpf (Fig. 8c). At larval stages (2–7 dpf), lateral line hair cells with and without FM 1–43 labeling can be detected, although by 5 dpf the majority of hair cells robustly label with FM 1–43 (Fig. 8d). These differences in labeling help to sort mature hair cells with functioning mechanotransduction from immature cells that have not yet become functional [47].

In addition to studying the onset of hair cell mechanotransduction, FM 1–43 labeling has been used extensively to characterize zebrafish with defects in mechanotransduction [19, 26, 71, 103, 104]. Many of these zebrafish mechanotransduction mutants have splayed hair bundles (e.g., *cdh23* (Fig. 7b, b')). Along with these morphological defects, the majority of these mutants also have impaired FM 1–43 labeling (e.g., *myo7aa* (Fig. 8e), *cdh23* (Fig. 7b, b'); also: *pcdh15a, myo6a, tmie, clarin1, ush1c, ush1ga*, [25, 29, 30, 32, 103, 104, 122, 124]). A subset of conserved zebrafish mechanotransduction mutants do not have splayed hair bundles but rather bundles that are thinner (*ap1b1, grxcr1*; [122, 130]), or do not impact hair bundle morphology at all (e.g., *tmc2b, tomt*; [19, 21, 131]). Despite relatively normal hair bundles, these mutations still disrupt FM 1–43 labeling in the lateral line.

If a zebrafish mutant has intact FM 1–43 labeling, this is a good indication that mechanotransduction channels are present and mechanotransduction is largely intact. It is important to point out that FM 1–43 is only a rough estimate of mechanotransduction. If a mutant has a subtle mechanotransduction phenotype, characterization of this defect will require a quantitative functional assessment (Subheading 4.1; but *see* **Note 6**). Even if a quantitative functional assessment reveals that mechanotransduction is intact in your zebrafish mutant, hair cell function could still be impaired.

Many mutations lead to defects that occur downstream of mechanotransduction, for example, at the hair cell synapse. Several conserved zebrafish mutants that disrupt hair cell neurotransmission have been characterized (*dmxl2, ca$_V$1.3a, vglut3, otofb, wrb*; [28, 31, 132–134]). In all of these mutants, FM 1–43 robustly labels lateral line hair cells. For example, in zebrafish mutants without the Ca$_V$1.3 channels required for presynaptic calcium influx prior to hair cell neurotransmission, have normal FM 1–43 labeling (Fig. 8f). Similar to mutants with disrupted hair cell neurotransmission, mutations that impair activity in afferent neurons or processes downstream of hair cells may also have intact FM 1–43 labeling. For example, in zebrafish *neurog1a* morphants that lack lateral line afferent neurons, FM 1–43 still labels lateral line hair cells (Fig. 6i; FM 4–64 is a red variant of FM 1–43 that labels hair cells in a *neurog1a* morphant).

One caveat of the FM 1–43 labeling outlined in this method is that it does not label hair cells in the zebrafish inner ear as they are not accessible to the external aqueous environment where FM 1–43 and other dyes are applied (Fig. 8a). However, in the majority of zebrafish mutants with hearing and balance impairments, both inner ear and lateral line hair cells are similarly affected due to homologous expression of genes in these two hair cell types. This makes the lateral line hair cells a good proxy for hair cells in the zebrafish inner ear. To assess FM 1–43 labeling in the inner ear of your mutant zebrafish, the dye can be microinjected into the otic capsule [21].

It is important to point out that there is a subset of mutants (*tmc1,2a,b, lhfpl5a,b*, and *otofa,b*) that have defects specific to the zebrafish inner ear or lateral line [21, 103, 132]. For example, *lhfpl5a* impairs inner ear mechanotransduction, while *lhfpl5b* impairs lateral line mechanotransduction [103]. These nonoverlapping expression patterns provide exciting new avenues to probe the functional and structural differences between these two important hair cell types. Overall, the methods described above together with previous work demonstrate that FM 1–43 is a rapid and powerful in vivo method to assess the onset and integrity of mechanotransduction in zebrafish hair cells.

4 Advanced In Vivo Methods: Moving Beyond the Methods Described Here

4.1 In Vivo Approaches to Study Hair Cell Activity

FM 1–43 is a fast and powerful way to assess the integrity and mechanosensitive properties of lateral line hair cells in zebrafish mutants. Several labs [135, 136] have developed additional methods to study the functional properties of zebrafish hair cell epithelia in vivo. These methods extend beyond what is covered in this chapter and are well-described in other published methods. The two main approaches to examine hair cell epithelia are (1) electrophysiology and (2) optogenetic-based imaging. In both of these approaches, whole larvae are commonly immobilized and paralyzed on a Sylgard-filled chamber and immersed in a specific extracellular solution [135–137]. To mechanically stimulate lateral line hair cells, either a stiff probe attached to a piezo-electric device or a fluid jet (a fluid-filled glass pipette attached to a device that applies positive or negative pressure) is used to deflect the hair bundles. This stimulation is then paired with either a recording electrode or optical imaging to measure activity in hair cell epithelia.

Electrophysiology-based methods were first applied to record extracellular microphonics to quantify the mechanosensitive properties among populations of zebrafish lateral line hair cells [27, 136]. Later, techniques to perform whole-cell recordings were established to isolate and measure whole-cell currents, specific currents, and exocytosis from individual zebrafish hair cells [137, 138]. In addition, downstream of sensory hair cells, methods were established to record spontaneous spikes from lateral line afferent neurons [51, 139]. Spontaneous spike rate has been linked to hair cell synapse sensitivity [50, 140, 141]. Alterations to spontaneous spike rate is associated with hair cell dysfunction and hearing loss in mice and zebrafish [133, 141–143]. Similar afferent recordings have also been applied to measure the timing and number of evoked spikes while stimulating an innervated neuromast using a fluid jet [136, 144, 145]. A striking feature of these electrophysiological recordings is that not only are they a powerful way to assess hair cell function in zebrafish, but all recordings are done in fully intact animals.

In addition to these electrophysiology-based measurements, optogenetics using genetically encoded indicators (GEIs) has provided a powerful, noninvasive way to optically study the function of zebrafish hair cells in vivo [146, 147]. GEIs are particularly advantageous for studies in zebrafish because zebrafish are genetically tractable and transparent as larvae, making optical imaging straightforward. The first GEIs applied to study activity in hair cell epithelia in zebrafish were calcium indicators [22, 47]. Calcium dynamics play a critical part in both mechanosensation and presynaptic activity in hair cells. In addition, calcium is an integral part of the postsynaptic response in the afferent processes beneath hair cells.

Table 4
Transgenic lines expressing genetically encoded indicators of activity in zebrafish hair cell epithelia

Transgenic line	Activity detected	References
$Tg(myo6b:GCaMP6scaax)^{idc1}$	Hair cell mechanotransduction presynaptic calcium	[24, 50]
$Tg(myo6b:SypHy)^{idc6}$	Hair cell exocytosis	[24]
$Tg(myo6b:mitoGCaMP3)^{w119}$; $Tg(myo6b: mitoRGECO1)^{idc12}$	Hair cell mitochondrial calcium	[148, 149]
$Tg(myo6b:GCaMP3)^{w78}$; $Tg(myo6b:RGECO1)^{vo10}$	Cytosolic hair cell calcium	[22, 71]
$Tg(hsp70l:GCaMP6s\text{-}SILL1)^{idc8}$; $Tg(neurod:GGECO)^{nl19}$	Afferent calcium	[24, 94]
$Tg(neurod:mitoRGECO)^{nl20}$	Afferent mitochondrial calcium	[94]
$Tg(myo6b:Rex\text{-}YFP)$	Hair cell $NAD^+/NADH$	[149]
$Tg[Sill2:Gal4\text{-}VP16^{bc4}, UAS:iGluSnFR^{uss2}]$	Afferent glutamate	[150]

Currently, there are established zebrafish transgenic lines that express calcium indicators specifically in lateral line hair cells or afferents (Table 4; [24]). These lines can measure activity in hair cells and afferent neurons and can also to examine how activity is impaired in various zebrafish mutants [130, 151]. One advantage of GEIs is that not only can they be expressed in specific cell types, but they can also be subcellularly localized. Taking advantage of this feature, researchers have localized variants of GCaMP, a GEI of calcium, to specific subcellular locations in zebrafish hair cells and afferents (Table 4). Transgenic zebrafish expressing localized GEIs has provided a more in-depth analysis of hair cell epithelia function and pathology.

For example, localizing GCaMP6s to the plasma membrane (myo6b:GCaMP6s-CAAX) has provided a robust and quantitative way to measure both mechanosensitive and presynaptic calcium signals in the same collection of zebrafish hair cells. As described earlier (Subheading 1), deflection of the mechanosensitive hair bundles opens the mechanoelectrical transduction (MET) channels at the tips of stereocilia, resulting in an influx of cations, including calcium. This results in hair cell depolarization, which opens voltage-gated calcium channels ($Ca_V1.3$) located basally at the hair cell synapse. This presynaptic calcium influx via $Ca_V1.3$ channels ultimately leads to vesicle fusion and neurotransmission. The localization of GCaMP6s in the hair cell plasma membrane (myo6b: GCaMP6s-CAAX) has allowed zebrafish researchers to detect calcium activity in both the hair bundles and at the presynapse simultaneously (Fig. 9; [24]). Methods to measure these calcium signals, along with the materials and equipment required have been documented thoroughly [135].

The ability to quantify activity at these two distinct levels using the *myo6b:GCaMP6s-CAAX* transgenic line has provided an important tool to parse out whether a mutant of interest impairs mechanotransduction at the apical end, or presynaptic function at the basal end of the hair cell. For example, during fluid-jet stimulation, robust GCaMP6s signals can be detected in the apical hair bundles in wildtype neuromast hair cells (Fig. 9c1, c1′, c3). Similarly, in wildtype animals, at the base of lateral line hair cells, presynaptic GCaMP6s signals can be detected during stimulation (Fig. 9c2, c2′, c3). In contrast, no GCaMP6s signals are detected in the hair bundles of *pcdh15a* mutants, which lack the tip links necessary to gate the MET channels in hair cells (Fig. 9d1, d3). As mechanotransduction is required to initiate opening of $Ca_V1.3$ channels at the presynapse, *pcdh15a* mutants subsequently have no presynaptic GCaMP6s signals (Fig. 9d2, d3). In contrast to *pcdh15a* mutants, the synaptic mutant *ca$_V$1.3a* has relatively normal GCaMP6s signals in hair bundles (Fig. 9e1, e3). As $Ca_V1.3$ channels are required for presynaptic calcium influx in hair cells, no GCaMP6s signals are detectable at the base of hair cells in *ca$_V$1.3a* mutants (Fig. 9e2, e3). After identifying potential defects in mechanotransduction via FM 1–43 labeling, GCaMP6s-based imaging is a powerful way to confirm and quantify the mechanosensitive properties of lateral line hair cells. In addition, this imaging can be used to quantify presynaptic calcium signals in response to mechanotransduction-based changes in hair cell receptor potential.

While GEIs of calcium have provided a powerful method to study activity in zebrafish, other GEIs have also been applied in studies of hair cell epithelia in zebrafish (Table 4). For example, the GEI SypHy has been used to measure hair cell exocytosis and the GEI GluSnFR has been used to measure glutamate released from hair cells onto lateral line afferents [24, 150]. Overall zebrafish provide a fertile testing ground to determine whether a GEI has potential to explore the function of hair cell epithelia. If a GEI is effective in zebrafish, then there is potential for its application to study hair cell epithelia in other model systems.

4.2 Advanced Behavioral and Optogenetic Approaches to Study Sensory System Function

Above we outlined several basic ways to behaviorally assess auditory and vestibular system function in a zebrafish mutant (tap-induced startle, posture, circling movement). To examine a behavioral phenotype in more depth, there are established methods for quantifying auditory, vestibular, and lateral line function [7, 8, 152]. These approaches rely on a camera and video-tracking software to capture the behavior of interest. The acoustic startle and escape response is an established reflex used to study the auditory portion of the inner ear as well as the escape response circuit (Fig. 10a, also *see* below) of zebrafish [1–4, 153, 154]. For acoustic startle behavior, a speaker, shaker, or tapper is used to stimulate zebrafish hair cells (*see* **Note 9**; [152, 155]). In response to the stimulus, larvae exhibit a robust

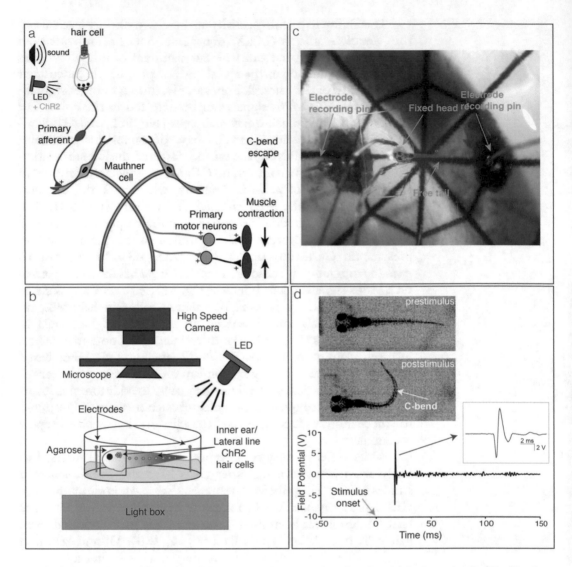

Fig. 10 Optogenetics and field potentials measure sensory system function. (**a**) Cartoon circuit of the Mauthner cell (M cell) escape response. In response to stimulus, hair cells activate downstream afferent neurons (blue). Afferent neurons synapse onto the M cell and coincident arrival of afferent excitatory potentials cause the M cell to reach threshold. The subsequent action potential travels along the contralaterally projecting primary axon and excites motor neurons (orange) that innervate the trunk and tail muscles (red). Simultaneous contraction of the musculature results in a "C"-shaped body bend and together with subsequent body bends the animal moves away from the source of the stimulus. (**b**) To capture behavioral videos together with field potentials during escape responses, a head-fixed larvae mounted in a recording chamber is placed under a dissecting microscope equipped with a camera. An infrared LED light box is used to illuminate the video recording outside the visual range of zebrafish. (**c**) Recording electrodes for measuring field potentials are placed precisely on either side of the larvae. Placement is established using a spiderweb template taped beneath the chamber. (**d**) A single frame from a video recording of a larval zebrafish (5 dpf) before (prestimulus, top panel) and during (poststimulus, middle panel; maximum C-bend shown) an evoked M-cell escape response. A field potential (bottom panel) from this M-cell evoked response is shown. Note the latency, amplitude, and other features of the field potential (onset)

escape response. Images of the response are captured at slow (10 frames/s) or high (1000 frames/s) frame rate. Slow-speed capture can be used to detect whether an escape occurred in response to a stimulus, while high-speed capture can be used to examine more subtle characteristics of the startle response including the latency to onset and the kinematics of the body during the behavior [156] as well as the duration and magnitude of the response behavior [157].

Another method used to measure vestibular behavior is the conserved vestibulo-ocular reflex (VOR) [7]. In this assay, larvae are head fixed and rotated in a vertical plane. A camera is used to capture the compensatory eye movements used to stabilize retinal images (saccades) during the rotations. The vestibulo-spinal reflex and other vestibular measurements are also powerful tools to examine mutant phenotypes [158]. To measure lateral line behavior, rheotaxis assays have been developed [8, 159, 160]. Rheotaxis is the innate behavior shared by many fishes including zebrafish to turn to face into an oncoming current and hold position [161]. The lateral line is important for this behavior. In zebrafish rheotaxis assays, a camera is used to capture the orientation of zebrafish relative to current flow [8, 46, 159]. Overall, behavioral assays have provided a quantitative way to assess the function of hair cell systems in wild-type and mutant zebrafish [133, 156, 162]. Cumulatively, these methods have demonstrated that the onset of these hair cell system functions in larval zebrafish is 4–5 dpf.

In addition to these robust behavioral approaches, optogenetics and electrophysiological methods have been combined to study the activity of hair cell sensory systems in zebrafish. For example, to aid in optogenetic studies, transgenic zebrafish expressing Channelrhodopsin-2 (ChR2) in hair cells have been created [23]. ChR2 is a light activated, nonspecific cation channel that is capable of inducing depolarization and action potentials in neurons [163]. In transgenic zebrafish expressing ChR2 in the ear and lateral line hair cells, blue light is able to activate and depolarize hair cells, leading to the excitation of downstream afferents, and ultimately a startle response (Fig. 10a; [23, 145, 164]). A transgenic zebrafish line expressing ChR2 in hair cells has been successfully paired with electrophysiological recordings of the lateral line afferents [23, 145]. In this way, blue light from an objective placed above a neuromast can be used to stimulate hair cells while recording the evoked spikes in connected afferent neurons. This approach has been used to examine important properties of the hair cell synapse including the latencies and jitter of evoked spikes [145].

Another way ChR2 has been used to investigate the integrity of the hair cell sensory systems is through measurements of extracellular field potentials during startle behavior (Fig. 10; [165, 166]. In zebrafish, the acoustic startle response is controlled by command neurons in the hindbrain [167, 168]. When stimulated, inner ear and lateral line hair cells activate afferent neurons that project to the Mauthner cell or non-Mauthner homolog neurons located in Rhombomere four of the hindbrain. Once excited, these command neurons propagate action potentials down the axons that project to motor neurons and ultimately muscles all along the trunk of the fish, resulting in contraction of the body away from the source of the stimulus [169]. This synchronized activation of the command neuron, motor neurons, and skeletal muscles generates a large extracellular field potential [167, 170]. Measuring the Mauthner cell-evoked field potential can provide a quantifiable way to measure the timing, probability, and amplitude of startle responses. Recently, detailed methods that use ChR2 to measure extracellular field potentials in zebrafish have been described [145, 164, 171].

Briefly, field potentials evoked from mechanically or optically triggering escape responses (e.g., from transgenic zebrafish expressing ChR2 in hair cells; Fig. 10a) can be recorded using a simple electrophysiology setup. For these experiments, larvae are head-embedded in low melting point (LMP) agarose with their dorsal side up (Fig. 10b) in reverse osmosis (RO) water (removal of all salts increases the resistance of the solution and the magnitude of the field potential). Two stainless steel insect pins are used as recording electrodes and are precisely placed 1 cm apart and equidistant to the rostro-caudal axis of the larva (Fig. 10c). Field potentials are then recorded using an AC/DC differential amplifier. If video capture of the escape response is desired, embedded larvae can be placed under a dissecting microscope (Subheading 2.3) equipped with a high-speed camera (IL5, Fastec Imaging, San Diego, CA) (Fig. 10c). To activate the startle response using ChR2-based optogenetics, a blue (approximately 470 nm) LED is used to activate hair cells in transgenic zebrafish and trigger a startle response. For specific details of this technique, the reader is directed to Ozdemir et al. [164].

After hair cells are activated (with sound, vibration, or blue light), a high-speed video of the escape response can be concomitantly acquired with field potentials in order to measure and compare the onset time (latency), probability, amplitude, or other characteristics of startle response field potentials and behaviors [145]. These types of measurements provide valuable insight into the functions and mechanisms of different sensory systems and can help to determine whether the startle response circuit is compromised in your zebrafish mutant (*see* **Note 10**).

4.3 Outlook and Future

Numerous additional in vivo approaches exist and can be pioneered to complement the approaches outlined in this chapter. The emergence of new GEIs, along with advances in microscopy methods, continues to create exciting ways to assay hair cell sensory systems.

For example, in addition to GEI of calcium, GPCR-activation-based (GRAB) optical indicators have been developed to detect a variety of neurotransmitters. GRABs open up the possibility of exploring the contribution of the dopaminergic and cholinergic efferents that innervate hair cell epithelia in zebrafish [172–174]. In addition to GRABs, real potential rests in GEIs of membrane voltage [175, 176]. Assaying in vivo electrical activity will be extremely useful to image activity within hair cell epithelia that is not driven by nor requires calcium.

The development and application of new GEIs have been greatly aided by advances in microscopy. For example, light-sheet fluorescence microscopy (LSFM) has enabled ultrafast, volumetric imaging with minimal phototoxicity [177–179]. LSFM offers great potential for in vivo imaging, particularly in dimly fluorescent samples or in robustly fluorescent samples, but over long durations and especially across populations of cells and large regions of the animal [180–182]. Overall, LSFM can be used to enhance many of the techniques outlined in the method. Recent work has used GEIs and LSFM to monitor activity at the systems level in zebrafish, during sensory-evoked behavior [160, 183, 184]. In the future, these types of imaging studies could be extended to understand how hair cell sensory systems respond after damage by noxious insults such as noise or chemical ototoxins. Overall, the future for zebrafish in hearing and balance research is bright and will continue to drive the research in the realm of in vivo imaging.

5 Notes

1. It is important to clean larvae well with fresh E3 media after they hatch from their chorions. The superficial location of the lateral line makes it excellent for imaging but also sensitive to the disposition of the E3 media. As a result, it is recommended to replace the E3 media daily. Shed chorions can enable the growth of bacteria and offer a food source for other microorganisms (e.g., coleps) that can negatively impact the health of lateral line hair cells. Clean E3 media is particularly important for zebrafish mutants with impaired hearing and balance as they often lack inflated swim bladders and rest on their sides in the dish.

2. Similar approaches can be used to create a multitude of transgenic zebrafish lines. For example, transgenic lines can be created to express genetically encoded indicators of interest in a specific cell type [185]. In addition, your gene of interest can be tagged with fluorescent proteins such as GFP to examine localization in living cells (Fig. 3d–g).

3. For time-lapse imaging wait approximately 20–30 min after adding the E3 media to allow the dish to stabilize (with regard to temperature and E3 fluid permeabilizing the agarose) to reduce sample drift (in X-Y and Z dimensions) during longer imaging sessions.

4. For camera-based imaging systems, a camera with a small pixel size (4–10 μm) is best for imaging zebrafish hair cells and subcellular structures. Some examples: QImaging Rolera EM-C2 CCD camera-8 μm pixel; Photometrics Dyno CCD camera-4.54 μm pixel; Hamamastu ORCA-FLASH 4.0 SCI CMOS V3–6.5 μm pixel. Alternatively, a camera with a larger pixel size can be used along with a 2× camera coupler (e.g., optovar) to magnify the sample.

5. In addition to FM 1–43, other cationic dyes can be used to visualize and assess the integrity and function of lateral line hair cells. For example: Yo-Pro, DAPI or Hoechst label hair cell nuclei, Mitotracker and TMRE label mitochondria, and FM 4–64, a red variant of FM 1–43 can be used to label hair cells. There are also fixable versions of some of these dyes that allow visualization alongside antibody labeling [22, 47, 49, 186, 187].

6. To assess mechanotransduction, it is important to use a brief FM 1–43 incubation time. Longer time spans (e.g., minutes) of dye incubations result in labeling that occurs via endocytosis, in addition to the rapid labeling that occurs via entry through the mechanotransduction channels [102]. Ideally, FM 1–43 labeling should be carried out in a rigorous and quantitative manner by using a timer to ensure all steps are done accurately and repeated consistently [188]. Precise timing of FM 1–43 labeling can be used to achieve quantitative results.

7. It is critical for fish to inflate their swim bladder and have an upright posture for proper vestibular function. Without vestibular function (e.g., gravity detection), larvae fail to swim and feed properly, and ultimately die (~10–14 dpf) after their yolk is depleted (7 dpf). Many zebrafish mutants with auditory and vestibular deficits are not homozygous viable as adults and are maintained as heterozygous stocks.

8. Circling behavior is an indication of severe vestibular dysfunction. Disruption to hair cell function, whether to the mechanotransduction apparatus or to the synaptic machinery, may result in fish that retain some level of crude auditory function, which is often sufficient to retain a startle response. However, the vestibular system is apparently not as resilient to the reduced hair cell function, and therefore mutants often display circling behavior and difficulty inflating their swim bladders [27].

9. One way to study acoustic startle responses is by placing larvae in a dish that is rested on a speaker for startle delivery, with a high-speed camera mounted (and video tracking software) [189]. The auditory stimulus (tone) can be controlled by computer interface software that can drive the speaker. Another related method involves placing the larva in a dish secured to a shaker or accelerometer (to measure acoustic particle motion levels of the auditory stimulus) that is driven by an amplifier and a Tucker-Davis Technologies (TDT Inc., Gainesville, FL, USA) System 3 [152, 155, 190]. The TDT System 3 can generate a variety of stimulus tones to the amplifier. An additional option is to provide "tap" stimuli, in which any small object or device may be used at predetermined times to produce acoustic taps on the dish or chamber holding the larvae [156, 191–193]. These methods are also useful in studying the effects of prepulse inhibition [152, 156].

10. By using an ototoxin (e.g., the aminoglycoside neomycin), one can induce hair cell death in the lateral line of zebrafish [194]. The inner ear hair cells will not be damaged as their internal location protects them from neomycin exposure. This enables assessment of how lateral versus inner ear hair cells contribute to behaviors or field potential measurements.

Acknowledgments

We acknowledge the following researchers for contributing images: Natalie Mosqueda (Fig. 6e–g), Katherine Pinter (Fig. 3a–c), and Candy Wong (Fig. 4c, d) and also Alisha Beirl for injecting the *neurog1a* morpholino (Fig. 6h, i). We thank Alma Jukic for her comments on the manuscript.

Funding: *This work was supported by a National Institute on Deafness and Other Communication Disorders (NIDCD) Intramural Research Program Grant 1ZIADC000085-01 (K.S.K) and by a NIDCD NIH R15DC014843 award (J.G.T).*

References

1. Montgomery JC, Coombs SL (2017) Lateral line neuroethology. In: Reference module in life sciences. Elsevier. https://doi.org/10.1016/B978-0-12-809633-8.03040-5

2. Haque KD, Pandey AK, Kelley MW, Puligilla C (2015) Culture of embryonic mouse cochlear explants and gene transfer by electroporation. J Vis Exp:e52260. https://doi.org/10.3791/52260

3. Ogier JM, Burt RA, Drury HR et al (2019) Organotypic culture of neonatal murine inner ear explants. Front Cell Neurosci 13. https://doi.org/10.3389/fncel.2019.00170

4. Kimmel CB (1989) Genetics and early development of zebrafish. Trends Genet 5: 283–288. https://doi.org/10.1016/0168-9525(89)90103-0

5. Vaz RL, Outeiro TF, Ferreira JJ (2018) Zebrafish as an animal model for drug discovery in Parkinson's disease and other movement disorders: a systematic review. Front Neurol 9: 347. https://doi.org/10.3389/fneur.2018. 00347

6. Kimmel CB, Patterson J, Kimmel RO (1974) The development and behavioral characteristics of the startle response in the zebra fish. Dev Psychobiol 7:47–60. https://doi.org/ 10.1002/dev.420070109

7. Mo W, Chen F, Nechiporuk A, Nicolson T (2010) Quantification of vestibular-induced eye movements in zebrafish larvae. BMC Neurosci 11:110. https://doi.org/10.1186/ 1471-2202-11-110

8. Suli A, Watson GM, Rubel EW, Raible DW (2012) Rheotaxis in larval zebrafish is mediated by lateral line mechanosensory hair cells. PLoS One 7. https://doi.org/10. 1371/journal.pone.0029727

9. Tanimoto M, Ota Y, Inoue M, Oda Y (2011) Origin of inner ear hair cells: morphological and functional differentiation from ciliary cells into hair cells in zebrafish inner ear. J Neurosci 31:3784–3794. https://doi.org/10.1523/ JNEUROSCI.5554-10.2011

10. Amsterdam A, Varshney G, Burgess S (2011) Retroviral-mediated insertional mutagenesis in zebrafish. Methods Cell Biol 104:59–82. https://doi.org/10.1016/B978-0-12-374814-0.00004-5

11. Bill BR, Petzold AM, Clark KJ et al (2009) A primer for morpholino use in zebrafish. Zebrafish 6:69–77. https://doi.org/10.1089/ zeb.2008.0555

12. Hoshijima K, Jurynec MJ, Klatt Shaw D et al (2019) Highly efficient CRISPR-Cas9-based methods for generating deletion mutations and F0 embryos that lack gene function in zebrafish. Dev Cell 51:645–657.e4. https:// doi.org/10.1016/j.devcel.2019.10.004

13. Irion U, Krauss J, Nüsslein-Volhard C (2014) Precise and efficient genome editing in zebrafish using the CRISPR/Cas9 system. Development 141:4827–4830. https://doi.org/ 10.1242/dev.115584

14. Kwan KM, Fujimoto E, Grabher C et al (2007) The Tol2kit: a multisite gateway-based construction kit for Tol2 transposon transgenesis constructs. Dev Dyn 236: 3088–3099. https://doi.org/10.1002/dvdy. 21343

15. Moens CB, Donn TM, Wolf-Saxon ER, Ma TP (2008) Reverse genetics in zebrafish by TILLING. Brief Funct Genomic Proteomic 7:454–459. https://doi.org/10.1093/ bfgp/eln046

16. Solnica-Krezel L, Schier AF, Driever W (1994) Efficient recovery of enu-induced mutations from the zebrafish germline. Genetics 136:1401–1420

17. Varshney GK, Carrington B, Pei W et al (2016) A high-throughput functional genomics workflow based on CRISPR/Cas9-mediated targeted mutagenesis in zebrafish. Nat Protoc 11:2357–2375. https://doi.org/ 10.1038/nprot.2016.141

18. Antonellis PJ, Pollock LM, Chou S-W et al (2014) ACF7 is a hair-bundle antecedent, positioned to integrate cuticular plate actin and somatic tubulin. J Neurosci 34: 305–312. https://doi.org/10.1523/ JNEUROSCI.1880-13.2014

19. Erickson T, Morgan CP, Olt J et al (2017) Integration of Tmc1/2 into the mechanotransduction complex in zebrafish hair cells is regulated by Transmembrane O-methyltransferase (Tomt). eLife 6: e28474. https://doi.org/10.7554/eLife. 28474

20. Graydon CW, Manor U, Kindt KS (2017) In vivo ribbon mobility and turnover of ribeye at zebrafish hair cell synapses. Sci Rep 7:7467. https://doi.org/10.1038/s41598-017-07940-z

21. Smith ET, Pacentine I, Shipman A et al (2020) Disruption of tmc1/2a/2b genes in zebrafish reveals subunit requirements in subtypes of inner ear hair cells. J Neurosci. https://doi.org/10.1523/JNEUROSCI. 0163-20.2020

22. Esterberg R, Hailey DW, Rubel EW, Raible DW (2014) ER–mitochondrial calcium flow underlies vulnerability of mechanosensory hair cells to damage. J Neurosci 34: 9703–9719. https://doi.org/10.1523/ JNEUROSCI.0281-14.2014

23. Monesson-Olson BD, Browning-Kamins J, Aziz-Bose R et al (2014) Optical stimulation of zebrafish hair cells expressing channelrhodopsin-2. PLoS One 9. https:// doi.org/10.1371/journal.pone.0096641

24. Zhang Q, Li S, Wong H-TC et al (2018) Synaptically silent sensory hair cells in zebrafish are recruited after damage. Nat Commun 9:1388. https://doi.org/10.1038/s41467-018-03806-8

25. Ernest S, Rauch GJ, Haffter P et al (2000) Mariner is defective in myosin VIIA: a zebrafish model for human hereditary deafness. Hum Mol Genet 9:2189–2196

26. Gleason MR, Nagiel A, Jamet S et al (2009) The transmembrane inner ear (Tmie) protein is essential for normal hearing and balance in the zebrafish. Proc Natl Acad Sci 106: 21347–21352. https://doi.org/10.1073/pnas.0911632106

27. Nicolson T, Rüsch A, Friedrich RW et al (1998) Genetic analysis of vertebrate sensory hair cell mechanosensation: the zebrafish circler mutants. Neuron 20:271–283

28. Obholzer N, Wolfson S, Trapani JG et al (2008) Vesicular glutamate transporter 3 is required for synaptic transmission in zebrafish hair cells. J Neurosci 28:2110–2118. https://doi.org/10.1523/JNEUROSCI.5230-07.2008

29. Seiler C, Finger-Baier KC, Rinner O et al (2005) Duplicated genes with split functions: independent roles of protocadherin15 orthologues in zebrafish hearing and vision. Dev Camb Engl 132:615–623. https://doi.org/10.1242/dev.01591

30. Seiler C, Ben-David O, Sidi S et al (2004) Myosin VI is required for structural integrity of the apical surface of sensory hair cells in zebrafish. Dev Biol 272:328–338. https://doi.org/10.1016/j.ydbio.2004.05.004

31. Sidi S, Busch-Nentwich E, Friedrich R et al (2004) gemini encodes a zebrafish L-type calcium channel that localizes at sensory hair cell ribbon synapses. J Neurosci 24:4213–4223. https://doi.org/10.1523/JNEUROSCI.0223-04.2004

32. Söllner C, Rauch G-J, Siemens J et al (2004) Mutations in cadherin 23 affect tip links in zebrafish sensory hair cells. Nature 428: 955–959. https://doi.org/10.1038/nature02484

33. Delmaghani S, Aghaie A, Bouyacoub Y et al (2016) Mutations in CDC14A, encoding a protein phosphatase involved in hair cell ciliogenesis, cause autosomal-recessive severe to profound deafness. Am J Hum Genet 98: 1266–1270. https://doi.org/10.1016/j.ajhg.2016.04.015

34. Riazuddin S, Belyantseva IA, Giese APJ et al (2012) Alterations of the CIB2 calcium- and integrin-binding protein cause Usher syndrome type 1J and nonsyndromic deafness DFNB48. Nat Genet 44:1265–1271. https://doi.org/10.1038/ng.2426

35. Jiang T, Kindt K, Wu DK (2017) Transcription factor Emx2 controls stereociliary bundle orientation of sensory hair cells. eLife 6: e23661. https://doi.org/10.7554/eLife.23661

36. Nicolson T (2005) The genetics of hearing and balance in zebrafish. Annu Rev Genet 39:9–22. https://doi.org/10.1146/annurev.genet.39.073003.105049

37. Yao Q, DeSmidt AA, Tekin M et al (2016) Hearing assessment in zebrafish during the first week postfertilization. Zebrafish 13: 79–86. https://doi.org/10.1089/zeb.2015.1166

38. Kwak S-J, Vemaraju S, Moorman SJ et al (2006) Zebrafish pax5 regulates development of the utricular macula and vestibular function. Dev Dyn 235:3026–3038. https://doi.org/10.1002/dvdy.20961

39. Riley BB, Moorman SJ (2000) Development of utricular otoliths, but not saccular otoliths, is necessary for vestibular function and survival in zebrafish. J Neurobiol 43:329–337. https://doi.org/10.1002/1097-4695(20000615)43:4<329::aid-neu2>3.0.co;2-h

40. Beck JC, Gilland E, Tank DW, Baker R (2004) Quantifying the ontogeny of optokinetic and vestibuloocular behaviors in zebrafish, medaka, and goldfish. J Neurophysiol 92: 3546–3561. https://doi.org/10.1152/jn.00311.2004

41. Fay RR, Popper AN (2000) Evolution of hearing in vertebrates: the inner ears and processing. Hear Res 149:1–10. https://doi.org/10.1016/S0378-5955(00)00168-4

42. Ladich F, Schulz-Mirbach T (2016) Diversity in fish auditory systems: one of the riddles of sensory biology. Front Ecol Evol 4. https://doi.org/10.3389/fevo.2016.00028

43. Faucher K, Parmentier E, Becco C et al (2010) Fish lateral system is required for accurate control of shoaling behaviour. Anim Behav 79:679–687. https://doi.org/10.1016/j.anbehav.2009.12.020

44. McHenry MJ, Feitl KE, Strother JA, Van Trump WJ (2009) Larval zebrafish rapidly sense the water flow of a predator's strike. Biol Lett 5:477–479. https://doi.org/10.1098/rsbl.2009.0048

45. Mekdara PJ, Schwalbe MAB, Coughlin LL, Tytell ED (2018) The effects of lateral line ablation and regeneration in schooling giant danios. J Exp Biol 221. https://doi.org/10.1242/jeb.175166

46. Olszewski J, Haehnel M, Taguchi M, Liao JC (2012) Zebrafish larvae exhibit rheotaxis and can escape a continuous suction source using their lateral line. PLoS One 7:e36661. https://doi.org/10.1371/journal.pone.0036661

47. Kindt KS, Finch G, Nicolson T (2012) Kinocilia mediate mechanosensitivity in developing zebrafish hair cells. Dev Cell 23:329–341. https://doi.org/10.1016/j.devcel.2012.05.022

48. Mackenzie SM, Raible DW (2012) Proliferative regeneration of zebrafish lateral line hair cells after different ototoxic insults. PLoS One 7:e47257. https://doi.org/10.1371/journal.pone.0047257

49. Pickett SB, Thomas ED, Sebe JY et al (2018) Cumulative mitochondrial activity correlates with ototoxin susceptibility in zebrafish mechanosensory hair cells. eLife 7. https://doi.org/10.7554/eLife.38062

50. Sheets L, He XJ, Olt J et al (2017) Enlargement of ribbons in zebrafish hair cells increases calcium currents but disrupts afferent spontaneous activity and timing of stimulus onset. J Neurosci 37:6299–6313. https://doi.org/10.1523/JNEUROSCI.2878-16.2017

51. Trapani JG, Nicolson T (2011) Mechanism of spontaneous activity in afferent neurons of the zebrafish lateral-line organ. J Neurosci 31:1614–1623. https://doi.org/10.1523/JNEUROSCI.3369-10.2011

52. Tilney LG, Tilney MS, DeRosier DJ (1992) Actin filaments, stereocilia, and hair cells: how cells count and measure. Annu Rev Cell Biol 8:257–274. https://doi.org/10.1146/annurev.cb.08.110192.001353

53. Hudspeth AJ, Corey DP (1977) Sensitivity, polarity, and conductance change in the response of vertebrate hair cells to controlled mechanical stimuli. Proc Natl Acad Sci U S A 74:2407–2411. https://doi.org/10.1073/pnas.74.6.2407

54. Fettiplace R, Kim KX (2014) The physiology of mechanoelectrical transduction channels in hearing. Physiol Rev 94:951–986. https://doi.org/10.1152/physrev.00038.2013

55. Puel J-L, Ladrech S, Chabert R et al (1991) Electrophysiological evidence for the presence of NMDA receptors in the guinea pig cochlea. Hear Res 51:255–264. https://doi.org/10.1016/0378-5955(91)90042-8

56. Ruel J, Emery S, Nouvian R et al (2008) Impairment of SLC17A8 encoding vesicular glutamate transporter-3, VGLUT3, underlies nonsyndromic deafness DFNA25 and inner hair cell dysfunction in null mice. Am J Hum Genet 83:278–292. https://doi.org/10.1016/j.ajhg.2008.07.008

57. Safieddine S, Wenthold RJ (1997) The glutamate receptor subunit δ1 is highly expressed in hair cells of the auditory and vestibular systems. J Neurosci 17:7523–7531. https://doi.org/10.1523/JNEUROSCI.17-19-07523.1997

58. Nicolson T (2015) Ribbon synapses in zebrafish hair cells. Hear Res 322. https://doi.org/10.1016/j.heares.2015.04.003

59. Sterling P, Matthews G (2005) Structure and function of ribbon synapses. Trends Neurosci 28:20–29. https://doi.org/10.1016/j.tins.2004.11.009

60. Eatock RA, Lysakowski A (2006) Mammalian vestibular hair cells. In: Vertebrate hair cells. Springer, New York, NY, pp 348–442

61. Burns JC, Kelly MC, Hoa M et al (2015) Single-cell RNA-Seq resolves cellular complexity in sensory organs from the neonatal inner ear. Nat Commun 6:8557. https://doi.org/10.1038/ncomms9557

62. Chen D-Y, Liu X-F, Lin X-J et al (2017) A dominant variant in DMXL2 is linked to nonsyndromic hearing loss. Genet Med 19:553–558. https://doi.org/10.1038/gim.2016.142

63. Ealy M, Ellwanger DC, Kosaric N et al (2016) Single-cell analysis delineates a trajectory toward the human early otic lineage. Proc Natl Acad Sci U S A 113:8508–8513. https://doi.org/10.1073/pnas.1605537113

64. Gao X, Dai P (2014) Impact of next-generation sequencing on molecular diagnosis of inherited non-syndromic hearing loss. J Otol 9:122–125. https://doi.org/10.1016/j.joto.2014.11.003

65. Kolla L, Kelly MC, Mann ZF et al (2020) Characterization of the development of the mouse cochlear epithelium at the single cell level. Nat Commun 11:2389. https://doi.org/10.1038/s41467-020-16113-y

66. Morgan A, Vuckovic D, Krishnamoorthy N et al (2019) Next-generation sequencing identified SPATC1L as a possible candidate gene for both early-onset and age-related hearing loss. Eur J Hum Genet 27:70–79. https://doi.org/10.1038/s41431-018-0229-9

67. Ryu N, Lee S, Park H-J et al (2017) Identification of a novel splicing mutation within SLC17A8 in a Korean family with hearing loss by whole-exome sequencing. Gene 627:233–238. https://doi.org/10.1016/j.gene.2017.06.040

68. Scheffer DI, Shen J, Corey DP, Chen Z-Y (2015) Gene expression by mouse inner ear hair cells during development. J Neurosci 35:6366–6380. https://doi.org/10.1523/JNEUROSCI.5126-14.2015

69. Vona B, Müller T, Nanda I et al (2014) Targeted next-generation sequencing of deafness genes in hearing-impaired individuals uncovers informative mutations. Genet Med 16:945–953. https://doi.org/10.1038/gim.2014.65

70. Westerfield M (1995) The zebrafish book: a guide for the laboratory use of zebrafish (Danio rerio), 3rd edn. M. Westerfield

71. Maeda R, Kindt KS, Mo W et al (2014) Tip-link protein protocadherin 15 interacts with transmembrane channel-like proteins TMC1 and TMC2. Proc Natl Acad Sci U S A 111:12907–12912. https://doi.org/10.1073/pnas.1402152111

72. Postlethwait JH, Woods IG, Ngo-Hazelett P et al (2000) Zebrafish comparative genomics and the origins of vertebrate chromosomes. Genome Res 10:1890–1902. https://doi.org/10.1101/gr.164800

73. Lush ME, Diaz DC, Koenecke N et al (2019) scRNA-Seq reveals distinct stem cell populations that drive hair cell regeneration after loss of Fgf and Notch signaling. eLife 8:e44431. https://doi.org/10.7554/eLife.44431

74. Créton R, Speksnijder JE, Jaffe LF (1998) Patterns of free calcium in zebrafish embryos. J Cell Sci 111(Pt 12):1613–1622

75. Rosen JN, Sweeney MF, Mably JD (2009) Microinjection of Zebrafish embryos to analyze gene function. J Vis Exp. https://doi.org/10.3791/1115

76. Xu Q (1999) Microinjection into Zebrafish embryos. In: Guille M (ed) Molecular methods in developmental biology: Xenopus and Zebrafish. Humana Press, Totowa, NJ, pp 125–132

77. Kok FO, Shin M, Ni C-W et al (2015) Reverse genetic screening reveals poor correlation between morpholino-induced and mutant phenotypes in zebrafish. Dev Cell 32: 97–108. https://doi.org/10.1016/j.devcel.2014.11.018

78. Stainier DYR, Raz E, Lawson ND et al (2017) Guidelines for morpholino use in zebrafish. PLoS Genet 13. https://doi.org/10.1371/journal.pgen.1007000

79. Hwang WY, Fu Y, Reyon D et al (2013) Heritable and precise zebrafish genome editing using a CRISPR-Cas system. PLoS One 8: e68708. https://doi.org/10.1371/journal.pone.0068708

80. Kim BH, Zhang G (2020) Generating stable knockout zebrafish lines by deleting large chromosomal fragments using multiple gRNAs. G3 10:1029–1037. https://doi.org/10.1534/g3.119.401035

81. Varshney GK, Pei W, LaFave MC et al (2015) High-throughput gene targeting and phenotyping in zebrafish using CRISPR/Cas9. Genome Res 25:1030–1042. https://doi.org/10.1101/gr.186379.114

82. Shah AN, Davey CF, Whitebirch AC et al (2015) Rapid reverse genetic screening using CRISPR in zebrafish. Nat Methods 12: 535–540. https://doi.org/10.1038/nmeth.3360

83. Buglo E, Sarmiento E, Martuscelli NB et al (2020) Genetic compensation in a stable slc25a46 mutant zebrafish: a case for using F0 CRISPR mutagenesis to study phenotypes caused by inherited disease. PLoS One 15: e0230566. https://doi.org/10.1371/journal.pone.0230566

84. El-Brolosy MA, Kontarakis Z, Rossi A et al (2019) Genetic compensation triggered by mutant mRNA degradation. Nature 568: 193–197. https://doi.org/10.1038/s41586-019-1064-z

85. Rossi A, Kontarakis Z, Gerri C et al (2015) Genetic compensation induced by deleterious mutations but not gene knockdowns. Nature 524:230–233. https://doi.org/10.1038/nature14580

86. Wilkie AO (1994) The molecular basis of genetic dominance. J Med Genet 31:89–98. https://doi.org/10.1136/jmg.31.2.89

87. Hyatt TM, Ekker SC (1998) Chapter 8 vectors and techniques for ectopic gene expression in zebrafish. In: Detrich HW, Westerfield M, Zon LI (eds) Methods in cell biology. Academic, pp 117–126

88. Distel M, Hocking JC, Volkmann K, Köster RW (2010) The centrosome neither persistently leads migration nor determines the site of axonogenesis in migrating neurons in vivo. J Cell Biol 191:875–890. https://doi.org/10.1083/jcb.201004154

89. Ernest S, Rosa FM (2015) A genomic region encompassing a newly identified exon provides enhancing activity sufficient for normal myo7aa expression in zebrafish sensory hair cells. Dev Neurobiol 75:961–983. https://doi.org/10.1002/dneu.22263

90. McDermott BM, Asai Y, Baucom JM et al (2010) Transgenic labeling of hair cells in the zebrafish acousticolateralis system. Gene Expr Patterns 10:113–118. https://doi.org/10.1016/j.gep.2010.01.001

91. Odermatt B, Nikolaev A, Lagnado L (2012) Encoding of luminance and contrast by linear and nonlinear synapses in the retina. Neuron 73:758–773. https://doi.org/10.1016/j.neuron.2011.12.023

92. Xiao T, Roeser T, Staub W, Baier H (2005) A GFP-based genetic screen reveals mutations that disrupt the architecture of the zebrafish retinotectal projection. Development 132: 2955–2967. https://doi.org/10.1242/dev.01861

93. Esain V, Postlethwait JH, Charnay P, Ghislain J (2010) FGF-receptor signalling controls neural cell diversity in the zebrafish hindbrain by regulating olig2 and sox9. Development 137:33–42. https://doi.org/10.1242/dev.038026

94. Mandal A, Pinter K, Mosqueda N et al (2021) Retrograde mitochondrial transport is essential for organelle distribution and health in zebrafish neurons. J Neurosci 41(7):1371–1392

95. Faucherre A, Pujol-Martí J, Kawakami K, López-Schier H (2009) Afferent neurons of the zebrafish lateral line are strict selectors of hair-cell orientation. PLoS One 4:e4477. https://doi.org/10.1371/journal.pone.0004477

96. Mo W, Nicolson T (2011) Both pre- and postsynaptic activity of Nsf prevents degeneration of hair-cell synapses. PLoS One 6: e27146. https://doi.org/10.1371/journal.pone.0027146

97. Baxendale S, Whitfield TT (2016) Chapter 6—methods to study the development, anatomy, and function of the zebrafish inner ear across the life course. In: Detrich HW, Westerfield M, Zon LI (eds) Methods in cell biology. Academic, pp 165–209

98. Swinburne IA, Mosaliganti KR, Upadhyayula S et al (2018) Lamellar projections in the endolymphatic sac act as a relief valve to regulate inner ear pressure. eLife 7:e37131. https://doi.org/10.7554/eLife.37131

99. Zecca A, Dyballa S, Voltes A et al (2015) The order and place of neuronal differentiation establish the topography of sensory projections and the entry points within the hindbrain. J Neurosci 35:7475–7486. https://doi.org/10.1523/JNEUROSCI.3743-14.2015

100. Gale JE, Marcotti W, Kennedy HJ et al (2001) FM1-43 dye behaves as a permeant blocker of the hair-cell mechanotransducer channel. J Neurosci 21:7013–7025

101. Meyers JR, MacDonald RB, Duggan A et al (2003) Lighting up the senses: FM1-43 loading of sensory cells through nonselective ion channels. J Neurosci 23:4054–4065

102. Seiler C, Nicolson T (1999) Defective calmodulin-dependent rapid apical endocytosis in zebrafish sensory hair cell mutants. J Neurobiol 41:424–434. https://doi.org/10.1002/(SICI)1097-4695(19991115)41:3<424::AID-NEU10>3.0.CO;2-G

103. Erickson T, Pacentine IV, Venuto A et al (2020) The lhfpl5 Ohnologs lhfpl5a and lhfpl5b are required for mechanotransduction in distinct populations of sensory hair cells in zebrafish. Front Mol Neurosci 12. https://doi.org/10.3389/fnmol.2019.00320

104. Pacentine IV, Nicolson T (2019) Subunits of the mechano-electrical transduction channel, Tmc1/2b, require Tmie to localize in zebrafish sensory hair cells. PLoS Genet 15: e1007635. https://doi.org/10.1371/journal.pgen.1007635

105. Hartwell RD, England SJ, Monk NAM et al (2019) Anteroposterior patterning of the zebrafish ear through Fgf- and Hh-dependent regulation of hmx3a expression. PLoS Genet 15:e1008051. https://doi.org/10.1371/journal.pgen.1008051

106. Monroe JD, Manning DP, Uribe PM et al (2016) Hearing sensitivity differs between zebrafish lines used in auditory research. Hear Res 341:220–231. https://doi.org/10.1016/j.heares.2016.09.004

107. Thomas ED, Raible DW (2019) Distinct progenitor populations mediate regeneration in the zebrafish lateral line. eLife 8:e43736. https://doi.org/10.7554/eLife.43736

108. Toro C, Trapani JG, Pacentine I et al (2015) Dopamine modulates the activity of sensory hair cells. J Neurosci 35:16494–16503. https://doi.org/10.1523/JNEUROSCI.1691-15.2015

109. Wada H, Ghysen A, Satou C et al (2010) Dermal morphogenesis controls lateral line patterning during postembryonic development of teleost fish. Dev Biol 340:583–594. https://doi.org/10.1016/j.ydbio.2010.02.017

110. Behra M, Gallardo VE, Bradsher J et al (2012) Transcriptional signature of accessory cells in the lateral line, using the Tnk1bp1: EGFP transgenic zebrafish line. BMC Dev Biol 12:6. https://doi.org/10.1186/1471-213X-12-6

111. Higashijima S, Hotta Y, Okamoto H (2000) Visualization of cranial motor neurons in live transgenic zebrafish expressing green fluorescent protein under the control of the islet-1 promoter/enhancer. J Neurosci 20:206–218

112. Nechiporuk A, Linbo T, Poss KD, Raible DW (2007) Specification of epibranchial placodes in zebrafish. Development 134:611–623. https://doi.org/10.1242/dev.02749

113. Haas P, Gilmour D (2006) Chemokine signaling mediates self-organizing tissue migration in the zebrafish lateral line. Dev Cell 10: 673–680. https://doi.org/10.1016/j.devcel.2006.02.019

114. Geng F-S, Abbas L, Baxendale S et al (2013) Semicircular canal morphogenesis in the zebrafish inner ear requires the function of gpr126 (lauscher), an adhesion class G protein-coupled receptor gene. Development 140:4362–4374. https://doi.org/10.1242/dev.098061

115. Whitfield TT, Granato M, van Eeden FJ et al (1996) Mutations affecting development of the zebrafish inner ear and lateral line. Dev Camb Engl 123:241–254

116. Riley BB, Chiang M, Farmer L, Heck R (1999) The deltaA gene of zebrafish mediates lateral inhibition of hair cells in the inner ear and is regulated by pax2.1. Dev Camb Engl 126:5669–5678

117. Kozlowski DJ, Whitfield TT, Hukriede NA et al (2005) The zebrafish dog-eared mutation disrupts eya1, a gene required for cell survival and differentiation in the inner ear and lateral line. Dev Biol 277:27–41. https://doi.org/10.1016/j.ydbio.2004.08.033

118. Andermann P, Ungos J, Raible DW (2002) Neurogenin1 defines zebrafish cranial sensory ganglia precursors. Dev Biol 251:45–58. https://doi.org/10.1006/dbio.2002.0820

119. Hwang P, Chou S-W, Chen Z, McDermott BM (2015) The stereociliary paracrystal is a dynamic cytoskeletal scaffold in vivo. Cell Rep 13:1287–1294. https://doi.org/10.1016/j.celrep.2015.10.003

120. Pelassa I, Zhao C, Pasche M et al (2014) Synaptic vesicles are "primed" for fast clathrin-mediated endocytosis at the ribbon synapse. Front Mol Neurosci 7:91. https://doi.org/10.3389/fnmol.2014.00091

121. Ohta S, Ji YR, Martin D, Wu DK (2020) Emx2 regulates hair cell rearrangement but not positional identity within neuromasts. eLife. https://elifesciences.org/articles/60432. Accessed 22 Jan 2021

122. Blanco-Sánchez B, Clément A, Fierro J et al (2018) Grxcr1 promotes hair bundle development by destabilizing the physical interaction between harmonin and sans usher syndrome proteins. Cell Rep 25:1281–1291.e4. https://doi.org/10.1016/j.celrep.2018.10.005

123. World Health Organization (2021) World report on hearing. https://apps.who.int/iris/handle/10665/339913. License: CCBY-NC-SA 3.0 IGO

124. Gopal SR, Chen DH-C, Chou S-W et al (2015) Zebrafish models for the mechanosensory hair cell dysfunction in usher syndrome 3 reveal that clarin-1 is an essential hair bundle protein. J Neurosci 35. 10188–10201. https://doi.org/10.1523/JNEUROSCI.1096-15.2015

125. Nagiel A, Andor-Ardó D, Hudspeth AJ (2008) Specificity of afferent synapses onto plane-polarized hair cells in the posterior lateral line of the zebrafish. J Neurosci 28: 8442–8453. https://doi.org/10.1523/JNEUROSCI.2425-08.2008

126. Dow E, Jacobo A, Hossain S et al (2018) Connectomics of the zebrafish's lateral-line neuromast reveals wiring and miswiring in a simple microcircuit. eLife 7:e33988. https://doi.org/10.7554/eLife.33988

127. Jacobo A, Dasgupta A, Erzberger A et al (2019) Notch-mediated determination of hair-bundle polarity in mechanosensory hair cells of the zebrafish lateral line. Curr Biol 29: 3579–3587.e7. https://doi.org/10.1016/j.cub.2019.08.060

128. Lozano-Ortega M, Valera G, Xiao Y et al (2018) Hair cell identity establishes labeled lines of directional mechanosensation. PLoS Biol 16:e2004404. https://doi.org/10.1371/journal.pbio.2004404

129. Navajas Acedo J, Voas MG, Alexander R et al (2019) PCP and Wnt pathway components act in parallel during zebrafish mechanosensory hair cell orientation. Nat Commun 10: 3993. https://doi.org/10.1038/s41467-019-12005-y

130. Grisham RC, Kindt K, Finger-Baier K et al (2013) Mutations in ap1b1 cause mistargeting of the Na+/K+-ATPase pump in sensory hair cells. PLoS One 8:e60866. https://doi.org/10.1371/journal.pone.0060866

131. Chou S-W, Chen Z, Zhu S et al (2017) A molecular basis for water motion detection by the mechanosensory lateral line of zebrafish. Nat Commun 8:1–16. https://doi.org/10.1038/s41467-017-01604-2

132. Chatterjee P, Padmanarayana M, Abdullah N et al (2015) Otoferlin deficiency in zebrafish results in defects in balance and hearing: rescue of the balance and hearing phenotype with full-length and truncated forms of mouse otoferlin. Mol Cell Biol 35: 1043–1054. https://doi.org/10.1128/MCB.01439-14

133. Einhorn Z, Trapani JG, Liu Q, Nicolson T (2012) Rabconnectin3α promotes stable activity of the H+ pump on synaptic vesicles in hair cells. J Neurosci 32:11144–11156. https://doi.org/10.1523/JNEUROSCI. 1705-12.2012

134. Lin S, Vollrath MA, Mangosing S et al (2016) The zebrafish pinball wizard gene encodes WRB, a tail-anchored-protein receptor essential for inner-ear hair cells and retinal photoreceptors. J Physiol 594:895–914. https://doi.org/10.1113/JP271437

135. Lukasz D, Kindt KS (2018) In vivo calcium imaging of lateral-line hair cells in larval zebrafish. J Vis Exp. https://doi.org/10.3791/58794

136. Trapani JG, Nicolson T (2010) Chapter 8—physiological recordings from zebrafish lateral-line hair cells and afferent neurons. In: William H, Detrich MW, Zon LI (eds) Methods in cell biology. Academic, pp 219–231

137. Olt J, Johnson SL, Marcotti W (2014) In vivo and in vitro biophysical properties of hair cells from the lateral line and inner ear of developing and adult zebrafish. J Physiol 592:2041–2058. https://doi.org/10.1113/jphysiol.2013.265108

138. Ricci AJ, Bai J-P, Song L et al (2013) Patch-clamp recordings from lateral line neuromast hair cells of the living zebrafish. J Neurosci 33:3131–3134. https://doi.org/10.1523/JNEUROSCI.4265-12.2013

139. Haehnel M, Taguchi M, Liao JC (2012) Heterogeneity and dynamics of lateral line afferent innervation during development in zebrafish (Danio rerio). J Comp Neurol 520:1376–1386. https://doi.org/10.1002/cne.22798

140. Ohn T-L, Rutherford MA, Jing Z et al (2016) Hair cells use active zones with different voltage dependence of Ca2+ influx to decompose sounds into complementary neural codes. Proc Natl Acad Sci U S A 113(32):E4716–E4725. https://doi.org/10.1073/pnas.1605737113

141. Song S, Lee JA, Kiselev I et al (2018) Mathematical modeling and analyses of interspike-intervals of spontaneous activity in afferent neurons of the zebrafish lateral line. Sci Rep 8:14851. https://doi.org/10.1038/s41598-018-33064-z

142. Trapani JG, Obholzer N, Mo W et al (2009) Synaptojanin1 is required for temporal fidelity of synaptic transmission in hair cells. PLoS Genet 5:e1000480. https://doi.org/10.1371/journal.pgen.1000480

143. Furman AC, Kujawa SG, Liberman MC (2013) Noise-induced cochlear neuropathy is selective for fibers with low spontaneous rates. J Neurophysiol 110:577–586. https://doi.org/10.1152/jn.00164.2013

144. Olt J, Ordoobadi AJ, Marcotti W, Trapani JG (2016) Physiological recordings from the zebrafish lateral line. Methods Cell Biol 133:253–279. https://doi.org/10.1016/bs.mcb.2016.02.004

145. Troconis EL, Ordoobadi AJ, Sommers TF et al (2017) Intensity-dependent timing and precision of startle response latency in larval zebrafish. J Physiol 595:265–282. https://doi.org/10.1113/JP272466

146. Zhang QX, He XJ, Wong HC, Kindt KS (2016) Functional calcium imaging in zebrafish lateral-line hair cells. Methods Cell Biol 133:229–252. https://doi.org/10.1016/bs.mcb.2015.12.002

147. Smedemark-Margulies N, Trapani JG (2013) Tools, methods, and applications for optophysiology in neuroscience. Front Mol Neurosci 6. https://doi.org/10.3389/fnmol.2013.00018

148. Esterberg R, Linbo T, Pickett SB et al (2016) Mitochondrial calcium uptake underlies ROS generation during aminoglycoside-induced hair cell death. J Clin Invest 126:3556–3566. https://doi.org/10.1172/JCI84939

149. Wong HC, Zhang Q, Beirl AJ et al (2019) Synaptic mitochondria regulate hair-cell synapse size and function. eLife 8:e48914. https://doi.org/10.7554/eLife.48914

150. Pichler P, Lagnado L (2019) The transfer characteristics of hair cells encoding mechanical stimuli in the lateral line of zebrafish. J Neurosci 39:112–124. https://doi.org/10.1523/JNEUROSCI.1472-18.2018

151. Sheets L, Kindt KS, Nicolson T (2012) Presynaptic CaV1.3 channels regulate synaptic ribbon size and are required for synaptic maintenance in sensory hair cells. J Neurosci 32:17273–17286. https://doi.org/10.1523/JNEUROSCI.3005-12.2012

152. Bhandiwad AA, Zeddies DG, Raible DW et al (2013) Auditory sensitivity of larval zebrafish (Danio rerio) measured using a behavioral prepulse inhibition assay. J Exp Biol 216:3504–3513. https://doi.org/10.1242/jeb.087635

153. Bang PI, Yelick PC, Malicki JJ, Sewell WF (2002) High-throughput behavioral screening method for detecting auditory response defects in zebrafish. J Neurosci Methods 118:

177–187. https://doi.org/10.1016/s0165-0270(02)00118-8

154. Zottoli SJ, Faber DS (2000) Review: The mauthner cell: what has it taught us? Neuroscientist 6(1):26–38. https://doi.org/10.1177/107385840000600111

155. Zeddies DG, Fay RR (2005) Development of the acoustically evoked behavioral response in zebrafish to pure tones. J Exp Biol 208:1363–1372. https://doi.org/10.1242/jeb.01534

156. Burgess HA, Granato M (2007) Sensorimotor gating in larval zebrafish. J Neurosci 27:4984–4994. https://doi.org/10.1523/JNEUROSCI.0615-07.2007

157. Friedrich T, Lambert AM, Masino MA, Downes GB (2012) Mutation of zebrafish dihydrolipoamide branched-chain transacylase E2 results in motor dysfunction and models maple syrup urine disease. Dis Model Mech 5:248–258. https://doi.org/10.1242/dmm.008383

158. Bagnall MW, Schoppik D (2018) Development of vestibular behaviors in zebrafish. Curr Opin Neurobiol 53:83–89. https://doi.org/10.1016/j.conb.2018.06.004

159. Oteiza P, Odstrcil I, Lauder G et al (2017) A novel mechanism for mechanosensory-based rheotaxis in larval zebrafish. Nature 547:445–448. https://doi.org/10.1038/nature23014

160. Vanwalleghem G, Schuster K, Taylor MA et al (2020) Brain-wide mapping of water flow perception in zebrafish. J Neurosci 40:4130–4144. https://doi.org/10.1523/JNEUROSCI.0049-20.2020

161. Arnold GP, Weihs D (1978) The hydrodynamics of rheotaxis in the plaice (Pleuronectes Platessa L.). J Exp Biol 75:147–169

162. Marsden KC, Jain RA, Wolman MA et al (2018) A Cyfip2-dependent excitatory interneuron pathway establishes the innate startle threshold. Cell Rep 23:878–887. https://doi.org/10.1016/j.celrep.2018.03.095

163. Lin JY (2011) A user's guide to channelrhodopsin variants: features, limitations and future developments. Exp Physiol 96:19–25. https://doi.org/10.1113/expphysiol.2009.051961

164. Ozdemir YI, Hansen CA, Ramy MA et al (2021) Recording channelrhodopsin-evoked field potentials and startle responses from larval zebrafish. Methods Mol Biol 2191:201–220. https://doi.org/10.1007/978-1-0716-0830-2_13

165. Eaton RC, Farley RD (1975) Mauthner neuron field potential in newly hatched larvae of the zebra fish. J Neurophysiol 38:502–512. https://doi.org/10.1152/jn.1975.38.3.502

166. Prugh JI, Kimmel CB, Metcalfe WK (1982) Noninvasive recording of the Mauthner neurone action potential in larval zebrafish. J Exp Biol 101:83–92

167. Zottoli SJ (1977) Correlation of the startle reflex and Mauthner cell auditory responses in unrestrained goldfish. J Exp Biol 66:243–254

168. Liu KS, Fetcho JR (1999) Laser ablations reveal functional relationships of segmental hindbrain neurons in zebrafish. Neuron 23:325–335. https://doi.org/10.1016/S0896-6273(00)80783-7

169. Fetcho JR (1991) Spinal network of the mauthner cell (part 1 of 2). Brain Behav Evol 37:298–306. https://doi.org/10.1159/000114367

170. Issa FA, O'Brien G, Kettunen P et al (2011) Neural circuit activity in freely behaving zebrafish (Danio rerio). J Exp Biol 214:1028–1038. https://doi.org/10.1242/jeb.048876

171. Monesson-Olson BD, Troconis EL, Trapani JG (2014) Recording field potentials from zebrafish larvae during escape responses. J Undergrad Neurosci Educ 13:A52–A58

172. Sun F, Zhou J, Dai B et al (2020) Next-generation GRAB sensors for monitoring dopaminergic activity in vivo. Nat Methods 17:1156–1166. https://doi.org/10.1038/s41592-020-00981-9

173. Jing M, Li Y, Zeng J et al (2020) An optimized acetylcholine sensor for monitoring in vivo cholinergic activity. Nat Methods 17:1139–1146. https://doi.org/10.1038/s41592-020-0953-2

174. Lunsford ET, Skandalis DA, Liao JC (2019) Efferent modulation of spontaneous lateral line activity during and after zebrafish motor commands. J Neurophysiol 122:2438–2448. https://doi.org/10.1152/jn.00594.2019

175. Panzera LC, Hoppa MB (2019) Genetically encoded voltage indicators are illuminating subcellular physiology of the axon. Front Cell Neurosci 13. https://doi.org/10.3389/fncel.2019.00052

176. Beck C, Zhang D, Gong Y (2019) Enhanced genetically encoded voltage indicators advance their applications in neuroscience. Curr Opin Biomed Eng 12:111–117. https://doi.org/10.1016/j.cobme.2019.10.010

177. Power RM, Huisken J (2017) A guide to light-sheet fluorescence microscopy for multiscale imaging. Nat Methods 14:360–373. https://doi.org/10.1038/nmeth.4224

178. Abu-Siniyeh A, Al-Zyoud W (2020) High-lights on selected microscopy techniques to study zebrafish developmental biology. Lab Anim Res 36:12. https://doi.org/10.1186/s42826-020-00044-2

179. Santi PA, Johnson SB, Hillenbrand M et al (2009) Thin-sheet laser imaging microscopy for optical sectioning of thick tissues. Bio-Techniques 46:287–294. https://doi.org/10.2144/000113087

180. Marques JC, Li M, Schaak D et al (2020) Internal state dynamics shape brainwide activity and foraging behaviour. Nature 577: 239–243. https://doi.org/10.1038/s41586-019-1858-z

181. Bahl A, Engert F (2020) Neural circuits for evidence accumulation and decision making in larval zebrafish. Nat Neurosci 23:94–102. https://doi.org/10.1038/s41593-019-0534-9

182. Vanwalleghem GC, Ahrens MB, Scott EK (2018) Integrative whole-brain neuroscience in larval zebrafish. Curr Opin Neurobiol 50: 136–145. https://doi.org/10.1016/j.conb.2018.02.004

183. Migault G, van der Plas TL, Trentesaux H et al (2018) Whole-brain calcium imaging during physiological vestibular stimulation in larval zebrafish. Curr Biol 28:3723–3735.e6. https://doi.org/10.1016/j.cub.2018.10.017

184. Favre-Bulle IA, Vanwalleghem G, Taylor MA et al (2018) Cellular-resolution imaging of vestibular processing across the larval zebra-fish brain. Curr Biol 28:3711–3722.e3. https://doi.org/10.1016/j.cub.2018.09.060

185. Kindt KS, Sheets L (2018) Transmission disrupted: modeling auditory synaptopathy in zebrafish. Front Cell Dev Biol 6. https://doi.org/10.3389/fcell.2018.00114

186. Owens KN, Cunningham DE, MacDonald G et al (2007) Ultrastructural analysis of aminoglycoside-induced hair cell death in the zebrafish lateral line reveals an early mito-chondrial response. J Comp Neurol 502: 522–543. https://doi.org/10.1002/cne.21345

187. Santos F, MacDonald G, Rubel EW, Raible DW (2006) Lateral line hair cell maturation is a determinant of aminoglycoside susceptibil-ity in zebrafish (Danio rerio). Hear Res 213: 25–33. https://doi.org/10.1016/j.heares.2005.12.009

188. Peterson HP, Troconis EL, Ordoobadi AJ et al (2018) Teaching dose-response relation-ships through aminoglycoside block of mechanotransduction channels in lateral line hair cells of larval zebrafish. J Undergrad Neurosci Educ 17:A40–A49

189. Kirshenbaum AP, Chabot E, Gibney N (2019) Startle, pre-pulse sensitization, and habituation in zebrafish. J Neurosci Methods 313:54–59. https://doi.org/10.1016/j.jneumeth.2018.12.017

190. Bhandiwad AA, Raible DW, Rubel EW, Sis-neros JA (2018) Noise-induced hypersensiti-zation of the acoustic startle response in larval zebrafish. J Assoc Res Otolaryngol 19: 741–752. https://doi.org/10.1007/s10162-018-00685-0

191. Best JD, Berghmans S, Hunt JJFG et al (2008) Non-associative learning in larval zeb-rafish. Neuropsychopharmacology 33: 1206–1215. https://doi.org/10.1038/sj.npp.1301489

192. Pantoja C, Hoagland A, Carroll E et al (2017) Measuring behavioral individuality in the acoustic startle behavior in zebrafish. Bio-Protoc 7. https://doi.org/10.21769/BioProtoc.2200

193. Faria M, Prats E, Novoa-Luna KA et al (2019) Development of a vibrational startle response assay for screening environmental pollutants and drugs impairing predator avoidance. Sci Total Environ 650:87–96. https://doi.org/10.1016/j.scitotenv.2018.08.421

194. Harris JA, Cheng AG, Cunningham LL et al (2003) Neomycin-induced hair cell death and rapid regeneration in the lateral line of zebra-fish (Danio rerio). J Assoc Res Otolaryngol 4: 219–234. https://doi.org/10.1007/s10162-002-3022-x

Chapter 10

Electrophysiological Recordings of Voltage-Dependent and Mechanosensitive Currents in Sensory Hair Cells of the Auditory and Vestibular Organs of the Mouse

Artur A. Indzhykulian, Stuart L. Johnson, and Gwenaëlle S. G. Géléoc

Abstract

Electrophysiological characterization of inner ear hair cell properties including assessment of voltage-dependent and mechanosensitive currents has provided invaluable information about their development and maturation in several animal models. Beyond the basic understanding of hair cell properties, electrophysiological investigations combined with the use of different mouse models, pharmacological tools, and exogenous gene expression systems such as those driven by viral vectors have been essential in providing insights into the functional role of various proteins expressed in hair cells, many of which are associated with deafness and/or balance deficits. This chapter provides detailed methods designed to optimize recordings of voltage-dependent and mechanosensitive currents in sensory hair cells of the auditory and vestibular organs of the mammal.

Key words Hair cells, Voltage-dependent currents, Mechanotransduction currents, Electrophysiology, Patch clamp, Mechanical stimulus, Stiff probe, Fluid jet, Mouse, Inner ear

1 Introduction

Sensory hair cells of the inner ear are primary receptors found in auditory and vestibular end organs of vertebrate animals. These cells mediate the senses of hearing and balance by transducing mechanical stimuli into electrical signals which are transmitted to the central nervous system. This process is called mechano-electrical transduction [1–5]. The sensitivity of the hair cells to displacement of their fluid environment, either during propagation of a sound wave or due to head movements, is provided by a highly specialized structure at the apex of the sensory hair cells: the *hair bundle* [2, 6]. The hair bundle comprises actin-filled microvilli, called *stereocilia*, organized in rows of graded height and

Artur A. Indzhykulian, Stuart L. Johnson and Gwenaëlle S. G. Géléoc contributed equally to this work.

Andrew K. Groves (ed.), *Developmental, Physiological, and Functional Neurobiology of the Inner Ear*, Neuromethods, vol. 176, https://doi.org/10.1007/978-1-0716-2022-9_10,

interconnected by several types of interciliary links [6, 7]. At the base of the stereocilia, a taper or narrowing of the ankle region forms a hinge where stereocilia pivot upon application of a displacement stimulus. Hair bundles can be displaced in several directions but only displacements towards or against the tallest row of stereocilia alter the hair cell *receptor potential* [4, 8]. The *mechanotransduction (MT) complex is the central component enabling the mechano-electrical transducer function and* is localized at the tips of stereocilia [9–11]. Adjacent stereocilia of graded height are interconnected at their apex by tip-links which convey the displacement force to the MT complex [12]. This complex includes transmembrane channel-like proteins, TMC1 and TMC2, which form the pore of the channel, and several other proteins that act as linker/modifiers of this complex [13–17]. The MT channel is permeable to cations, thereby allowing positive charges to flow into the cell [4]. *In vivo*, the hair bundle bathes in a potassium-rich media called *endolymph*, while the cell body bathes in a sodium-rich media called *perilymph* [18]. Upon opening of the channels, an influx of positive ions depolarizes the cell and results in the generation of a *receptor potential* [4, 8]. The hair cell receptor potential does not generate action potentials but activates voltage-dependent calcium channels allowing entry of calcium which activates the release of neurotransmitter, glutamate, at the base of the sensory cells. Glutamate, in turn, stimulates afferent nerve fibers connected to the sensory hair cells.

As many deafness genes encode proteins essential for hair bundle development, structure, and function, different techniques have been used to assess hair cells and hair bundle function in animal models of deafness. In particular, electrophysiological recordings of mechano-electrical signals in single hair cells, initially developed by Hudspeth and Corey [4] to study the frog sacculus, have now been widely used and adapted to study hair cells from different species. For this chapter we will focus on the study of sensory hair cells in the mammalian inner ear.

The auditory organ comprises two types of *cochlear hair cells. Outer hair cells* (OHCs) organized along the organ of Corti in three longitudinal rows play an important role in frequency tuning. They express a protein called prestin [19], a motor protein which provides mechanical energy that manifests itself in the form of somatic electromotility upon depolarization of the cell membrane. While OHCs are contacted by thin unmyelinated type II afferent fibers, the function of this afferent signal remains poorly understood. Each type II afferent fiber synapses with ten or more OHCs. *Inner hair cells* (IHCs), organized in a single row along the organ, are innervated by type I cochlear afferents which are large diameter myelinated neurons that each connect to a single ribbon synapse of a single hair cell. IHCs are considered "true" sensory receptors of the inner ear as these cells are responsible for the transmission of auditory information to the central nervous system. These two cell types express different sets of proteins [20, 21] and in particular various types of ion channels. Changes in the hair cell's complement of ion channels and their expression levels have been observed during development and along the organ from the basal high-frequency end to the apical low frequency end [17, 22–27].

The vestibular organs of mammals include two otoliths (utricle and saccule), which sense linear acceleration and gravity, and three semicircular canals, which sense angular acceleration. The vestibular apparatus possesses two major sensory hair cell types: the piriform *type I* hair cells and the cylindrical *type II* hair cells. Type I hair cells are innervated by calyx ending irregularly firing afferent fibers. Type II hair cells are innervated by bouton ending regularly firing afferent fibers. These two hair cell types express different sets of ionic conductances which determine their electrophysiological properties.

Hair cell function largely depends upon the expression of a specific set of voltage-dependent ion channels and their modulation upon a specific context, may this be an ionic gradient, expression of a different channel subunit, or alteration of the protein via biochemical modulation (phosphorylation, etc.). With the advancement of patch-clamp technology, numerous studies have been conducted to assess the functional relevance of these conductances and determine how genetic manipulations affect hair cell function leading to deafness or balance deficit.

This chapter aims to provide an overview of the methods used to record voltage-dependent and mechanosensitive currents in sensory hair cells of the auditory and vestibular organs of the neonatal and adult mouse or gerbil. Specific methods describing tissue preparation and recording procedures are depicted as well as a description of a typical setup used for these experiments.

2 Methods

2.1 Tissue Preparation (Acute and Culture)

The inner ear organs are excised from postnatal (P0–P6) mice. The microdissection is similar to that previously described [28, 29] with a few optimizations designed to preserve the hair bundles.

2.1.1 Microdissection of the Neonatal Utricle for Electrophysiological Recordings

- In brief, neonatal mice (P0–P8) are euthanized by decapitation according to the appropriate and approved institutional animal care and use committee protocol. The head is excised by cutting low in the neckline well below the pinna (Fig. 1a), rinsed briefly in 75% ethanol and placed in MEM (Invitrogen, Carlsbad, CA). The head is then bisected along two cuts: one from the foramen magnum towards the top of the skull up to the nose and the other one from the foramen magnum to the middle of the mouth roof and up to the nose (Fig. 1b). The brain tissue is

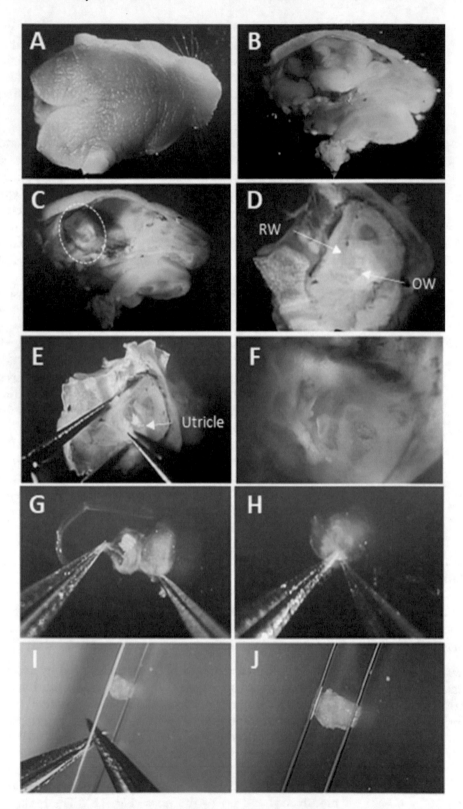

Fig. 1 Dissection of the neonatal mouse utricle. (**a**) Head from a neonatal mouse as seen after cut below the neck line. (**b**) Hemi-skull before and (**c**) after removal of the brain in fresh extracellular solution. The bony

removed (Fig. 1c) and the preparation is quickly placed in fresh MEM supplemented with HEPES and Ampicillin (*see* Subheading "Solutions").

- *Temporal bone extraction*: The temporal bone is removed by cutting cartilage and bone around it with fine scissors. After transfer to a new dish of MEM, all overlying tissue, including the surface skin and mesenchymal tissue is removed to expose the tympanic ring (skull side). Once the tympanic ring is removed, the cochlear portion of the labyrinth becomes visible. The middle ear bones are typically removed along with the tympanic ring. As a result, both the oval and round windows are exposed (Fig. 1d).

  Note: A key to preserving healthy tissues is to keep fresh media during the entire dissection. For this reason, it is important to regularly transfer the preparation to a new petri dish. One indication that a new fresh media is needed is when the pH indicator changes (use of phenol red in the dissecting media is helpful) or when the media becomes cloudy.

- *Preparation of the tissue for enzyme treatment*: The otolithic membrane includes a gelatinous extracellular layer above which lay otoconia, small crystals composed of organic and inorganic elements. This structure couples the stereocilia at the surface of the utricle (and saccule). Removal of this matrix without enzymatic treatment is possible but will lead to hair bundle damage. If the goal is to obtain a pristine structure, a short protease treatment is advised that will loosen the connection between the extracellular layer and the hair bundles. The approach described below involves the opening of the bony labyrinth and preservation of the epithelium within the temporal bone.

  As the utricle is just beneath the bony medial surface, caution should be used when opening the bony labyrinth to expose the utricle. In older animals the bone is more calcified and harder. In young mice it is advised to use forceps to pop up a portion of the bone. When working on older mice (P10 or more), a small scalpel may be used to gently scrape away the bone layer (Fig. 1d, e). When working on pigmented mice the location of the utricle is obvious due to the presence of a

---

Fig. 1 (continued) labyrinth can be seen towards the back of the head (dotted line). (**d**) The skull is trimmed away with small scissors to extract the bony labyrinth. (**e**) The isolated bony labyrinth is held by with forceps placed on each side of the superior semicircular canal. The bone is removed to make a small hole close to the oval window and above the utricle. (**f**) The membranous labyrinth is trimmed away above the utricle to expose the otolith. (**g**) After enzymatic treatment and removal of the otolithic membrane, the utricle is excised from the temporal bone with one of the ampulla. (**h**) Ampullas and nerve fibers are trimmed away to have a clean and thin preparation. (**i**) The utricle is placed under thin glass fibers. (**j**) Example of a flat and trimmed utricle mounted under the thin glass fibers

pigmented membranous labyrinth. Conversely, this membrane is translucent in albino mice making it a little more challenging at first. Nevertheless, once the bones and cartilages are removed the presence of a bright white otoconial mass should be obvious to the experimenter (Fig. 1e).

To allow solutions to access the epithelium, the upper part of the membranous labyrinth is cut open on one side of the utricle. It is better to use fine forceps to expose the entire surface while preserving the contacts with the nearby ampullas, so the sensory epithelium stays in place in the temporal bone.

The temporal bone serves as a convenient scoop to bring the tissue through to the enzymatic bath and after that back in the rinse solution. This will preserve the stereocilia as the epithelium will not be handled directly and will not pass through a meniscus.

The temporal bones are then bathed in MEM solution (Invitrogen, Carlsbad, CA) containing 0.1 mg/ml protease XXIV (Sigma, St. Louis, MO) for 15–20 min. The bones are placed now faced down so the otoconia (also called otoliths) may fall naturally during the enzymatic treatment. The right concentration of enzyme will ensure success. Too much enzymatic treatment will result in a sticky epithelium, too little may pose a problem while removing the otolithic membrane.

- *Final steps*: After the otolithic membrane has been removed (either during the enzymatic treatment or using a fine fiber or eyelash), the utricle is excised using fine forceps. Since it is extremely important to avoid touching the apical surface of the epithelium, one option is to grab it by the nerve or otherwise by one side. Sometimes, it is better to pull it out via the anterior ampulla. The extra tissues and nerves are then trimmed away. The older the prep, the thicker the nerve. The fibers can be grabbed from underneath with two forceps and slowly trimmed away to obtain a thin sensory layer.

- *Mounting option*: Utricles are mounted onto a glass coverslip to allow access to the hair cells for performing electrophysiological recordings. A successful approach in our lab has been to mount the tissue onto a round glass coverslip, holding it in a flat position with two glass fibers glued to the coverslip with a small drop of sylgard (Fig. 1i, j). Other approaches are described in Subheading 2.1.2.

Organotypic Cultures

To produce organotypic cultures, tissues are maintained at 37 °C in MEM (Invitrogen) supplemented with 10 mM HEPES (pH 7.4), 5 mg/100 ml Ampicillin, 1 mg/100 ml Cipro, and up to 3% FBS. The culturing medium is replaced every 2–3 days.

Materials   Dissecting tools; Cell culture dish (35 and 60 mm); Glass fibers chambers (Poly-L-Lysin or Cell-Tak, Corning, may be used instead of fibers); Glass coverslips 18 mm; Glass pipettes for chamber: VWR Calibrated pipettes 100 µl; Sylgard kit Corning; Fine eyelash or equivalent.

Solutions   Dissections are performed in standard MEM solution (Invitrogen, Carlsbad, CA) supplemented with 10 mM HEPES (pH 7.4) and 5 mg/100 ml Ampicillin. Use of solution with phenol red is suggested. For organotypic cultures, use MEM or equivalent (with Earle's salts and glutamax)—Invitrogen, supplemented with 10 mM HEPES (pH 7.4), 5 mg/100 ml Ampicillin, and 1 mg/ 100 ml Ciprofloxacin, warmed to room temperature; Fetal Bovine Serum (FBS) is added to the culture media (Invitrogen); Protease XXIV Sigma at ~1 mg/10 ml.

2.1.2 Microdissection of the Organ of Corti for Electrophysiological Recordings

Electrophysiological studies of developing and mature mammalian auditory hair cells have provided a better understanding of normal hair cell function. Such studies have determined tonotopic differences in the properties of hair cells along the organ of Corti [17, 22–26, 30–35] as well as developmental patterns leading to a fully functional mature sensory cell [34, 36–39]. Hair cell electrophysiology has also been fundamental for investigating the function of specific proteins in hair cells using knockout or mutant models [40, 41]. Such studies, too many to cite, have all relied on ex vivo preparations of the organ of Corti at various ages.

While the overall goal and general approaches to dissecting out the mammalian organ of Corti are similar to what has been described before for adult mouse tissue [42], different limitations and considerations arise depending on the age of the animal, the region of the organ (more basal or more apical), and the cell type the experimenter wants to record from. In the following sections we will focus on these different approaches. In all cases, the key to successful electrophysiological recording is to have a good dissection that optimizes access to healthy viable cells.

Dissection of the Neonatal Organ of Corti

Up to about 9 days of postnatal development (P9), the mammalian cochlea (mouse, gerbil, and rat) can be dissected using the approach detailed below. The neonatal organ of Corti, unlike that at mature stages, can be dissected as one piece and can be used in its entirety for recording, or cut into discrete sections. As the temporal bones and the cochleae undergo substantial change after P10, in part due to the calcification of the bony structure, a slightly modified approach is advised as outlined in the next section (Subheading "Dissection of the Mature/Adult Organ of Corti (>P10)").

Note: It is important to use a very cold extracellular solution while performing the tissue dissection. Solutions are used at almost freezing point, kept on ice throughout the dissection. A frozen cold pack may also be used to keep the tissue cold during the dissection.

- *Temporal Bone Extraction*: The first stages of the dissection are similar to those described above for the utricle (Subheading 2.1.1). Briefly, the head is bisected with scissors, the brain removed with large forceps, and the half heads are placed into petri dishes with cold solution. The bony labyrinth of the cochlea and vestibular apparatus is visible inside the skull towards the back of the head (Fig. 2a) and is carefully taken out of the skull by removing the connective tissue and bone surrounding it with medium forceps (Fig. 2b, c). Care should be taken to not break the cochlea from the vestibular apparatus since these are important for holding the cochlea in the next stage.

- *Exposing the Organ of Corti*: The isolated bony labyrinth is placed into a petri dish with fresh cold solution. Medium forceps are used to hold the cochlea via the vestibular apparatus so that the tip of the cochlear apex, or helicotrema, is facing up and forward (Fig. 2d). Fine forceps are used to make a hole in the bone around the middle coil of the cochlea to expose the tissue below (Fig. 2e). More of the bone is removed from the front face of the cochlea up to the apex by breaking pieces off with fine forceps (Fig. 2f). The cochlea is then turned over to remove the bone from the back surface (Fig. 2g). At this stage, the entire cochlea should be exposed, coiled around the central modiolus (Fig. 2h).

- *Dissecting Out the Organ of Corti*: Neonatal tissue (up to ~P5) can be pulled and removed from the modiolus (Fig. 2i). For older preparations (P5–P9) a modified dissection is advised as the hair cells can become detached from the organ of Corti when the modiolus is removed. Instead, the cochlea is turned upside down and the modiolus separated in stages progressing up towards the apex. The stria vascularis is then removed with fine forceps, one holding the stria and the other one holding the basal end of the organ of Corti. Up to P5, the stria can simply be pulled away from the rest of the organ of Corti (Fig. 2j, k). For older preparations (P5–P9), the stria is pulled carefully and incrementally from the base to prevent hair cells from detaching and coming along with the stria. At this point, the organ of Corti is ready to be moved to the recording chamber using a transfer spoon (Fig. 2k, l).

- *Mounting the Tissue*: As mentioned in Subheading 2.1.1, there are several techniques that may be considered to hold the cochlea down in the recording chamber including:
  - using glass fibers or tungsten pins glued to a coverslip (Subheading 2.1.1) to hold the tissue down.

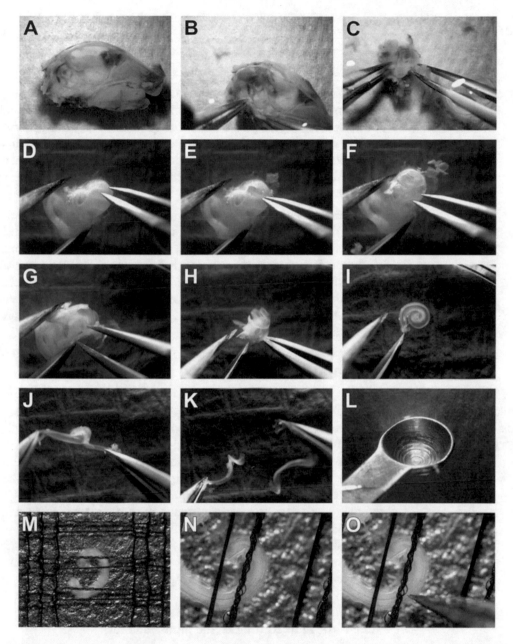

Fig. 2 Dissection of the neonatal mammalian cochlea. (**a**) The bisected skull from a P4 mouse after removal of the brain in cold extracellular solution. The bony labyrinth of the cochlea and vestibular apparatus can be seen towards the back of the head. (**b, c**) The bony labyrinth is removed from the skull with forceps. (**d**) The isolated bony labyrinth held by the vestibular bones with forceps. The bone covering the cochlea is cracked below the very apical turn. (**e**) The bone is removed to make a small hole. (**f**) More of the bone is removed from the top side of the cochlea, revealing the membranous labyrinth of the cochlear duct below. (**g**) The bony labyrinth is turned over and bone is removed from the back of the cochlea. (**h**) The spiral of the isolated cochlear duct attached to the modiolus. (**i**) The cochlear duct after it has been separated from the modiolus. (**j**) The stria vascularis is removed from the organ of Corti by pulling them apart with forceps. (**k**) The isolated organ of Corti and the removed stria vascularis. (**l**) The organ of Corti in the spoon ready to be transferred to the recording chamber. (**m**) The whole neonatal organ of Corti held down under the nylon fibers of the grid in the recording chamber. The apical and basal turns are positioned such that the hair cells are accessible for recording. (**n**)

    – treating the glass coverslip with a tissue adhesive such as Cell-Tak (Corning™), a specially formulated protein solution extracted from marine mussels.

    – using a nylon mesh fixed to a stainless steel ring [43].

- In this case the nylon mesh is used and put into the recording chamber with fresh cold solution with the dissected cochlea placed on top. A pair of fine forceps is used to hold up one of the nylon fibers and the cochlea is placed below such that the section that is to be recorded from is held securely. For apical and basal coil recordings, the cochleae can be placed as shown (Fig. 2m), each section held by a different nylon strand.

- *Removal of the Tectorial Membrane*: An important final step in the dissection is the removal of the tectorial membrane that lies above the hair cells. This step can be difficult because the membrane is transparent and closely juxtaposed onto the sensory patch. Starting near the nylon fiber, the tip of the forceps is used to go below the tectorial membrane and lift it so its edge can be seen. It is then held and peeled away from the organ of Corti.

- The recording chamber is then mounted onto the experimental microscope ready for patching.

Dissection of the Mature/Adult Organ of Corti (>P10)

The following section describes a procedure used to obtain healthy hair cells from the mature or adult mammalian cochlea, from around P10 onwards. The procedure is different from that used for neonatal animals due to the much harder cochlear bone and because some of the cochlear tissue is removed with it. Unlike at neonatal stages, the mature organ of Corti cannot be isolated as a whole. Therefore, the best approach is to decide which region is required, usually a half to one turn of the cochlea.

Isolation of the bony labyrinth from the skull is simpler than for the neonatal cochlea because the bone is less fragile, and it can be removed by eye using large forceps. The bony labyrinth is then placed into a petri dish containing very cold solution (Fig. 3a). The following description focuses on the gerbil cochlea, but the procedure is the same for that of the mouse and rat.

Dissection of the Mature Apical Coil

- The bone is removed from the apical coil by making a small crack below the apex and a hole made by removing a piece of bone (Fig. 3b, c). As more bone is removed from the apex the stria vascularis may become detached (Fig. 3d). The bone around the rest of the apex and middle is then removed (Fig. 3e) and the

Fig. 2 (continued) The apical turn of the organ of Corti. (**o**) Very fine forceps are used to remove the tectorial membrane from the apical turn of the organ of Corti to allow access to the hair cells with a recording electrode

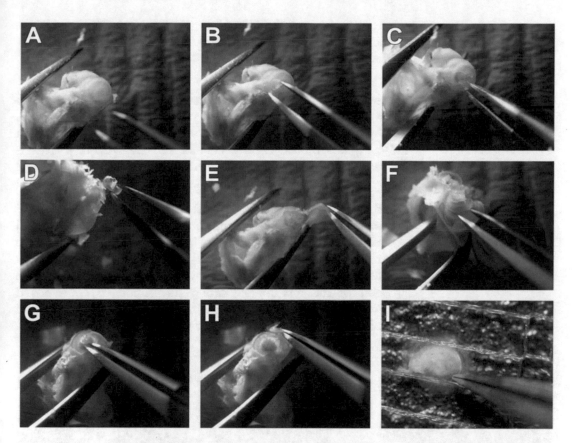

Fig. 3 Dissection of the apical turn of the mature/adult mammalian cochlea. (a) The intact gerbil bony labyrinth showing the cochlea and vestibular apparatus that is held with forceps. (b) The bone is cracked below the apical turn. (c) The bone is removed where the crack was to make a small hole. (d) The stria vascularis may become detached when the bone is removed and can be carefully taken away. (e) A large piece of bone is removed from the side and back of the cochlea. (f) Most of the bone has been removed from the circumference of the cochlea, leaving the spiral of the organ of Corti attached to the modiolus. (g) The modiolus below the apical coil is pinched with fine forceps to break it. (h) The lower end of the apical coil is clamped with forceps and the cochlear apical turn gently removed. (i) The apical turn of the mature cochlea held down under the nylon grid in the recording chamber. Very fine forceps are used to remove the tectorial membrane from the region to be recorded from, which in this case is that between the two nylon fibers

cochlea turned to remove the bone from the back. The apical coil of the organ of Corti is now visible (Fig. 3f). The modiolus under the apical coil is now pinched with forceps to break it (Fig. 3g). The lower end of the apical coil is then clamped with forceps and the cochlear apex is gently removed (Fig. 3h). The tissue can now be transferred to the recording chamber with the spoon.

- The apical coil is positioned below the nylon grid so it is held down securely with a region of hair cells accessible for recording. Figure 3i shows a region of the apical coil secured between two nylon strands. The apical coil can also be held with a single strand as shown for the neonatal dissection (Fig. 1n). The tectorial membrane is then removed (Fig. 3i) as described for the neonatal preparation.

Dissection of the Mature Basal/Middle Coil

The basal and middle coils of the mammalian cochlea have proven more difficult to obtain viable hair cells from than the apical coil. Only very few studies show electrophysiological recordings from these regions of the mature cochlea [22–24, 32, 33]. This is because the hair cells tend to remain attached to the stria vascularis and are removed when the bone is taken off.

- After isolating the bony labyrinth (Fig. 4a), the outer bone of the basal or middle turn of the cochlea is removed (the basal coil is shown in Fig. 4b). Removal of the outer bone will remove the stria vascularis and expose the organ of Corti. The bone below the organ of Corti can then be weakened, or broken, with the point of the fine forceps along the region to be removed (Fig. 4c). The organ of Corti is then cut at the beginning and end of this region by pinching it with fine forceps (Fig. 4d, e).

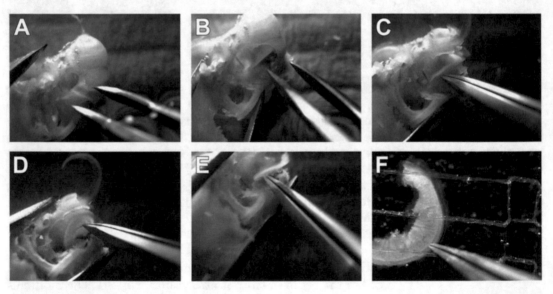

Fig. 4 Dissection of the basal turn of the mature/adult mammalian cochlea. (**a**) The gerbil bony labyrinth with the cochlea and vestibular apparatus held between forceps. The cochlea is oriented so that the basal turn can be seen. (**b**) The bone of the cochlea is broken and removed from the beginning of the basal turn from the oval window. (**c**) The bone has been removed from the basal turn leaving the spiral of the organ of Corti visible attached to the modiolus. The point of the forceps is used to weaken and break the bone below the organ of Corti that attaches it to the modiolus. (**d**) The upper end of the basal turn is broken or pinched with fine forceps. (**e**) The lower end of the basal turn is broken or pinched and the turn becomes detached from the modiolus. The basal turn is then removed with the forceps. (**f**) The basal turn of the mature cochlea is held down under the grid in the recording chamber. Very fine forceps are used to remove the tectorial membrane from the region to be recorded from, which is between the two nylon fibers

The region of the organ of Corti is then pulled away from the modiolus (Fig. 4e).

- The basal or middle coil of the cochlea is then transferred to the recording chamber and secured under the nylon grid (Fig. 4f). It is better to place the section of organ of Corti to be recorded from between nylon strands, as shown in Fig. 3f, rather than using one stand as shown for the neonatal cochlea (Fig. 1n). This is because the hard bone below the tissue can become pushed upward making the hair cells difficult to see if the latter approach is used. The tectorial membrane is then removed (Fig. 4f) as described for the neonatal preparation.

2.2 Equipment

2.2.1 Electrophysiology

The following paragraphs outline the equipment and techniques that are required to successfully pursue these experiments. Additional information is available in previously published reviews [44, 45], books [46–48] along with an excellent video [49] and the Axon Guide from Molecular Devices (the "patch-clamp bible" available for free download https://www.moleculardevices.com/en/assets/user-guide/dd/cns/axon-guide-to-electrophysiology-and-biophysics-laboratory-techniques-gref).

Vibration Isolation Table

Two main goals when constructing and troubleshooting an electrophysiology recording system are to isolate the system from all potential mechanical vibrations and the electrical noise. Having a properly installed, leveled, and adjusted air table is the first step towards having a successful vibration isolation. The air table should be large enough to comfortably accommodate the microscope along with any additional equipment the user might need for their work. Typically, a standalone electrophysiology system without any additional equipment needs an air table of $30'' \times 36''$ (750×900 mm). The tables are often ordered with casters to allow for easy and safe positioning within the room, as well as arm rests and side shelves to keep all minor equipment and controllers off the tabletop, further reducing unintentional mechanical vibrations. A Faraday cage, although optional, is a great way to mitigate the electrical noise.

Microscope

The overwhelming majority of hair cell physiology work is done using fixed stage upright microscope systems equipped with differential interference contrast (DIC) optics. The fixed stage option with a focusing nosepiece is a combination in which the objective (and not the microscope stage) is moved in Z-plane to focus. This ensures that the tissue, the recording, and stimulating pipettes remain stationary when the operator adjusts the focus. A lot of attention is paid to reducing any possible sources of mechanical vibration and drift. Some commonly used models are Olympus BX-51WI/BX-61WI, Nikon FN-1, Zeiss Examiner D1, or similar systems (Fig. 5). The microscope is equipped with a water-immersion lens, with a long working distance (1.0–2.0 mm), a numerical aperture (NA) often ranging between 0.9 and 1.1 for good lenses, and magnification of $60\times$ or $63\times$, but sometimes as low as $40\times$ and as high as $100\times$. In addition, a lower magnification objective ($4\times$ to $10\times$, dry or water immersion) is often used to position the pipettes and approach the tissue before switching to a higher magnification dipping lens. Most manufacturers also offer electrophysiology lens turrets, also known as the microscope 'nosepiece', that hold two objectives and allow to switch between the objectives clear of any pipettes inserted to the recording chamber.

In some cases, an inverted microscope system could be used for hair cell electrophysiology recordings (Fig. 5). Since the tissue will be observed from the bottom through the coverslip or a glass bottom dish, this approach is limited by the working distance of the lens and can only be used in thinner tissue preparations, usually after culturing the explant for a few days in vitro. Among the benefits is the ability to use high NA oil immersion lenses to visualize the tissue, albeit at a much shorter working distance of ~200 µm. And since there is no longer a massive lens dipping into the recording chamber, there is ample of space to approach the tissue with as many pipettes as needed, and more flexibility with pipette approach angles.

Recording Chamber

There is a huge variety of recording chambers available to meet any specific needs (temperature control, perfusion, low approach angle for the pipette, etc.). They are especially useful for acutely dissected tissue samples but could also be used for cultured explants attached to a coverslip. Alternatively, although more costly, glass bottom dishes can be used. The use of plastic dishes is not advisable: due to their depolarizing effect, any plastic introduced in the light path of the microscope dramatically decreases the contrast and resolution. Recording chambers can also be 3D-printed in-house and fitted into a microscope stage using an adapter. Some examples are shown in Fig. 5, illustrating a white, 3D-printed recording chamber that can hold a coverslip of up to 18 mm in diameter. The chamber is then mounted on a microscope stage using either a transparent plexiglass adapter manufactured using laser-cutting technology or a machined adapter made from a 6 mm aluminum sheet.

Camera

The choice of the camera is often defined by a few parameters, and there is a large selection of them on the market for any budget. If there is no intention of imaging in low-light conditions or performing any fluorescence imaging, the main points of consideration would be the camera's sensor size, measured diagonally in millimeters or inches (defines the area of a lenses' field of view available for imaging); the camera resolution, i.e. the number of

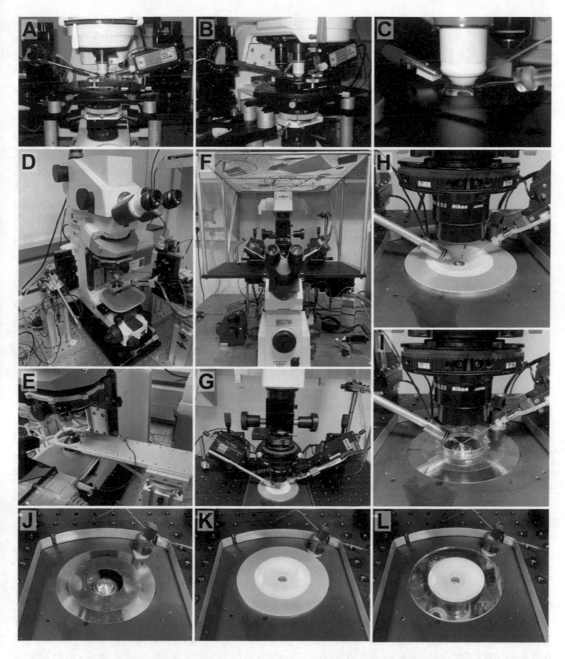

Fig. 5 An upright and inverted microscope set-up for electrophysiology recording. (**a–c**) An upright Axioskop FS Zeiss microscope equipped with a 63× water immersion objective and two MP-285 micromanipulators (Sutter Instruments). (**c**) An experimental arrangements showing the recording (patch) pipette on the right, and the stiff probe mounted on a piezo actuator assembly on the left. The tissue is mounted in a recording chamber and observed with an objective lens. (**d**) An upright Zeiss Examiner D1 equipped with a 63× 0.95NA water dipping objective with DIC optics and a physiology nosepiece. The microscope is mounted on a translation stage and can be manually moved in *X* and *Y* for micropositioning. Two manipulator stands are set up at 90° holding two MP-285 manipulators (Sutter Instruments). The piezo probe is mounted on the left manipulator, while the patch pipette is on the right side of the image. (**e**) An additional manipulator stand gantry assembly is used to hold the tissue preparation (stationary), while the microscope is moved around the preparation. (**f–i**) An inverted Nikon TE2000 equipped with a 100× 1.3NA oil immersion objective, DIC optics, and 0.52 NA long

pixels within the pixel array (defines the maximum allowed resolution for the acquired image), and the frame rate (30 fps or more). It is advisable to use a monochrome camera without a fan or any other moving parts to minimize any potential vibrations. Lower magnification objective lens users (40×) might benefit from a smaller chip size (1/2″ or 1/3″) or may consider adding an additional magnification lens (1.5×, 1.8×, or 2.0×) before the camera.

Note: Some camera manufacturers include an infrared (IR) filter in front of the camera's sensor, which will block any incoming IR light. This may be an issue if the operator is set up to use an IR DIC system. Conversely, it is advisable to use an IR filter in the light path *before* the light is delivered to the tissue to decrease any potential overheating the light may cause. This is especially important when using halogen light sources and is less of an issue with LED light sources.

Patch-Clamp Amplifiers

The most commonly used amplifier is the Axon Instruments Axopatch 200B or MultiClamp 700B, although other brands like HEKA instruments, Warner Instruments, AM Systems, and dPatch (by Sutter Instruments) are also used. Some amplifiers record one channel only, while others may enable an additional functionality of multi-channel recording. Some amplifiers are bundled with a software license, while in other cases the software is offered for an additional cost.

Analog-to-Digital Signal Converter

Most patch-clamp amplifiers use a standalone analog-to-digital (A/D) converter, like Axon Digidata 1550B, while some patch-clamp amplifiers include a built-in A/D converter and connect to a PC via a USB cable.

Software

Some A/D converters would only work with a specific software, like *pClamp Software Suite* by Axon Instruments, or *SutterPatch®* by Sutter Instruments, while other A/D converters can be used with third party software choices (*jClamp* and others).

Fig. 5 (continued) working distance condenser. (**f**) The microscope stage is enclosed in a small Faraday cage. Mounted on the right side of the vibration isolation table are a digital manometer (Fisher Scientific, 0–5 psi) and a microinjector (Narishige, IM-9C) to control the pressure at the back of the pipette. (**g**) A closer view of the two Burleigh micromanipulators holding the piezo actuator and the patch pipette on a headstage, mounted at 180°. (**h**) A closer view featuring a Piezosystem Jena PA8-12 piezo actuator holding a stiff probe (left), and Warner Instruments QSW-T10P pipette holder (right) with a 1.0 mm OD patch pipette. The image also features a 3D printed recording chamber that holds 18 mm coverslips. (**i**) Same as in (**h**), but with a 50 mm glass bottom dish (WPI Inc., FD5040-100) often used to culture Organ of Corti explants. The dish is mounted on a microscope stage plate. (**j–l**) A closer view of the microscope stage inserts used to hold the tissue preparations in a glass bottom dish (**j**), in a 3D-printed recording chamber using a specially machined aluminum adaptor (**k**), and the same recording chamber mounted using a laser cut ring made from a 6 mm Plexiglas (**l**). Also visible is the ground electrode on a holder with a ball joint (ALA Scientific)

Filters, Power Supplies, and Oscilloscope

It is advisable to have a low-pass or a bandpass filter to have a better control of the stimulus parameters. A low-noise analog power supply (0–12 V or 0–15 V is a typical choice) and an oscilloscope are commonly used to power small equipment and troubleshoot the system. A great attention is given to grounding all equipment to minimize the electrical noise.

Micromanipulators

Having a set of stable and precise micromanipulators is important to achieve good outcomes in an experiment. The most commonly used micromanipulators are the variety offered by Sutter Instruments and Scientifica brands. One can evaluate the mechanical stability of the entire system by positioning the tip of the pipette in the center of the field of view (or the screen), leaving it for an extended period of time, as long as the experiment might last. Such an exercise would expose any slow drifts originating from the microscope stage, microscope objective (drifting down), as well as any drifts originating within the micromanipulators. Once identified, such drifts can be mitigated and reduced when possible. Special attention should be given to securing the cables, making sure they do not pull on any of the equipment. Some systems benefit from having a motorized microscope stage for X–Y translation, while other users may choose to move the entire microscope around the specimen, as shown in Fig. 5d.

Perfusion Heating Devices and Heating and Cooling Platforms

Different temperature controllers are available commercially that can be used to control the temperature of the chamber and maintain the samples at a specific temperature. Such devices can be used to perform recordings in physiological conditions at body temperatures. Typically, these are coupled to perfusion chambers which bring heating and cooling to the recording chambers. Available, as well, are heating and cooling platforms which control the temperature of the sample.

2.2.2 Mechanical Stimulus: Stiff Glass Probes

Piezo Stack

Two of the most used manufactured piezo stacks are PI (Physik Instrumente) and Piezosystem Jena which are designed to deliver the desired range of bundle displacement. One such miniature multilayer piezo actuator, P885-31, is a 5 × 5 mm stack with a travel range of 13.5 µm when driven 0. . .100 V. A 10 V step sent directly from an A/D converter to this piezo would deliver a displacement of ~1 µm. The disadvantage of piezo stacks with a large travel range is a lower precision and low reproducibility of the displacement step, as they tend to have a larger hysteresis. Another unit, PL033 by *PI*, is a 3 × 3 mm stack with a travel range of 2.2 µm when driven from −20 to 120 V (Fig. 6). A commonly used displacement of ~1 µm would require a driver voltage of ~50 V, which can be delivered by a high voltage amplifier. Alternatively, PA8–12 by PiezoJena can be used (Fig. 5h). Although more massive and of much higher capacitance, this piezo stack travels for

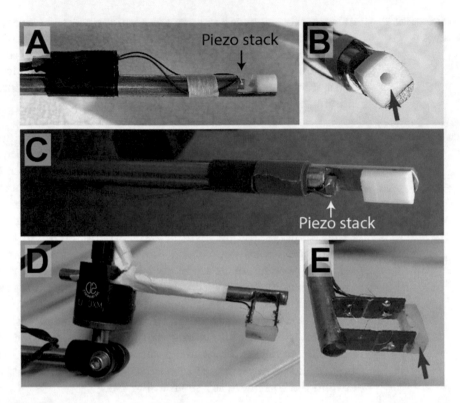

Fig. 6 Piezoelectrical devices for stiff glass probe mechanical stimuli. (**a–c**) A miniature multilayer piezo actuator (PL033 by PI) mounted on a 6 mm aluminum rod. The stiff probe is inserted through the white plastic piece with a concentric hole matching the pipette diameter (red arrow in **b**). The pipette directly contacts the piezo stack and is often secured with wax. (**d, e**) Example of a dual piezoelectrical bending transducer assembled for stimulation of vestibular hair cells. The pipette is inserted in the plastic mount through a concentric hole (red arrow) and secured with a side screw (not shown)

9.5 µm when driven from −20 to 130 V. A major benefit of using this piezo stack is its built-in strain gauge sensor reporting the axial displacement of the probe [50]. Prior to the actual experiment, the probe's linear displacement in the imaging plane (µm/V) can be calibrated by imaging the probe position at different voltage steps. Note that the calibration will be affected by the angle with which the probe approaches the tissue.

Piezoelectric Bending Transducers

Another commonly used transducer described originally by Corey and Hudspeth [51] involves assembly of piezoelectric bimorph elements [8] (Fig. 6d, e; Piezo.com). One of the difficulties encountered with this system is that there is typically substantial resonance which can be observed at the onset of the square step stimuli [51]. This resonance can be reduced by removing high-frequency components from the driving voltage using a low-pass filter and by adjusting the stimulus protocol to impose a sigmoidal ramp. Other factors such as the use of a short and stiff mounting arm, limiting the weight burden on the piezo ceramics using light solder flux (Piezo.com), use of light and flexible wires as well as a light stiff probe holder will all contribute to limiting resonance in the system. The stimulator shown in Fig. 6d, e uses a parallel mount of two layers of piezoelectric actuators. This configuration improves the linearity of the motion. We have used this stimulator to record mechanotransduction currents in vestibular hair cells [52]. This design allows access to the back of the stimulus pipette to place a small tube used to control the pressure at the back of the pipette to apply suction to the kinocilium.

Calibration

Setups are typically equipped with a CCD camera which is used to help position the probes and monitor the cell and pipettes during the experiments. The extent and linearity of the stimulus are assessed prior to the experiment by placing a pipette onto the piezo stimulator, applying voltage drives leading to probe displacements in the range of 1–2 mm. Images are acquired for each of the sequential steps and calibration of the stimulus is done by analyzing the probe displacement. Calibration of the camera is done by imaging a micrometer scale observed on the microscope to determine the effective pixel size of the resulting image. This calibration step is also crucial to determine that a straight motion is observed. Any damage to the piezo system, loss of a connection can lead to alteration of the motion. It is also important to confirm that there is no drift in the stimulus probe. A drift may arise, for example, if there is tension applied to the cable connected to the piezo system.

Stiff Glass Probe

Stimulus pipettes are often pulled from the same glass as the patch pipettes, then fire polished with a microforge until the tip is melted and rounded to fit the bundle V-shape (Fig. 7a–c). The approximate diameter of the tip would range ~5–8 μm to fit OHC bundles and 7–10 μm for straighter IHC stereocilia bundles. The shorter the stimulus pipette measured from its mounting point to the tip, the higher its resonance frequency is likely to be. For an upright microscope, the stimulus pipette length is largely defined by the diameter of the microscope's objective lens and its approach angle. For objectives with shorter working distance, 1.0 mm diameter glass capillaries can be used to manufacture probes, although they are likely to resonate at a lower frequency when compared to an otherwise similar 1.5 mm glass probe. In some cases, the microforge is also used to bend the tip of the stimulus probe to point it further down to prevent the side of the pipette from touching the spiral limbus (Fig. 7d–j). This can be considered if motion artifacts are observed at the time of bundle deflection with a stimulus probe, likely due to the side of the pipette coming in contact with the epithelium outside the field of view. In some cases, glass probes could also be pulled to manufacture much longer, flexible fibers and be used for force application and measurements

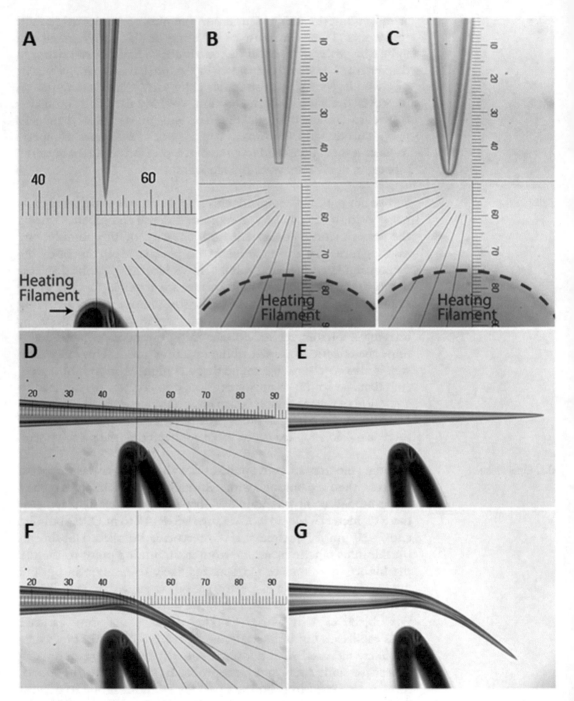

Fig. 7 Fabrication of the stiff probe stimulus pipette. (**a–c**) A pulled pipette is approached to the heating filament of the microforge (Narishige MF-830) and rounded to the desired diameter. (**d–g**) The pipette is rounded using a microforge to permit access to the hair bundle and bypass the spiral limbus

[53–57] [PMCID: PMC2267152, PMCID: PMC6757600, PMCID: PMC6772778].

High Voltage Amplifiers

Depending on the piezo stack's capacitance, the travel range (V/µm), and required stimulus rise time, use of a high voltage amplifier may be required. Both PI and Piezosystem Jena offer commercially available amplifiers to drive the piezo stacks.

Note: Another detailed chapter on this topic was previously published which includes further details about the glass probe stimulations [58].

2.2.3 Mechanical Stimulus: Fluid Jet

The fluid jet, also called pressure clamp, was first used to deflect the hair bundles of hair cells in vitro almost three decades ago [59]. It has been used since then for many studies on hair cell mechano-transduction [30, 60–69] (*Note that this reference list is not exhaustive as there are too many to cite here).

Fluid jets provide a targeted steam of fluid to push the hair bundle and pull it back in the opposite direction. This method offers a few advantages: (1) it mimics the physiological stimulus that inner hair cells normally receive in vivo, (2) the pipette does not come into direct contact with the hair bundle, (3) the fluid jet can be filled with different solutions such as low Ca^{2+} to mimic the composition of the endolymph.

Fluid Jet Design

The fluid jet shown in Fig. 8a is custom-made and was initially designed for use with hair cells by Corné Kros [59]. Made from acrylic material, it consists of a chamber with a piezoelectric disk inside. The chamber is wide enough to fit the piezo disk and is made of two parts that screw together to securely clamp the disk inside. The chamber is sealed with a silicone washer (Fig. 8b). When the fluid jet is assembled the metal surface of the piezo disk is covered with a layer of Sylgard 184 (Sigma Aldrich) which prevents electrical contact between the piezo disk and the solution inside the chamber.

Setting Up the Fluid Jet

A 1.5 mm diameter borosilicate glass pipette is pulled to a tip diameter of between 5–10 µm, depending on the width of the hair bundles to be stimulated, and inserted into the pipette holder of the fluid jet. The fluid jet chamber and the glass pipette are filled with extracellular solution. The piezo disk works in a similar way to a speaker by moving inward or outwards in response to voltage changes. This movement pushes extracellular solution out from the tip of the glass pipette or sucks it back in. The fluid jet is mounted on a manipulator with coarse and fine movement controls so that the tip of the pipette can be positioned behind the hair bundle of the hair cell so it can be deflected by the jet of fluid (*see* Subheading 2.3.3 for an additional description of the positioning).

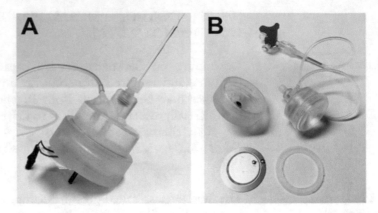

Fig. 8 Fluid jet used to deflect the hair bundles of hair cells. (**a**) Fluid jet with a glass pipette in the pipette holder. The fluid jet is not connected to the power supply and is not mounted on the experimental setup. (**b**) The fluid jet in (**a**) disassembled to reveal its constituent parts. The fluid jet chamber consists of two acrylic halves that have been unscrewed. The half with the pipette holder also has a side-arm tube that ends in a 3-way tap. The piezoelectric disk shown bottom left is situated inside the chamber when it is assembled. The chamber is sealed with a silicone washer, bottom right. The red and black wires shown in (**a**) are soldered onto the piezo disk, but have been removed in (**b**)

Positioning the Fluid Jet

To eliminate a steady flow of fluid into or out of the pipette in the absence of a stimulus, we use a fluid-filled tube that emanates from the chamber and ends in a 3-way tap attached to a post (Fig. 9a, b). The resting pressure can be adjusted simply by opening the tap and altering the position (height) of the tap (Fig. 9b). Increasing the height raises the pressure, lowering it reduces the pressure. This can be controlled by placing the stimulus pipette near a piece of debris away from the sensory cells. When the correct resting pressure has been found, the tap is closed before any stimulus is applied. The fluid jet pipette is then positioned behind the desired hair bundle. The fluid jet resting pressure is zeroed regularly during an experiment, even if done by simply opening and closing the tap.

Delivery of the Stimulus

The fluid jet is controlled by a custom-built power supply (Fig. 9c) that can be driven by a standard patch-clamp DA/AD interface and incorporated in the protocol used to stimulate and record the sensory cell. The analog command voltage from the digitizer is amplified ten times before going to the fluid jet, up to a maximum voltage of ±50 V.

Calibration

The force generated by fluid motion delivered by the fluid jet requires calibration. This is done by measuring the movement of a carbon or flexible fiber of known stiffness placed within the stream of the fluid jet, at a given distance [59]. The fluid jet system shown in Fig. 8 can generate a constant pressure of ~5 pN/μm^2 per volt of driver voltage [59]. Such information can be used to determine the stiffness of the hair bundle of different hair cell types.

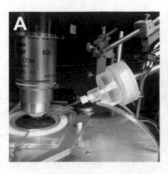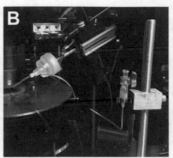

Fig. 9 The fluid jet mounted on the experimental setup. (**a**) The fluid jet mounted on a manipulator with the pipette tip positioned in the recording chamber on the experimental microscope. (**b**) A wider view of the fluid jet showing the tap of the fluid-filled side-arm clipped to a clamp, the height of which can be adjusted. (**c**) The rack-mounted 8-pole Bessel filters (top left) and power supply (bottom right) that drive the fluid jet. The command input comes from the patch-clamp DA/AD converter

Bundle Displacement

To determine the extent and kinetics of the bundle displacement, one can project the DIC image of the hair bundle onto a pair of photodiodes. These photodiodes sense the changing light intensity from the moving dark and light bands for a small region of the bundle image and transform it into a photovoltage [67, 70–74]. The photovoltage is converted into distance by moving the photodiodes over the stationary hair bundle a known amount to create a calibration curve. The calibration curve is then used to report on the bundle movement. A laser spot has also been used to illuminate the top of the hair bundle that is then projected onto a photodiode pair to create a photovoltage [67, 70–72]. The laser illumination has the benefit of increasing the contrast of the bundle image and improve the measurement accuracy. With the development of affordable high speed, high resolution, cameras it is also possible to measure bundle displacement with a video image [62, 75–77]. Note that not all experiments require the measurement of hair bundle displacement as it can be just as informative to analyze transduction current amplitude against incrementing voltage steps applied to the piezo system, or piezo's strain gauge feedback signal reporting its axial displacement position, until mechanotransduction current saturation is reached.

Rise Time

Fluid jet stimulators are typically associated with longer rise times than stiff probes [59, 62, 78, 79]. The input signal to the fluid jet is usually filtered below 1 kHz to avoid activating the internal resonance of the piezo disk. An input signal 10–90% rise time of 460 μs has been reported [59]. However, neither method of bundle stimulation has been rapid enough to measure the activation kinetics of the transducer channel itself, without lowering the temperature in the recording chamber to 4 C [80]. Only with the development of faster probe actuators, or different methods of bundle deflection, will this become possible.

Commercial Pressure Clamp

As an alternative to the custom-made fluid jet described above, there are commercially available pressure clamp systems that can be adapted to stimulate hair cell bundles. The HSPC-2-SB (ALA Scientific; https://alascience.com/products/hspc-2sb/) is a closed loop pressure-control servo, optimized to generate reproducible and rapid pressure/vacuum steps. The headstage can be connected to a pipette holder with a side port to apply the pressure or suction steps to the hair cells in a similar way as described above. The pressure steps are controlled via a command voltage that can be supplied through the patch-clamp software. The speed of the pressure response is amplitude dependent, with the rise time reported at 12 ms for a 0–100 mmHg step (measured as 0–100% settling time). Since the pressure applied at the back of the pipette that would be sufficient to deflect a bundle is of a much lower amplitude (5–20 mmHg), a rise time of less than 1 ms can be achieved for such stimuli.

2.2.4 Mechanical Stimulus: Kinocilium Coupled Probes

This third method of stimulation has been used to assess mechanotransduction in utricular hair cells [4, 52, 80–83]. This method can only be applied to cells that have a kinocilia firmly coupled to the rest of the bundle. Neonatal cochlear hair cells possess a small kinocilia which is weakly coupled to the hair bundle. The kinocilia in these cells also disappears as cells mature. Vestibular hair cells, on the other end, possess a prominent and robust kinocilia that can be coupled to a stiff probe to displace the hair bundle. In this case a piezoelectrically driven glass probe is either attached to the kinociliary bulb or the upper portion of the kinocilium can be inserted into a small pipette (diameter ~ 600 μm) with negative pressure that is maintained constant during the recordings. The piezoelectrical system used in this case is similar to what is described in the previous section. The only difference is that the pipette is coupled to a tube that is used to apply negative pressure. This can be done simply with a small mouth suction and valves that allow to close the system and maintain the pressure during the experiment. In some cases, the probe can be directly attached to the kinocilium using an acid-treated glass probe.

2.3 Electrophysiological Recording Procedures

2.3.1 General Approach

The purpose of this chapter is to provide some specific tips to *optimize* successful electrophysiological recordings from hair cells. The general approach is otherwise similar to procedures previously described for other preparations in particular brain slices [84].

- Sensory epithelia are placed onto a microscope chamber and initially viewed at low magnification (10×) to adjust orientation of the sample and electrode placements. High magnification 40× or 63× water-immersion lenses with differential interference contrast optics are used for the experimental recording portion of the work. Tissues are placed in extracellular recording sodium-rich solutions.

- Recording pipettes are pulled from capillary glass to resistances ranging 2–5 MΩ. Different types of glass can be used, either borosilicate or soda (soft) glass typically with a diameter of 1.5–1.0 mm (matching the pipette holder), with or without a filament. When using borosilicate glass, it is recommended to fire polish the tip of the pipette beforehand to improve the seal.

- Recording pipettes are filled with an intracellular solution, the composition of which depends on the purpose of the experiment (typically KCl or CsCl based). The pipette is visually inspected to make sure no bubbles are present. A ground electrode is also placed in the recording chamber.

- The apical surface of the epithelium is viewed from above as the pipette is advanced using the low magnification lens. Initially a larger "cleaning pipette" is approached near the tissue to clear out an area of interest. The cleaning pipette is a 1 or 1.5 mm borosilicate glass pipette pulled to a tip diameter of 3–5 μm. The pipette is filled with extracellular solution and used to remove cells and debris around the hair cell of interest. Alternatively, some choose to skip this step instead using positive pressure while advancing the recording pipette through the tissue. Both techniques were shown to work well and result in high success rates. But when applying positive pressure, it is important to keep in mind that should the low calcium intracellular solution exiting the pipette be released near the hair bundles, it will lead to loss of tip-links, thereby abolishing mechanotransduction.

- The filled recording pipette is lowered into the recording chamber with slight positive pressure applied at the back of the pipette to avoid clogging the tip. The pressure is decreased, once in the bath, to reduce the amount of internal solution exiting the pipette.

- The resistance of the recording pipette is checked by applying a 10 mV voltage command step (*see* Axon guide), also confirming that the pipette tip is not clogged with a bubble or debris.

- When the hair cell of interest is identified, the recording pipette is positioned ~10–15 μm away from the targeted cell and slowly approached to the basolateral side while keeping slight positive pressure.

- Holding current is now adjusted to zero to compensate for liquid junction potential that results at the interface between intracellular and extracellular solutions.

- The patch pipette is, then, advanced towards the hair cell until a dimple is seen on the cell body resulting from the positive pressure in the pipette. When the positive pressure is released the hair cell membrane touches the pipette and a tight-seal may be formed. *A tight giga-ohm seal is required before attempting to rupture the membrane for whole-cell recording.* A giga-ohm seal may occur as soon as the positive pressure is released or may take more time and require gentle suction (by mouth or syringe). Gradually adjusting the voltage clamp holding potential to a negative value, close to the expected cell resting potential (−60 to −80 mV), can aid the sealing process. If a giga-seal cannot be obtained, the pipette should be removed, discarded, and the procedure repeated with a new clean pipette.

- Following the formation of a giga-seal, any fast current transients that result from the stray capacitance of the recording pipette can be nulled with the fast capacitance compensation circuitry of the amplifier. Before attempting to *rupture the membrane for* whole-cell recording the holding potential should be set to the desired value according to the experimental needs.

- The whole-cell configuration is achieved by applying slight suction to the patch pipette, usually by mouth to give more control, which breaks the patch of the membrane separating the inside of the pipette from the cell. An indication that the whole-cell configuration has been reached is the sudden appearance of capacitive transients on the test pulse and a large decrease of the membrane resistance. At this instance the mouth suction pressure to the recording pipette is released. These slower transients are caused by the voltage across the series resistance (Rs), between the pipette tip and the inside of the cell, and cell membrane capacitance (Cm). The transients are nulled using the Rs and Cm transient compensation circuit on the patch-clamp amplifier. These important parameters must be recorded or noted for each cell. Cm reflects the size of the cell (chargeable cell surface area). Rs should be as low as possible since it affects the voltage the cell sees compared to the command, the higher the Rs the greater the voltage drop to the cell. The voltage drop due to Rs can, and should, be compensated by the dedicated compensation circuitry of the amplifier in voltage clamp mode, but a low Rs to begin with is always better (usually between 5 and 10 MΩ). From this point the cell is ready to be stimulated as desired.

- Recordings are typically obtained at room temperature (22–24 °C) with regular perfusion of the chamber with a peristaltic pump. Neonatal preparations can last a few hours under these conditions. Recordings can also be obtained at around body temperature (35–37 °C), which is important when investigating how the cells would respond in vivo, such as when monitoring the cell's voltage responses in current clamp or the properties of synaptic transmission that are very temperature dependent. In fact, any study on mammalian hair cells investigating their biophysical properties and how they would function in vivo should be done at body temperature, or at least this should be taken into account when drawing conclusions. Neonatal cells can remain viable for a few hours at body temperature if adequate cellular perfusion is maintained. Adult cells tend not to last as long, showing signs of deterioration, such as swelling or weak membranes that do not seal well, after an hour or so in these conditions.

- In the whole-cell tight-seal configuration described above, data can be acquired in either voltage- or current-clamp mode using the patch-clamp amplifier (Axopatch 200B, Axon multiclamp 700B or Cairn Optopatch in our case, *see* Subheading 2.2.1). The voltage clamp mode is used to measure ionic currents in biological membrane. In voltage clamp mode, the cell membrane potential is clamped at a desired voltage by injection of a mirror current. As the membrane potential changes, resulting voltage-gated currents activated at different potentials, are recorded. Usually, incremental voltage steps from the holding potential, or below, are applied to see the progressive activation of voltage-gated channels. Different voltage stimulus protocols can be used to assess voltage-dependent currents (*see* Subheading 2.3.2) or assess the voltage-dependence of mechanosensitive currents (*see* Subheading 2.3.3). The current clamp mode is used to measure transmembrane potential resulting from injection of a hyperpolarizing or depolarizing current into the cell. These recordings are useful for investigating how the cell would respond to stimulation in vivo, such as to depolarizing currents through the MT channels, where the MT current is mimicked by a depolarizing current step or waveform. Current clamp allows investigation of how the voltage and ion-gated channels interact. For example, neonatal auditory hair cells fire Ca-dependent action potentials [37, 85–89], whereas mature hair cells respond to current injection with sustained and graded changes in membrane potential [37, 90] (Fig. 12).

- The current or voltage recordings are filtered at an appropriate frequency for the protocol (usually at 1 or 5 kHz) with a low-pass Bessel filter, digitized at 5 kHz (or greater) with a data acquisition interface (Axon Digidata 1440, Molecular devices, CA) and collected using pClamp software (Axon Instruments; alternatives such as jClamp can be used). Data are stored appropriately for off-line analysis using Clampfit or other data analysis and processing software.

General Materials and Solutions Required

Dissecting tools; Cell culture dish (35 and 60 mm); Glass coverslips, Ø18 mm; Glass pipettes for electrophysiology: R6 glass (size according to pipette holder); WPI 1B100F-4; Glass fibers for stimulus probe; Syringe filter with a 0.22 μm pore and Ø33mm and Ø13mm; Vacuum grease Corning; DMEM/F-12, GlutaMAX™ supplement (Invitrogen # 10565018); alternatively, MEM (with Earle's salts and Glutamax, Invitrogen #41090-101) can also be used; Ampicillin (powder—Invitrogen); Ciprofloxacin; Various Salts + HEPES; Geltrex (ready to use solution that is easier to work with, can be stored at 4 °C, Invitrogen #A1569601) or Matrigel (should be stored at −20 °C and must be diluted with the culturing media before use); Glass bottom dishes for organotypic cultures (optional) can be replaced with glass coverslips for culturing in plastic dishes; Syringes.

2.3.2 Voltage-Dependent Currents

Whole-cell patch-clamp is used to study voltage-dependent currents in hair cells. It is applied to activate the cell's voltage-gated ion channels and record either the resulting membrane currents using voltage clamp, or the change in membrane potential using current clamp.

The following section describes typical stimulus protocols and hair cell responses and demonstrates how different portions of the recordings can be analyzed to obtain specific information. It also includes information about the use different solutions to study specific currents or block currents that may obscure those that are being studied.

Voltage Clamp

The whole-cell voltage clamp mode, as its name implies, allows the experimenter to clamp the cell membrane potential at a given value and record the currents that results from a change in the voltage applied to the cell. Due to their shape and small size, hair cells are particularly suited to this approach. This technique has revolutionized the field of neurobiology [45].

Examples of whole-cell voltage clamp recordings are shown in Fig. 10 to illustrate the type of recordings that have been obtained from immature and mature IHCs using different protocols. These recordings were performed using a standard KCl-based intracellular solution and a normal NaCl-based extracellular solution. This condition allows the cells complete set of voltage-gated currents to be investigated with voltage protocols. A common voltage clamp protocol is shown in Fig. 10a. Typical examples of the current responses from an immature and mature IHC to this protocol are shown in Fig. 10b, c, respectively.

Steady-state current–voltage (I–V) curves (Fig. 10d) were obtained by plotting the steady-state current at the end of the first voltage step (indicated by the thick blue line) against the membrane potential the cell is stepped to. This analysis shows much larger outward currents in mature cells. The outward current has been shown to be the delayed-rectifier $I_{K,neo}$ in immature cells and the rapidly activating $I_{K,f}$ together with the delayed $I_{K,s}$ in

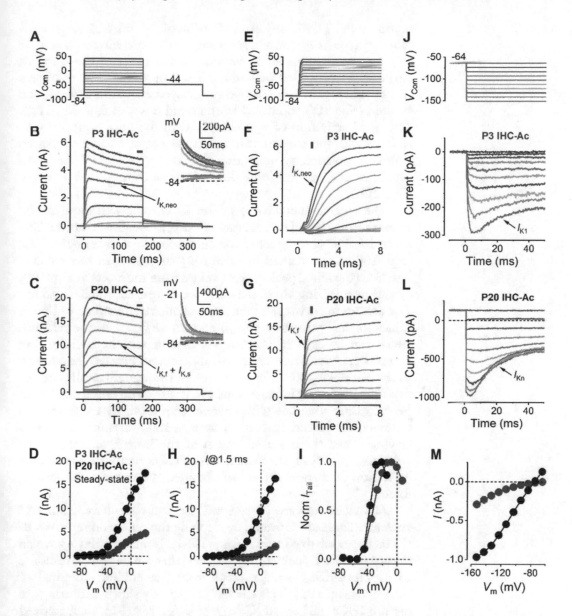

Fig. 10 Whole-cell voltage clamp recordings from IHCs. (**a**) A voltage clamp protocol used to stimulate the cells consisting of the holding potential of −84 mV followed by 160 ms voltage step to potentials in 10 mV increments. Each of these is followed by a 160 ms step to −44 mV before returning to the holding potential. (**b, c**) Outward K⁺ current responses from an immature P3 apical coil IHC (**b**) and a mature P20 apical coil IHC (**c**). The responses are color coded to match the corresponding voltage stimulus. Arrows indicate the currents present at the different ages, responsible for all currents in the recording, not just the labeled response. The thick blue horizontal line above the traces delineates the region used to measure steady-state currents. The insets show the tail current responses to the −44 mV step on an expanded scale with the membrane potential of the previous step indicated next to the largest and smallest responses. (**d**) Steady-state *I*–*V* curves obtained from the K⁺ current responses in (**b**) and (**c**). (**e**–**g**) The same responses as in (**a**–**c**) but on an expanded time scale to show the activation time course of the K⁺ currents. The thick vertical blue line indicates where the currents were measured at 1.5 ms. (**h**) *I*–*V* curves for the currents measured at 1.5 ms. (**i**) Activation curves for the K⁺ currents in (**b, c**), obtained by normalizing the size of the peak instantaneous tail currents, elicited at −44 mV, and plotting this against the membrane potential of the preceding voltage step. (**j**) A voltage protocol

mature cells [22, 37, 39]. An indication of the different K^+ channel types can be seen from the shape of the current responses. It becomes more apparent if we look at the onset of the outward currents. An expanded view of the first 8 ms of the same responses is shown in Fig. 10e–g. The slower delayed activation of $I_{K,neo}$ can be seen (Fig. 10f) compared to the rapid onset of $I_{K,f}$ (Fig. 10g). The rapid activation of $I_{K,f}$, compared to the other outward currents, allows the current through these channel to be measured almost in isolation by measuring the current size at 1.5 ms (indicated by the vertical blue line) and plotting the I–V curve (Fig. 10h).

The voltage-dependence of the K^+ current activation can be assessed from the tail currents at -44 mV, as mentioned above. The normalized K^+ current activation curves for the two example IHCs (Fig. 10i) were obtained by measuring the instantaneous tail current size (the initial peak current values from the insets in Fig. 10b, c) and normalizing these between the lowest and highest values. The activation curves can be fit with sigmoidal functions to obtain values for half activation (V_{Half}) and the slope (s) of the relation, which can be used to characterize or compare the current. Tail currents are used to generate activation curves because they ensure a constant driving force for K^+ is present throughout. If the current from the preceding voltage steps was used the driving force would be increasing together with channel activation. By using the tail currents, the preceding voltage steps activate varying numbers of channels, and these are still open at the instant of stepping to -44 mV. Therefore, the instantaneous tail currents are an accurate indication of channel activation obtained at a constant K^+ driving force.

As well as outward K^+ currents, hair cells also have inward K^+ currents. These are activated by stepping the cell membrane potential to hyperpolarized values (Fig. 10j–l). Immature IHCs have an inward rectifier K^+ current I_{K1} [38, 39] that can be seen in response to hyperpolarizing voltage steps from the holding potential of -64 mV (Fig. 10k). This current is relatively slowly activating. In mature IHCs the inward K^+ current is $I_{K,n}$ [39, 91], characterized by a very negative activation range. Because of this, $I_{K,n}$ is substantially open at the holding potential of -64 mV (Fig. 10l), which causes the instantaneous inward currents upon hyperpolarization. The I–V curves for I_{K1} and $I_{K,n}$ (Fig. 10m) were obtained by plotting the peak inward current against membrane potential.

Fig. 10 (continued) used to activate inward K^+ currents, consisting of hyperpolarizing voltage steps in 10 mV decrements from the holding potential of -64 mV. (**k** and **l**) Inward K^+ currents in response to the protocol (**j**), from an immature and mature IHC. The arrows indicate the different currents present at the two ages. (**m**) I–V curves for the peak inward K^+ currents in **k** and **l**

These examples show that a number of different K⁺ channels can be evaluated in hair cells at different stages of development using relatively simple voltage clamp protocols. Knowledge of the channel properties allows them to be measured almost independently without the need for specific channel blockers. However, not all voltage-dependent currents can be studied as easily. Calcium currents, for example, are masked by the comparatively large size of the K⁺ currents. Studies of hair cell function and synaptic transmission often require an accurate measure of the Ca^{2+} current. In order to do this a combination of K⁺ channel blockers is required to obtain I_{Ca} in relative isolation. The example Ca^{2+} currents (Fig. 11) were obtained from immature and adult basal-coil gerbil IHCs using both intracellular and extracellular K⁺ channel blockers. Caesium was used as the main intracellular ion, which does not move as easily through K⁺ channels, and K⁺ channel blockers TEA (30 mM) and 4-AP (15 mM) were used for both recordings. The SK channel blocker apamin (300 nM) was additionally applied to the immature cell to specifically block the SK2 channels in these cells [92], and linopirdine (100 μM) was applied to the adult cell to block $I_{K,n}$ that tends to be resistant to the other blockers. The recordings were obtained at body temperature since I_{Ca} is much smaller at room temperature [32, 93].

Using these recording conditions, a well isolated I_{Ca} can be recorded in response to depolarizing voltage steps from the holding potential of −81 mV (Fig. 11a, b; the liquid junction potential for the solutions used was −11 mV). For these recordings the baseline

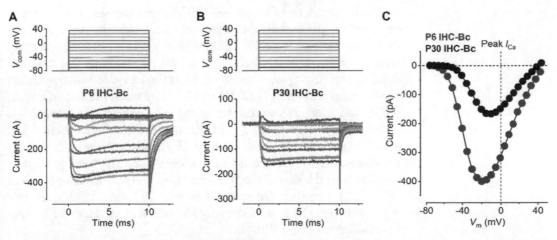

Fig. 11 Inward Ca^{2+} currents isolated in immature and adult IHCs. (**a**) and (**b**) Upper panels show the voltage protocol consisting of 10 ms depolarizing voltage steps in 10 mV increments from the holding potential of −81 mV, followed by a return to the holding level. The lower panels show the resulting I_{Ca} recorded from immature P6 and adult P30 basal-coil gerbil IHCs, respectively. The color coding matches the corresponding voltage command. (**c**) *I–V* curves for I_{Ca} obtained from measuring the peak inward current during the voltage steps for the cells in **a** and **b**. The *I–V* curves show current values in response to 5 mV voltage increments that are not all shown in **a** and **b**. The recordings were previously published [32]

current has been subtracted to zero and the leak current subtracted to isolate I_{Ca} as much as possible. The leak current represents the passive leak of current though the cell membrane and is identified from its linearity at potentials where few voltage-gated channels are active. In this case the leak current was measured around -80 mV since most of the K^+ currents were blocked. Leak subtraction was done using Clampfit software of pClamp by subtracting the command waveform scaled-down according to the cell membrane resistance. Note that leak subtraction was not done for the K^+ currents (Fig. 10) since there were currents active at all potentials, making it difficult to accurately measure the real passive leak.

The I–V curves for I_{Ca} (Fig. 11c) were made by plotting the peak current during the 10 ms voltage step against cell membrane potential. Steps of 5 mV increments were used to generate the I–V curves for improved resolution. The I–V curves have the characteristic bell-shaped relation of a Ca^{2+} current, reaching a maximum at around -20 to -10 mV and reversing at positive potentials. The I_{Ca} is about two to three times smaller in mature compared to immature cells [36, 94].

Current Clamp

The functional interaction of the cells compliment of voltage and ion-gated channels can be seen by performing current clamp recordings. In this condition, hyperpolarizing or depolarizing currents are injected into the cell and the resulting membrane potential is recorded. It is advisable to perform these at body temperature to get the most physiological representation of how the underlying currents would be interacting in vivo.

A characteristic of immature IHCs is that they fire Ca^{2+}-dependent action potentials spontaneously that increase in frequency with depolarization (Fig. 12a). This action potential activity comes about because of the interaction of the large I_{Ca} and the delayed-rectifier $I_{K,neo}$ that were identified in the voltage clamp experiments shown in Figs. 10 and 11. The combination of larger, faster, and more negatively activating K^+ currents ($I_{K,f}$ and $I_{K,n}$), together with the smaller I_{Ca}, in mature IHCs prevents them firing action potentials. Instead, mature IHCs show sustained voltage responses, for the duration of the current step, that are graded in amplitude to the size of the stimulus (Fig. 12b). The example traces (Fig. 12b), obtained from an apical coil P21 gerbil IHC, show a slight voltage relaxation following the initial peak response to current injections up to 900 pA.

The use of whole-cell patch-clamp recording remains a powerful method to scrutinize the functional properties of hair cells. Depending on the question being addressed, it can be used to assess the cells complement of ion channels in total or in isolation. It can also be used to observe the functional interaction of the identified ion channels and how these combine to generate the characteristic voltage responses of the cells.

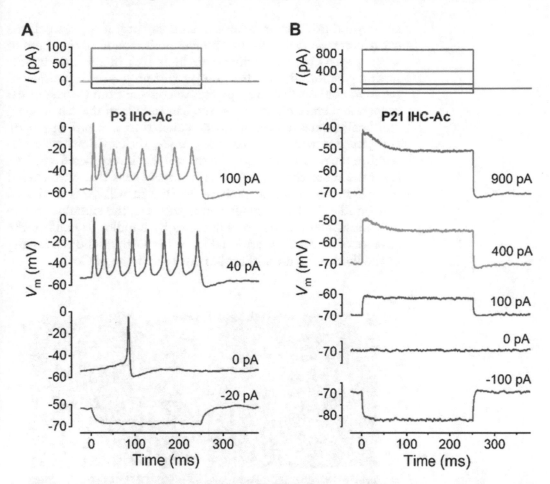

Fig. 12 Voltage responses of immature and mature IHCs. (**a**) Upper panel shows the current clamp protocol applied to an immature IHC. Hyperpolarizing and depolarizing currents were applied in 10 pA increments, only some of the steps are shown. The lower panels show the voltage responses of a P3 apical coil mouse IHC to the current steps, with the current level indicated to the right. The cell fired a spontaneous action potential and these increased in frequency with depolarization. (**b**) Larger current steps (upper panel) were applied to a mature P21 apical-coil gerbil IHC that showed graded voltage responses up to 900 pA depolarizing current injection (lower panels). The color coding matches the corresponding command current

2.3.3 Mechanosensitive Currents

As described above, mechanosensitive currents can be evoked with the use of a stiff probe or fluid jet. The protocols that are typically applied with a stiff probe consist of square step stimuli, while sine wave stimuli are more often applied when using a fluid jet.

We discuss below the different aspects that need to be considered when performing such recordings.

Positioning the Stimulus Pipette

The first step when performing these experiments consists of approaching and positioning the stimulus pipette. There are different aspects that need to be considered: first the pipette should be positioned in such a way that bundle displacements are optimized and reproducible upon repeating stimulus protocols; second, the experimenter should monitor bundle position and resting current when approaching the pipette to limit bias in the hair bundle resting position; third, when using a stiff probe, the experimenter needs to confirm that the pipette is uncoupled from the rest of the preparation and only touching the hair bundle of the cell studied; and finally, once the cell has been coupled to a recording pipette and placed in whole-cell recording mode, the experimenter has to confirm the pipette is coupled to the cell that is recorded from. This is a little more challenging when studying cells that have altered mechanotransduction as perhaps no current will be evoked with hair bundle displacement. If that is the case, the stimulus pipette may be moved around to deflect nearby bundles to confirm the absence of mechano-sensation. Figure 13 illustrates the positioning of the fluid jet or the stiff probes onto the organ of Corti.

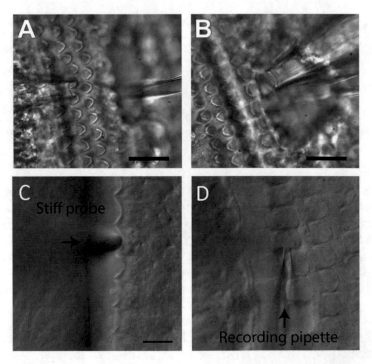

Fig. 13 Stimulus pipette and recording electrode placement for mechanotransduction current recordings. (**a**, **b**) Fluid jet stimulation of neonatal mouse OHCs (**a**) and IHCs (**b**). The stimulus probe is shown positioned near the hair bundle, while the patch electrode, appearing out of focus, is sealed onto the hair cell body below. Scale bar 20 μm. (**c**, **d**) Stiff probe stimulation of neonatal mouse IHCs. The stimulus probe is positioned onto the hair bundle (**c**) and the recording patch pipette is approached perpendicularly onto the cell body of the IHC. Scale Bar 10 μm

Resting Open Probability

Mechanosensitive channels are not all closed when hair bundles sit in their resting/unstimulated position. Instead, a fraction of hair cell mechanosensitive channels are open at any time as they fluctuate between their open and closed state at rest and generate a standing inward current which keeps the cell further depolarized. The resting open probability (P_{open}) of the channel is regulated by Ca^{2+} entering through the channel [63]. In vitro, about 8–10% of the channels are open at rest in perilymphatic Ca^{2+} conditions, but under endolymphatic Ca^{2+} concentrations, P_{open} increases to ~20–35% [63, 64, 66]. In vivo, the electrochemical driving force set by the hair cells resting potential, the endocochlear potential and the composition of the endolymph all contribute to setting P_{open} [63]. In vitro, P_{open} can be estimated by displacing the bundle in the inhibitory direction (away from the tallest row), thereby closing all the channels, and comparing the resting current measured with the pipette placed away from the hair bundle, with the currents measured when all the channels are closed (upon inhibitory stimulation).

Maximum Transduction Current and Operating Range

Mechanosensitive currents are evoked by gradually increasing step sizes until reaching saturation (i.e. maximum current) and before risking hair bundle damage. The range of the stimuli used will depend on the hair cell type, age and its position along the organ. Maximum, saturating displacements are typically in the range of 1–1.5 µm for in vitro experiments, although they were estimated to be an order of magnitude smaller in vivo.

For fluid jets, the stimulus protocols can be in the form of a sinewave that moves the hair bundle back and forth from its resting position, similar to the motion normally experienced in vivo (Fig. 14). Square steps stimuli, also used with fluid jet, can remain at a constant pressure for at least 1 s (Fig. 14) [59]. When displacement steps are used, it is important to alternate positive and negative steps to avoid build-up of positive or negative pressure inside the fluid jet [59]. In the case of stiff probe, square step stimuli are applied to the hair bundle (Fig. 14).

Bundle displacement can be visualized with a digital camera (*see* Subheading 2.2.3) or can be assessed by projecting the bundle image onto a pair of photodiodes [67, 74].

Kinetics of Activation and Adaptation

The kinetics of the transduction current recorded from a stimulated hair cell largely depends on the mode of hair bundle stimulation used for the experiments (*see* Subheading 2.2.3). Improvements to stiff probe technology now produces rise times as short as 11 µs [77] which has allowed more accurate measurements of the kinetics of activation and adaptation of the transduction current. Adaptation is characterized by a slow and a fast component which has been described in mammalian hair cells [62, 63, 79, 82, 95–99]. The fast component of adaptation occurs within the first few milliseconds,

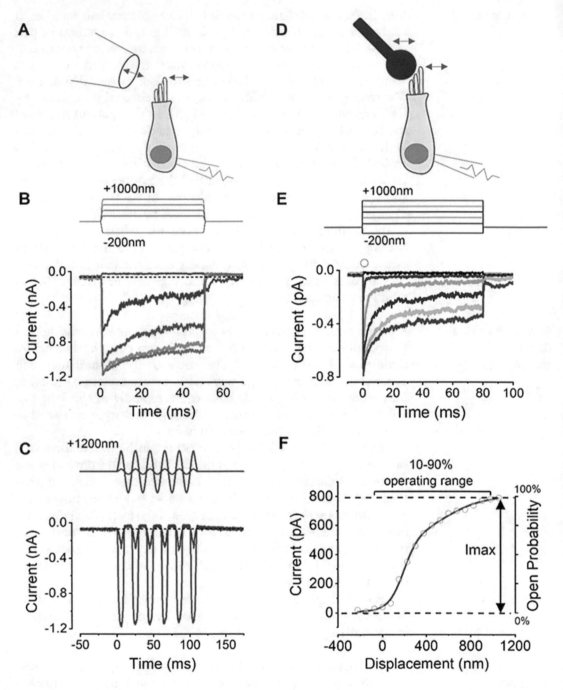

Fig. 14 Mechanotransduction currents evoked using fluid jet or stiff probe stimulation. (**a–c**) Currents evoked by fluid jet stimulation (**a**) recorded from neonatal OHCs using square step stimuli (**b**) or sinewave stimuli (**c**). (**d–f**) Currents evoked by stiff probe stimulation (**d**) recorded from neonatal OHCs using square step stimuli (**e**) and current/displacement curve fitted with a double Boltzmann curve illustrating the increased open probability of the channels with bundle displacement, the maximum transduction current (I_{max}) and the 10–90% operating range. P7 OHCs, holding potential −84 mV, 1.3 mM external Ca^{2+}, internal Cs^{+} solution

while the slow component occurs over the subsequent 20–50 ms. Several variables affect the study of adaptation, including which type of stimulus is used, its rise time and how well the probe is coupled to the hair bundle [62].

2.4 Additional Technical Tips

2.4.1 Dissection Tools

The main dissection tools are Dumont #5 forceps that are sharpened to different extents using a wet stone. We use two pairs that are sharpened to a medium point (Fig. 1b) for the removal of hard bone. An additional two pairs are sharpened to a fine point for the dissection of the soft cochlear tissue after the removal of the bony casing. Finally, we use one very fine pair of forceps to remove the tectorial membrane from the surface of the organ of Corti in the final step of the dissection. Another important dissection tool is a small spoon used for transferring the dissected cochlea to the recording chamber (Fig. 2l), which is approximately 6 mm wide by 2 mm deep. Other standard tools are used for the initial part of the dissection such as a scalpel, a small pair of scissors for bisecting the head, and large forceps for removing the brain or isolating the bony labyrinth for mature/adult dissections.

2.4.2 Perfusion

Perfusion systems are commonly used to replace the bath solution either with a fresh one, a solution with a different ionic composition, or containing a drug. They are especially important during longer recording sessions to make sure the composition, hence the osmolarity of the bath solution does not change due to evaporation. Both gravity fed systems and peristaltic pumps are often used.

2.4.3 Tubing

Tygon tubing of various diameters is commonly used to modulate the pressure at the back of the patch pipette. The tubing with an inner diameter of ~1.0–1.5 mm and an outer diameter of ~3.0 mm is a common choice. Some headstage pipette holders have a metal pressure port connector (like Axon Instruments), while others are plastic (like Warner Instruments Cat #QSW-T10P). Sometimes the tygon tubing does not easily come off the plastic connector, and some users choose to extend the tygon tubing with a small piece of a more flexible silicon tubing.

2.4.4 Glass Fibers and Fire Polishing

There is a great variety of glass filaments used for physiology. As a personal preference, authors reference borosilicate glass capillaries Cat#1B100F-4 (WPI Inc.)/1 mm outer diameter recording pipettes, and R-6 glass (King Precision, Claremont, CA)/1.5 mm outer diameter recording pipette. Please refer to the latest version of the Pipette Cookbook by Adair Oesterle and Sutter Instruments, available for download from sutter.com for details on patch pipette manufacturing. A microforge such as the MF2 (Narishige) is a common choice that fits most pipette fabrication needs.

| 2.4.5 Other Considerations | Although optional, a digital manometer (0–30 psi) can be used in-line with the pressure port to monitor the pressure at the back of the patch pipette. |

3 Conclusions

The 1991 Nobel prize in Physiology or Medicine was recognized to Erwin Neher and Bert Sakmann for their "discovery of the function of single ion channels." Together they developed the patch-clamp technique in the late 1970s and refined this technique to allow determination of single channel properties. This technology has revolutionized the study of ionic currents within excitable cells, including auditory and vestibular hair cells. The patch-clamp technique has been crucial in providing an understanding of how sensory hair cells of the inner ear transform an initial mechanical input into an electrochemical output capable of stimulating afferent neuronal fibers that, in turn, carry information to the central nervous system. Over 30 years later, it remains an essential tool for understanding the functional role of molecules within the hearing and balance organs and is ideal for integration with optical techniques such as confocal imaging and optogenetics. While this chapter provides detailed guidance on how to record mechanotransduction and voltage-dependent currents from sensory hair cells, practical aspects of the experiments will differ for each setup and user. There are many published protocols available, yet there is nothing better than assembling one's own setup and being prepared for trial and error, tweaking the assembly until consistent and stable recordings are obtained. A successful electrophysiologist learns that every failure brings them one step closer to success and that perseverance is key.

Acknowledgments

The authors would like to thank their many mentors who have trained them in these different techniques and have shared several technical tips highlighted in this review. To cite a few, the authors wish to thank Dr. Corné Kros (PhD co-mentor for Dr. Geleoc and PhD and postdoctoral mentor for Dr. Johnson), Dr. Walter Marcotti (postdoctoral mentor of Dr. Johnson), Dr. Gregory Frolenkov (postdoctoral mentor of Dr. Indzhykulian) as well as Dr. David Corey (postdoctoral mentor for Drs. Geleoc and Indzhykulian) and Dr. Jonathan Ashmore (postdoctoral mentor for Dr. Geleoc). The authors also would like to thank Dr. Corne Kros for initially developing the fluid jet some 30 years ago as well as Drs. David Corey and Jim Hudspeth for the development of hair bundle micromanipulation with piezoelectric bimorph elements. We also thank

Dr. David Corey for providing images for this publication and Dr. Walter Marcotti who contributed to the development of the dissection techniques of the organ of Corti as described in this chapter. We thank Dr. Corne Kros for his feedback on a previous version of this manuscript. A.A.I. is supported by the NIH grant R01DC017166 and the Bertarelli Program in Translational Neuroscience and Neuroengineering. G.S.G work is supported by the NIH grant RO1DC008853, the Barber Research fund for Gene therapy for Genetic Deafness and the US-Israel Binational Science foundation.

References

1. Fettiplace R, Crawford AC, Evans MG (1992) The hair cell's mechanoelectrical transducer channel. Ann N Y Acad Sci 656:1–11. https://doi.org/10.1111/j.1749-6632.1992.tb25196.x

2. Hackney CM, Furness DN (1995) Mechanotransduction in vertebrate hair cells: structure and function of the stereociliary bundle. Am J Phys 268(1 Pt 1):C1–C13. https://doi.org/10.1152/ajpcell.1995.268.1.C1

3. Howard J, Roberts WM, Hudspeth AJ (1988) Mechanoelectrical transduction by hair cells. Annu Rev Biophys Biophys Chem 17:99–124. https://doi.org/10.1146/annurev.bb.17.060188.000531

4. Hudspeth AJ, Corey DP (1977) Sensitivity, polarity, and conductance change in the response of vertebrate hair cells to controlled mechanical stimuli. Proc Natl Acad Sci U S A 74(6):2407–2411. https://doi.org/10.1073/pnas.74.6.2407

5. Roberts WM, Howard J, Hudspeth AJ (1988) Hair cells: transduction, tuning, and transmission in the inner ear. Annu Rev Cell Biol 4:63–92. https://doi.org/10.1146/annurev.cb.04.110188.000431

6. Hudspeth AJ, Jacobs R (1979) Stereocilia mediate transduction in vertebrate hair cells (auditory system/cilium/vestibular system). Proc Natl Acad Sci U S A 76(3):1506–1509. https://doi.org/10.1073/pnas.76.3.1506

7. Goodyear RJ, Marcotti W, Kros CJ, Richardson GP (2005) Development and properties of stereociliary link types in hair cells of the mouse cochlea. J Comp Neurol 485(1):75–85. https://doi.org/10.1002/cne.20513

8. Corey DP, Hudspeth AJ (1979) Ionic basis of the receptor potential in a vertebrate hair cell. Nature 281(5733):675–677. https://doi.org/10.1038/281675a0

9. Denk W, Holt JR, Shepherd GM, Corey DP (1995) Calcium imaging of single stereocilia in hair cells: localization of transduction channels at both ends of tip links. Neuron 15(6):1311–1321. https://doi.org/10.1016/0896-6273(95)90010-1

10. PMCID: PMC6564293 (Hudspeth AJ. Extracellular current flow and the site of transduction by vertebrate hair cells. J Neurosci. 1982 Jan;2(1):1–10. https://doi.org/10.1523/JNEUROSCI.02-01-00001.1982. PMID: 6275046; PMCID: PMC6564293.)

11. PMCID: PMC2712647 (Beurg M, Fettiplace R, Nam JH, Ricci AJ. Localization of inner hair cell mechanotransducer channels using high-speed calcium imaging. Nat Neurosci. 2009 May;12(5):553–8. https://doi.org/10.1038/nn.2295. Epub 2009 Mar 29. PMID: 19330002; PMCID: PMC2712647.)

12. Assad JA, Shepherd GM, Corey DP (1991) Tip-link integrity and mechanical transduction in vertebrate hair cells. Neuron 7(6):985–994. https://doi.org/10.1016/0896-6273(91)90343-x

13. Holt JR, Pan B, Koussa MA, Asai Y (2014) TMC function in hair cell transduction. Hear Res 311:17–24. https://doi.org/10.1016/j.heares.2014.01.001

14. Kawashima Y, Geleoc GS, Kurima K, Labay V, Lelli A, Asai Y, Makishima T, Wu DK, Della Santina CC, Holt JR, Griffith AJ (2011) Mechanotransduction in mouse inner ear hair cells requires transmembrane channel-like genes. J Clin Invest 121(12):4796–4809. https://doi.org/10.1172/JCI60405

15. Kurima K, Ebrahim S, Pan B, Sedlacek M, Sengupta P, Millis BA, Cui R, Nakanishi H, Fujikawa T, Kawashima Y, Choi BY, Monahan K, Holt JR, Griffith AJ, Kachar B (2015) TMC1 and TMC2 localize at the site

of mechanotransduction in mammalian inner ear hair cell stereocilia. Cell Rep 12(10):1606–1617. https://doi.org/10.1016/j.celrep.2015.07.058

16. Pan B, Akyuz N, Liu XP, Asai Y, Nist-Lund C, Kurima K, Derfler BH, Gyorgy B, Limapichat W, Walujkar S, Wimalasena LN, Sotomayor M, Corey DP, Holt JR (2018) TMC1 forms the pore of mechanosensory transduction channels in vertebrate inner ear hair cells. Neuron 99(4):736–753.e736. https://doi.org/10.1016/j.neuron.2018.07.033

17. Pan B, Geleoc GS, Asai Y, Horwitz GC, Kurima K, Ishikawa K, Kawashima Y, Griffith AJ, Holt JR (2013) TMC1 and TMC2 are components of the mechanotransduction channel in hair cells of the mammalian inner ear. Neuron 79(3):504–515. https://doi.org/10.1016/j.neuron.2013.06.019

18. Waltner JG, Raymond S (1950) On the chemical composition of the human perilymph and endolymph. Laryngoscope 60(9):912–918. https://doi.org/10.1288/00005537-195009000-00003

19. Dallos P, Zheng J, Cheatham MA (2006) Prestin and the cochlear amplifier. J Physiol 576 (Pt 1):37–42. https://doi.org/10.1113/jphysiol.2006.114652

20. Li Y, Liu H, Giffen KP, Chen L, Beisel KW, He DZZ (2018) Transcriptomes of cochlear inner and outer hair cells from adult mice. Sci Data 5: 180199. https://doi.org/10.1038/sdata.2018.199

21. Liu H, Pecka JL, Zhang Q, Soukup GA, Beisel KW, He DZ (2014) Characterization of transcriptomes of cochlear inner and outer hair cells. J Neurosci 34(33):11085–11095. https://doi.org/10.1523/JNEUROSCI.1690-14.2014

22. Jeng JY, Ceriani F, Hendry A, Johnson SL, Yen P, Simmons DD, Kros CJ, Marcotti W (2020) Hair cell maturation is differentially regulated along the tonotopic axis of the mammalian cochlea. J Physiol 598(1):151–170. https://doi.org/10.1113/JP279012

23. Johnson SL (2015) Membrane properties specialize mammalian inner hair cells for frequency or intensity encoding. elife 4. https://doi.org/10.7554/eLife.08177

24. Johnson SL, Forge A, Knipper M, Munkner S, Marcotti W (2008) Tonotopic variation in the calcium dependence of neurotransmitter release and vesicle pool replenishment at mammalian auditory ribbon synapses. J Neurosci 28(30):7670–7678. https://doi.org/10.1523/JNEUROSCI.0785-08.2008

25. Johnson SL, Olt J, Cho S, von Gersdorff H, Marcotti W (2017) The coupling between Ca (2+) channels and the exocytotic Ca(2+) sensor at hair cell ribbon synapses varies tonotopically along the mature cochlea. J Neurosci 37(9):2471–2484. https://doi.org/10.1523/JNEUROSCI.2867-16.2017

26. Lelli A, Asai Y, Forge A, Holt JR, Geleoc GS (2009) Tonotopic gradient in the developmental acquisition of sensory transduction in outer hair cells of the mouse cochlea. J Neurophysiol 101(6):2961–2973. https://doi.org/10.1152/jn.00136.2009

27. Tang F, Chen X, Jia L, Li H, Li J, Yuan W (2019) Differential gene expression patterns between apical and basal inner hair cells revealed by RNA-Seq. Front Mol Neurosci 12:332. https://doi.org/10.3389/fnmol.2019.00332

28. Brandon CS, Voelkel-Johnson C, May LA, Cunningham LL (2012) Dissection of adult mouse utricle and adenovirus-mediated supporting-cell infection. J Vis Exp 61. https://doi.org/10.3791/3734

29. Cunningham LL (2006) The adult mouse utricle as an in vitro preparation for studies of ototoxic-drug-induced sensory hair cell death. Brain Res 1091(1):277–281. https://doi.org/10.1016/j.brainres.2006.01.128

30. Beurg M, Barlow A, Furness DN, Fettiplace R (2019) A Tmc1 mutation reduces calcium permeability and expression of mechanoelectrical transduction channels in cochlear hair cells. Proc Natl Acad Sci U S A 116(41):20743–20749. https://doi.org/10.1073/pnas.1908058116

31. Fettiplace R, Fuchs PA (1999) Mechanisms of hair cell tuning. Annu Rev Physiol 61: 809–834. https://doi.org/10.1146/annurev.physiol.61.1.809

32. Johnson SL, Marcotti W (2008) Biophysical properties of CaV1.3 calcium channels in gerbil inner hair cells. J Physiol 586(4):1029–1042. https://doi.org/10.1113/jphysiol.2007.145219

33. Meyer AC, Frank T, Khimich D, Hoch G, Riedel D, Chapochnikov NM, Yarin YM, Harke B, Hell SW, Egner A, Moser T (2009) Tuning of synapse number, structure and function in the cochlea. Nat Neurosci 12(4):444–453. https://doi.org/10.1038/nn.2293

34. Moser T, Beutner D (2000) Kinetics of exocytosis and endocytosis at the cochlear inner hair cell afferent synapse of the mouse. Proc Natl Acad Sci U S A 97(2):883–888. https://doi.org/10.1073/pnas.97.2.883

35. Waguespack J, Salles FT, Kachar B, Ricci AJ (2007) Stepwise morphological and functional maturation of mechanotransduction in rat outer hair cells. J Neurosci 27(50):13890–13902. https://doi.org/10.1523/JNEUROSCI.2159-07.2007

36. Johnson SL, Marcotti W, Kros CJ (2005) Increase in efficiency and reduction in Ca2+ dependence of exocytosis during development of mouse inner hair cells. J Physiol 563 (Pt 1):177–191. https://doi.org/10.1113/jphysiol.2004.074740

37. Kros CJ, Ruppersberg JP, Rusch A (1998) Expression of a potassium current in inner hair cells during development of hearing in mice. Nature 394(6690):281–284. https://doi.org/10.1038/28401

38. Marcotti W, Geleoc GS, Lennan GW, Kros CJ (1999) Transient expression of an inwardly rectifying potassium conductance in developing inner and outer hair cells along the mouse cochlea. Pflugers Arch 439(1–2):113–122. https://doi.org/10.1007/s004249900157

39. Marcotti W, Johnson SL, Holley MC, Kros CJ (2003) Developmental changes in the expression of potassium currents of embryonic, neonatal and mature mouse inner hair cells. J Physiol 548(Pt 2):383–400. https://doi.org/10.1113/jphysiol.2002.034801

40. Ahituv N, Avraham KB (2000) Auditory and vestibular mouse mutants: models for human deafness. J Basic Clin Physiol Pharmacol 11(3):181–191. https://doi.org/10.1515/jbcpp.2000.11.3.181

41. Ohlemiller KK (2019) Mouse methods and models for studies in hearing. J Acoust Soc Am 146(5):3668. https://doi.org/10.1121/1.5132550

42. Fang QJ, Wu F, Chai R, Sha SH (2019) Cochlear surface preparation in the adult mouse. J Vis Exp 153. https://doi.org/10.3791/60299

43. Marcotti W, Kros CJ (1999) Developmental expression of the potassium current IK,n contributes to maturation of mouse outer hair cells. J Physiol 520(Pt 3):653–660. https://doi.org/10.1111/j.1469-7793.1999.00653.x

44. Goutman JD, Pyott SJ (2016) Whole-cell patch-clamp recording of mouse and rat inner hair cells in the intact organ of Corti. Methods Mol Biol 1427:471–485. https://doi.org/10.1007/978-1-4939-3615-1_26

45. Neher E, Sakmann B (1992) The patch clamp technique. Sci Am 266(3):44–51. https://doi.org/10.1038/scientificamerican0392-44

46. Gamper N (2013) Ion channels: methods and protocols, Methods in molecular biology, vol 998, 2nd edn. Humana Press, New York

47. Martina M, Taverna S (2014) Patch-clamp methods and protocols, Methods in molecular biology, vol 1183, 2nd edn. Humana Press, New York

48. Okada Y (2012) Patch clamp techniques: from beginning to advanced protocols, Springer protocols handbooks. Springer, New York

49. Grant L, Yi E, Goutman JD, Glowatzki E (2011) Postsynaptic recordings at afferent dendrites contacting cochlear inner hair cells: monitoring multivesicular release at a ribbon synapse. J Vis Exp 48. https://doi.org/10.3791/2442

50. Indzhykulian AA, Stepanyan R, Nelina A, Spinelli KJ, Ahmed ZM, Belyantseva IA, Friedman TB, Barr-Gillespie PG, Frolenkov GI (2013) Molecular remodeling of tip links underlies mechanosensory regeneration in auditory hair cells. PLoS Biol 11(6):e1001583. https://doi.org/10.1371/journal.pbio.1001583

51. Corey DP, Hudspeth AJ (1980) Mechanical stimulation and micromanipulation with piezoelectric bimorph elements. J Neurosci Methods 3(2):183–202. https://doi.org/10.1016/0165-0270(80)90025-4

52. Geleoc GS, Holt JR (2003) Developmental acquisition of sensory transduction in hair cells of the mouse inner ear. Nat Neurosci 6(10):1019–1020. https://doi.org/10.1038/nn1120

53. PMCID: PMC2267152 (Beurg M, Nam JH, Crawford A, Fettiplace R. The actions of calcium on hair bundle mechanics in mammalian cochlear hair cells. Biophys J. 2008 Apr 1;94 (7):2639–53. https://doi.org/10.1529/biophysj.107.123257. Epub 2008 Jan 4. PMID: 18178649; PMCID: PMC2267152)

54. PMCID: PMC6757600 (Ricci AJ, Crawford AC, Fettiplace R. Mechanisms of active hair bundle motion in auditory hair cells. J Neurosci. 2002 Jan 1;22(1):44–52. https://doi.org/10.1523/JNEUROSCI.22-01-00044.2002. PMID: 11756487; PMCID: PMC6757600.)

55. PMCID: PMC6772778 (Ricci AJ, Crawford AC, Fettiplace R. Active hair bundle motion linked to fast transducer adaptation in auditory hair cells. J Neurosci. 2000 Oct 1;20 (19):7131–42. https://doi.org/10.1523/JNEUROSCI.20-19-07131.2000. PMID: 11007868; PMCID: PMC6772778.)

56. PMCID: PMC1288017 (Le Goff L, Bozovic D, Hudspeth AJ. Adaptive shift in the domain of negative stiffness during spontaneous

oscillation by hair bundles from the internal ear. Proc Natl Acad Sci U S A. 2005 Nov 22;102(47):16996–7001. https://doi.org/10.1073/pnas.0508731102. Epub 2005 Nov 15. PMID: 16287969; PMCID: PMC1288017.)

57. Howard J, Hudspeth AJ (1988) Compliance of the hair bundle associated with gating of mechanoelectrical transduction channels in the bullfrog's saccular hair cell. Neuron 1(3):189–199. https://doi.org/10.1016/0896-6273(88)90139-0

58. Peng AW, Ricci AJ (2016) Glass probe stimulation of hair cell stereocilia. Methods Mol Biol 1427:487–500. https://doi.org/10.1007/978-1-4939-3615-1_27

59. Kros CJ, Rusch A, Richardson GP (1992) Mechano-electrical transducer currents in hair cells of the cultured neonatal mouse cochlea. Proc Biol Sci 249(1325):185–193. https://doi.org/10.1098/rspb.1992.0102

60. Alagramam KN, Goodyear RJ, Geng R, Furness DN, van Aken AF, Marcotti W, Kros CJ, Richardson GP (2011) Mutations in protocadherin 15 and cadherin 23 affect tip links and mechanotransduction in mammalian sensory hair cells. PLoS One 6(4):e19183. https://doi.org/10.1371/journal.pone.0019183

61. Beurg M, Xiong W, Zhao B, Muller U, Fettiplace R (2015) Subunit determination of the conductance of hair-cell mechanotransducer channels. Proc Natl Acad Sci U S A 112(5):1589–1594. https://doi.org/10.1073/pnas.1420906112

62. Caprara GA, Mecca AA, Wang Y, Ricci AJ, Peng AW (2019) Hair bundle stimulation mode modifies manifestations of mechanotransduction adaptation. J Neurosci 39(46):9098–9106. https://doi.org/10.1523/JNEUROSCI.1408-19.2019

63. Corns LF, Johnson SL, Kros CJ, Marcotti W (2014) Calcium entry into stereocilia drives adaptation of the mechanoelectrical transducer current of mammalian cochlear hair cells. Proc Natl Acad Sci U S A 111(41):14918–14923. https://doi.org/10.1073/pnas.1409920111

64. Corns LF, Johnson SL, Kros CJ, Marcotti W (2016) Tmc1 point mutation affects Ca2+ sensitivity and block by dihydrostreptomycin of the mechanoelectrical transducer current of mouse outer hair cells. J Neurosci 36(2):336–349. https://doi.org/10.1523/JNEUROSCI.2439-15.2016

65. Corns LF, Johnson SL, Roberts T, Ranatunga KM, Hendry A, Ceriani F, Safieddine S, Steel KP, Forge A, Petit C, Furness DN, Kros CJ, Marcotti W (2018) Mechanotransduction is required for establishing and maintaining

mature inner hair cells and regulating efferent innervation. Nat Commun 9(1):4015. https://doi.org/10.1038/s41467-018-06307-w

66. Fettiplace R, Kim KX (2014) The physiology of mechanoelectrical transduction channels in hearing. Physiol Rev 94(3):951–986. https://doi.org/10.1152/physrev.00038.2013

67. Geleoc GS, Lennan GW, Richardson GP, Kros CJ (1997) A quantitative comparison of mechanoelectrical transduction in vestibular and auditory hair cells of neonatal mice. Proc Biol Sci 264(1381):611–621. https://doi.org/10.1098/rspb.1997.0087

68. Marcotti W, Corns LF, Goodyear RJ, Rzadzinska AK, Avraham KB, Steel KP, Richardson GP, Kros CJ (2016) The acquisition of mechano-electrical transducer current adaptation in auditory hair cells requires myosin VI. J Physiol 594(13):3667–3681. https://doi.org/10.1113/JP272220

69. Marcotti W, van Netten SM, Kros CJ (2005) The aminoglycoside antibiotic dihydrostreptomycin rapidly enters mouse outer hair cells through the mechano-electrical transducer channels. J Physiol 567(Pt 2):505–521. https://doi.org/10.1113/jphysiol.2005.085951

70. Crawford AC, Fettiplace R (1985) The mechanical properties of ciliary bundles of turtle cochlear hair cells. J Physiol 364:359–379. https://doi.org/10.1113/jphysiol.1985.sp015750

71. Kros CJ, Lennan GWT, Richardson GP (1995) Transducer currents and bundle movements in outer hair cells of neonatal mice. In: Flock A (ed) Active hearing. Elsevier Science, Oxford, pp 113–125

72. Kros CJ, Rüsch A, Lennan GWR, Richardson GP (1993) Voltage dependence of transducer currents in outer hair cells of neonatal mice. In: GDuifuis H, Horst JW, van Dijk P, van Netten SM (eds) Biophysics of hair cell sensory system. World Scientific, Singapore, pp 141–150

73. Ohmori H (1985) Mechano-electrical transduction currents in isolated vestibular hair cells of the chick. J Physiol 359:189–217. https://doi.org/10.1113/jphysiol.1985.sp015581

74. Ricci AJ, Crawford AC, Fettiplace R (2000) Active hair bundle motion linked to fast transducer adaptation in auditory hair cells. J Neurosci 20(19):7131–7142

75. Caprara GA, Mecca AA, Peng AW (2020) Decades-old model of slow adaptation in sensory hair cells is not supported in mammals. Sci Adv 6(33):eabb4922. https://doi.org/10.1126/sciadv.abb4922

76. Fridberger A, Tomo I, Ulfendahl M, Boutet de Monvel J (2006) Imaging hair cell transduction at the speed of sound: dynamic behavior of mammalian stereocilia. Proc Natl Acad Sci U S A 103(6):1918–1923. https://doi.org/10.1073/pnas.0507231103

77. Peng AW, Effertz T, Ricci AJ (2013) Adaptation of mammalian auditory hair cell mechanotransduction is independent of calcium entry. Neuron 80(4):960–972. https://doi.org/10.1016/j.neuron.2013.08.025

78. Dinklo T, Meulenberg CJ, van Netten SM (2007) Frequency-dependent properties of a fluid jet stimulus: calibration, modeling, and application to cochlear hair cell bundles. J Assoc Res Otolaryngol 8(2):167–182. https://doi.org/10.1007/s10162-007-0080-0

79. Vollrath MA, Eatock RA (2003) Time course and extent of mechanotransducer adaptation in mouse utricular hair cells: comparison with frog saccular hair cells. J Neurophysiol 90(4):2676–2689. https://doi.org/10.1152/jn.00893.2002

80. Corey DP, Hudspeth AJ (1983) Kinetics of the receptor current in bullfrog saccular hair cells. J Neurosci 3(5):962–976

81. Eatock RA, Corey DP, Hudspeth AJ (1987) Adaptation of mechanoelectrical transduction in hair cells of the bullfrog's sacculus. J Neurosci 7(9):2821–2836

82. Holt JR, Corey DP, Eatock RA (1997) Mechanoelectrical transduction and adaptation in hair cells of the mouse utricle, a low-frequency vestibular organ. J Neurosci 17(22):8739–8748

83. Holton T, Hudspeth AJ (1986) The transduction channel of hair cells from the bull-frog characterized by noise analysis. J Physiol 375:195–227. https://doi.org/10.1113/jphysiol.1986.sp016113

84. Segev A, Garcia-Oscos F, Kourrich S (2016) Whole-cell patch-clamp recordings in brain slices. J Vis Exp 112. https://doi.org/10.3791/54024

85. Beutner D, Moser T (2001) The presynaptic function of mouse cochlear inner hair cells during development of hearing. J Neurosci 21(13):4593–4599

86. Glowatzki E, Fuchs PA (2002) Transmitter release at the hair cell ribbon synapse. Nat Neurosci 5(2):147–154. https://doi.org/10.1038/nn796

87. Johnson SL, Eckrich T, Kuhn S, Zampini V, Franz C, Ranatunga KM, Roberts TP, Masetto S, Knipper M, Kros CJ, Marcotti W (2011) Position-dependent patterning of spontaneous action potentials in immature cochlear inner hair cells. Nat Neurosci 14(6):711–717. https://doi.org/10.1038/nn.2803

88. Johnson SL, Kennedy HJ, Holley MC, Fettiplace R, Marcotti W (2012) The resting transducer current drives spontaneous activity in prehearing mammalian cochlear inner hair cells. J Neurosci 32(31):10479–10483. https://doi.org/10.1523/JNEUROSCI.0803-12.2012

89. Marcotti W, Johnson SL, Rusch A, Kros CJ (2003) Sodium and calcium currents shape action potentials in immature mouse inner hair cells. J Physiol 552(Pt 3):743–761. https://doi.org/10.1113/jphysiol.2003.043612

90. Jeng JY, Carlton AJ, Johnson SL, Brown SDM, Holley MC, Bowl MR, Marcotti W (2021) Biophysical and morphological changes in inner hair cells and their efferent innervation in the ageing mouse cochlea. J Physiol 599(1):269–287. https://doi.org/10.1113/JP280256

91. Oliver D, Knipper M, Derst C, Fakler B (2003) Resting potential and submembrane calcium concentration of inner hair cells in the isolated mouse cochlea are set by KCNQ-type potassium channels. J Neurosci 23(6):2141–2149

92. Marcotti W, Johnson SL, Kros CJ. A transiently expressed SK current sustains and modulates action potential activity in immature mouse inner hair cells. J Physiol. 2004 Nov 1;560(Pt 3):691–708. https://doi.org/10.1113/jphysiol.2004.072868. Epub 2004 Aug 26. PMID: 15331671; PMCID: PMC1665291.

93. Nouvian R (2007) Temperature enhances exocytosis efficiency at the mouse inner hair cell ribbon synapse. J Physiol 584(Pt 2):535–542. https://doi.org/10.1113/jphysiol.2007.139675

94. Brandt A, Khimich D, Moser T (2005) Few CaV1.3 channels regulate the exocytosis of a synaptic vesicle at the hair cell ribbon synapse. J Neurosci 25(50):11577–11585. https://doi.org/10.1523/JNEUROSCI.3411-05.2005

95. Colclasure JC, Holt JR (2003) Transduction and adaptation in sensory hair cells of the mammalian vestibular system. Gravit Space Biol Bull 16(2):61–70

96. Kennedy HJ, Evans MG, Crawford AC, Fettiplace R (2003) Fast adaptation of mechanoelectrical transducer channels in mammalian cochlear hair cells. Nat Neurosci 6(8):832–836. https://doi.org/10.1038/nn1089

97. Kros CJ, Marcotti W, van Netten SM, Self TJ, Libby RT, Brown SD, Richardson GP, Steel KP (2002) Reduced climbing and increased slipping adaptation in cochlear hair cells of mice with Myo7a mutations. Nat Neurosci 5(1):41–47. https://doi.org/10.1038/nn784

98. Stauffer EA, Holt JR (2007) Sensory transduction and adaptation in inner and outer hair cells of the mouse auditory system. J Neurophysiol 98(6):3360–3369. https://doi.org/10.1152/jn.00914.2007

99. Stepanyan R, Frolenkov GI (2009) Fast adaptation and Ca2+ sensitivity of the mechanotransducer require myosin-XVa in inner but not outer cochlear hair cells. J Neurosci 29(13):4023–4034. https://doi.org/10.1523/JNEUROSCI.4566-08.2009

Chapter 11

Biophysical Recording from Adult Hair Cells

Antonio Miguel Garcia de Diego and Jonathan F. Ashmore

Abstract

Many biophysical recordings from mammalian cochlear hair cell are made using neonatal or relatively juvenile systems. These tissues are often more experimentally more robust. This chapter describes two methods which provide insight into how the adult hair cells perform. The first is the isolation method for guinea pig outer hair cells. This preparation allows recording of ionic currents and the study of the prestin/ SLC26A5 outer hair cell motility. The second system described is the preparation of in situ hair cells from the mouse temporal bone. Some examples are shown where electrical patch clamp recordings can be made. The methods of preparation also allow access to the organ of Corti for imaging.

Key words Hair cell, Cochlea, Mouse, Guinea pig, In situ recording, Outer hair cells

1 Introduction

Recording from individual adult mammalian hair cells has always presented particular challenges as the cells are generally more difficult to dissect and survive less well, if at all, for any length of time in culture. For this reason the more popular systems have been based on early stage tissues. In rodents this means within the first two postnatal weeks (P0–P15). There are certainly results from later stage animals (for a recent example, *see* ref. 1), but in general the data is quite sparse. Recording from adult cells obviates issues about emergence of novel features of the cell characteristic of particular stages of development. However, knowledge of the properties of adult hair cells (and other cells in the inner ear) is critical so that the data can be matched to the extensive data known from adult auditory nerve fiber recording or adult auditory brainstem recordings (ABRs), where the animal has normally reached maturity. This short chapter will describe two systems that can achieve this end.

The first method is recording from hair cells of the guinea pig; this historically was the first system to be used for hair cells recoding as even in newborn animals the cells have reached maturity. This is the system used from the early 1980s onwards to study outer hair

Andrew K. Groves (ed.), *Developmental, Physiological, and Functional Neurobiology of the Inner Ear*, Neuromethods, vol. 176, https://doi.org/10.1007/978-1-0716-2022-9_11,

cell motility [2–4] as well as a characterization of some of the OHC ionic currents [5].

The second method describes how recording from early stage rodent cochleas can be extended to adult systems by resorting to an in situ system where the damage to hair cells due to dissection is minimized.

2   Methods

For details of patch clamp recording, *see* the chapter by Geleoc et al. (this volume) and [6].

2.1   Solutions and Recording Conditions

Solution choice is important for any sort of physiological recording experiment. They need to be carefully prepared to ensure reproducibility and interpretation of the data. The following are some standard solutions that can be used in the experiments described here.

2.1.1   External Solution (in mM)

NaCl, 142; KCl, 4; $CaCl_2$, 1.5; $MgCl_2$, 1; Na_2HPO_4, 8; NaH_2PO_4, 2; D-glucose, 10.

(10 mM Hepes can be used instead of the phosphates. In this case the solution pH has to be titrated back to 7.4 with NaOH. Check the pH of all solutions before experimenting, as left on the bench the solution pH can drift. The solution osmotic strength should be between 300 and 320 mOsm/kg.)

2.1.2   Internal Solution for the Patch Pipette Recording (in mM)

KCl, 144; $MgCl_2$, 2; Na_2HPO_4, 8; NaH_2PO_4, 2 D-glucose, 10; ATP, 1; GTP, 1; EGTA,0.5.

(As with the external solution Hepes can be used instead of the phosphates. Since only a small volume is required for pipette filling, when prepared the internal solution can be divided into 1 mL aliquots and stored at −20 °C. Since the ATP and GTP degrade at room temperature it is a good idea to keep each aliquot when used on ice; it should be usable for several hours.)

2.2   Further Notes

2.2.1   pH Buffering

In vivo, fluids surrounding cells are buffered against pH changes by bicarbonate. In practice this presents technical problems as the solution has to be bubbled continuously with 5% CO_2, a difficult situation to maintain in the small chambers used in these experiments. As a result the pH buffer is usually chosen to be a phosphate buffer or Hepes buffer. In a phosphate buffer (i.e. a defined ratio of mono- and bi-phosphate salts) there is a limit to how much external calcium can be used in the medium as calcium phosphate salts precipitate. Outer hair cells (OHCs) survive for longer periods with a phosphate buffer, however (for reasons related to the properties of the OHC motor prestin [7]). As a convenient default, we recommend using conventional phosphate buffer saline, either acquired for tissue culture (without indicator) or made up from tablets. This will be suitable for short term studies.

2.2.2 Recording Temperature

In general dissections are carried out at room temperature. This often applies to the recording conditions as well, although there may be particular reasons for running at 37 °C (e.g. estimation of the in vivo current kinetics), it is more convenient to work at lower temperatures and the tissue survives longer. Working at higher temperatures, particularly with small chamber volumes, requires additional solution superfusion as evaporation becomes a problem.

2.2.3 Dye Loading with Patch Pipettes

When loading a cell with a calcium indicator (e.g. fluo3, OGB-1, etc.) the volume of dye used to fill a cell is very small (typically 10 pL). For this reason it is not necessary to backfill the entire pipette and small aliquots can be made up and stored frozen for each experiment. We put the intracellular solution and the dye with a total volume of 100 μL into a small PCR vial. Using pipettes with a fiber, put the pipette open, un-pulled end into the vial and wait for 30–60 s. You should be able to see just the tip filling up on holding the pipette to the light. The pipette can then be backfilled in the normal way. This economically provides more than enough dye for each cell.

2.2.4 Choice of Microscope

The viewing and recording of both isolated and in situ hair cells can be carried out on an inverted microscope or an upright microscope. For recording from cells a fixed stage microscope is required.

An *upright microscope* equipped with a 40× or higher power water immersion objective can be used for both isolated or in situ systems. There is typically 1–2 mm clearance under the objective to insert a micropipette. Practice is required for the use of an upright microscope with isolated cells as cells can be disturbed as the objective is moved and focused.

An *inverted microscope* is probably better for isolated cells. In this case larger numerical aperture and even oil immersion objectives can be used (e.g. 40, 60, or 63× 1.3 NA, depending on the make of microscope). In such cases it is necessary to use 35 mm or 60 mm diameter glass bottom dishes (Matek or WPI) as the chamber. Note that as the numerical aperture (NA) increases the working distance decreases, so practice is required to align both cell and pipette together in the same plane without breaking the glass pipette against the dish.

3 Dissection Protocols

3.1 Isolation of Hair Cells from the Organ of Corti of the Guinea Pig

This protocol is, in fact, applicable to all cochleas, although it is most easily implemented for a large cochlea such as that of the guinea pig. The anatomy of the structures should be studied to enable suitable landmarks to be identified easily. For an example guide, *see* [8].

The isolation of cells followed a number of published protocols [9].

1. Guinea pigs are killed by procedures which conform to local and international standards of animal health and welfare.

2. The tympanic bulla (the middle ear cavity) is removed from the head and opened up to reveal the cochlea lying on the medial surface.

3. A small score mark with a sharp scalpel (No 11) is made longitudinally down the side of the cochlea allowing the cochlear bone to be opened up and the coils of the cochlea revealed. The cochlear coils can be removed by cutting through the base with the scalpel and then grasping the coil base to lift it free into dissecting solution.

4. Unpeel and pick the organ of Corti off with a fine pin (e.g. a 25 gauge hypodermic needle). It is helpful if it is broken up into small sections 1–2 mm in length.

5. Suck the pieces up into a total of 40 μL droplet in a Gilson pipette. It is helpful if the pipette tip has previously sucked up some solution and then expelled to reduce the tissue sticking to the fresh plastic.

6. Transfer to a further drop of 40 μL of solution containing 1 mg/mL of Trypsin (Sigma type IV). Other enzymes can also be used but the length of time required may have to be explored. Leave for 15 min at room temperature. We find it convenient to carry out this incubation by keeping the drop and the subsequent trituration in the corner of a petri dish.

7. Suck up into a 100 μL volume in the same Gilson tip and expel 2–3 times reasonably vigorously. The bubbles which form provide a surface tension force to dissociate the cells. Depending on this step and the incubation time cells may have adhering cells of the organ of Corti or in some cases undissociated synaptic endings may sometimes be present.

8. Place the triturated cell solution down onto the chamber surface. Gently add a volume of extracellular solution to make up the chamber volume. A typical volume might be 1 mL.

The cells settle quite slowly, with the OHCs dropping down first within about 5 min. They should then be visible (Fig. 1). Some cells stick, some do not. For the purposes of electrophysiological

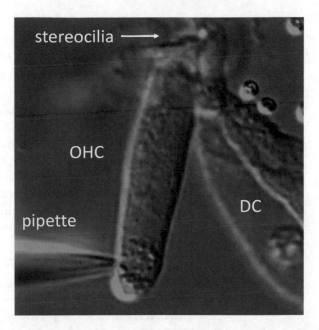

Fig. 1 Isolated guinea pig OHC recording. Based on its length, this cell is from the apical cochlea. The patch pipette forms a seal most readily, prior to going whole cell, at the pole of the OHC near the nucleus. The cell attached to the OHC is a Dieters cell (DC), only partially separated during isolation. DIC optics, 40× 1.30 NA objective, Zeiss Axiovert microscope. Recording: MBL BIE workshop 2015

recording it is immaterial as the cells can be approached with a patch pipette even if they are not stuck down. Most cells will have some point of attachment however. It is often found that approaching the membrane at the basal pole produces good seals most easily.

4   In Situ Recording from the Mouse Cochlea

1. Remove the cochlea from the mouse. We do this by first killing the mouse by procedures which conform local and international standards of animal health and welfare. We then remove the head and peel back the skin. Bisect the head and scoop out the brain leaving just the lateral portions of the skull. We do the whole operation just wearing surgical gloves as this allows identification of the ear canal and a feel for the bulla nearby—it should feel just like a small grain of rice. Once located, gently tease it out.

2. Place the bulla in a small dish of extracellular solution. With two sets of forceps, gently tease the bulla open to reveal the cochlea on the medial surface of the bone. It should be possible to identify the exit site of the auditory nerve into the brain cavity. Clear away as much of the residual connective tissue and muscle leaving clear surfaces of the bone.

3. Identify the semi-circular canals. A portion of the cerebellum is sometimes found in the volume delineated by the canals. It is best for subsequent optical reasons to pick this out and to discard it as the illuminating light will be projected through this cavity.

4. We use the top of a 35 mm plastic petri dish as the chamber. An alternative is to use 13 mm coverslips which can then be transferred to a recording chamber although sometimes the glue (see below) detaches. Under a dissecting microscope put a small drop of cyanoacrylate glue (we use Loctite standard superglue) onto the bottom of a 35 mm petri dish. Carefully pick up the bone with the cochlea, touch against a tissue to dry it, and lower onto the glue droplet with the cochlea uppermost. Orientate the bone as the glue dries (you have 1–2 min) so that as much of the apical cochlea as possible is horizontal (Fig. 2).

5. Gently add extracellular solution to cover the bone. It can be left for 5 min for the glue—which will go white—and will harden better for the further dissection.

6. Identify the helicotrema. You may have to adjust the lighting angle and the focus to get a good view of the bone. The magnification also needs to be adjusted so the 1 mm diameter of the cochlea fills a reasonable fraction of the field of view.

7. Now, with no 5 forceps, make a very small hole in the bone near the helicotrema and work toward the base about half a turn, picking away the overlying bone. Be careful not to break the cochlear modiolus. Also be careful not to go too deep with the forceps or else you will also pick away the cochlear partition as well. Sometimes, having opened up half a turn apex to base, it is helpful to work back toward the apex to remove the stria as this allows better access for the patch pipette.

8. Place the dissected cochlea under the objective of an *upright* microscope with a water immersion objective. Use bright field illumination. If observed with a $10 \times$ WI objective, the cochlea coil should appear as a bright ring. Having centered the area of interest, switch to a higher power water immersion objective.

9. The angle of approach for pipettes is critical: too steep and the pipette will hit against the objective; too shallow and the pipette tip will not reach down to the tissue. Some experimentation and familiarity with the microscope and the manipulation are required. We also find that using the smaller diameter 1.2 mm OD thin wall glass for the pipettes can be an advantage to provide extra clearance (Fig. 3).

10. With practice it will be found that a new pipette can be inserted under the higher power WI objective without first lining the tip up with a low power (e.g. $10 \times$) objective. This decreases the

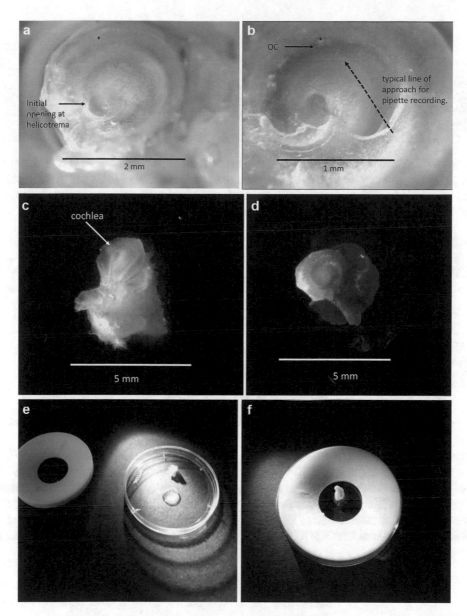

Fig. 2 Dissecting the mouse cochlea for in situ recording. 1. The temporal bone in dissecting medium with the bulla removed and placed with medial side underneath exposing the cochlea (**a, b**). 2. The temporal bone is placed on a cyanoacrylate drop in the dish so that the cochlear axis is vertical. 3. Lower magnification showing the relative size of the cochlea on a 35 mm dish. A circular PTFE ring is used to reduce the volume of solution, but at the same time allows a shallow angle of approach for patch pipette manipulations (**c, d**). 4. A small opening is made at the helicotrema to start the removal of bone over the cochlear partition. The line of the stria vascularis can be seen winding around the cochlear coil. (**e**). 5. When the bone is removed, the lines of the organ of Corti can just be made out with a dissection microscope. A typical line of approach for pipette recording is shown (**f**)

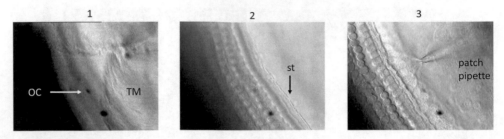

Fig. 3 Preparing the organ of Corti for recording. 1. Peel back the tectorial membrane (TM) by using a 10 μm diameter pipette to hold it by suction. 2. On focusing down, the three rows of OHCs and the stereocilia (st) of the IHCs are revealed. 3. Using a patch pipette to attach and to pull away the inner border cells from around the IHCs, leaving the membrane clean for forming a gigaseal with the patch pipette

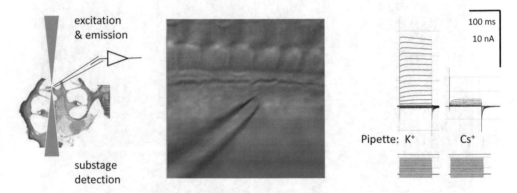

Fig. 4 Recording from the mouse hair cells. 1. The cells can be visualized in various ways. In Fig. 3, the approach can be seen in bright field illumination from above. Alternatively, the pipette and the tissue can be visualized when trans-illuminated from above in a multiphoton microscope. The 930 nm excitation beam penetrates bone and is collected by a substage photodetector during scanning. 2. The currents recorded from an IHC during a series of command steps (a) with K^+ as the dominant ion in the recording pipette and (b) with Cs^+ replacing K^+ to suppress the large K^+ currents. Under such condition a small inward current due to Ca^{2+} can be measured

turnaround time when there can be inevitable pipette exchanges and so increases the information from one dissection.

11. In our hands the mouse in situ preparation lasts for up to 2–3 h. The second cochlea, if kept at 4 °C, is still usable at his point although the quality of the recordings is noticeably worse.

12. Recordings can be made from cells from P18 upwards. The oldest mouse recorded successfully was 12 months old.

An example of a recorded current from an IHC is shown in Fig. 4. With K^+ as the major ion in the patch pipette the currents can be too large to obtain good clamp control. Cs^+ included to block K^+ channels, and under these circumstances the calcium currents can be observed and measured.

5   Subsidiary Recording Techniques

5.1   Measurement of Movements in Adult Cochlear Structures (OHCs, etc.)

There are several methods of measuring movement of cells. The most commonly used methods use optical measurement techniques. A feature of the cell boundary is lined up with a movement detector in the microscope optical path and the signal can be recorded with a bandwidth of up to 100 kHz, for example [10]. Modern video cameras now have high frame rates and these have also been employed successfully to identify movement of structures well into the acoustic range of frequencies [11]. Coupled with image tracking software (e.g. using ImageJ) but sometime requiring custom scripts, the responses can be identified.

5.2   Intracellular Calcium Measurements from IHCs

There are a wide range of indicator dyes to measure intracellular calcium in cells. Such dyes can either be added in the membrane permeant form as the acetoxy-methyl ester or directly introduced during the recording process. An example of dye filling of an isolated cell through the patch pipette is shown in Fig. 5. This particular measurement was carried out on a multiphoton upright microscope to reduce long term photobleaching of the cell, although comparable images are obtainable with a conventional confocal microscope. The cell shows a characteristic IHC morphology; none of the still intact surrounding cells is labeled. In some cases, however, cells have been found to be coupled together [12].

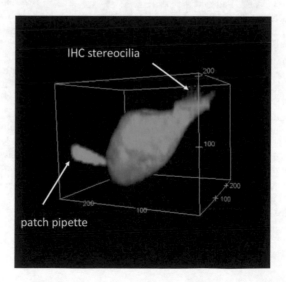

Fig. 5 Optical imaging of mouse inner hair cells. z-stack of a mouse IHC recorded in situ loaded with the indicator dye OGB-1 to detect Ca^{2+}. The multiphoton recording produces minimal photobleaching and cells can be recorded continuously for over 30 min, in some cases showing evidence of cell–cell coupling [12]. The recording pipette has been sealed to the base of the cell and can be seen projecting as it contains the fluorescent dye

6   Conclusions

The above provides a brief summary of two main techniques which have been used to probe the properties of mammalian cochlear hair cells. The same methods are in principle applicable to cells of the vestibular system, although the access for in situ recording has yet to be fully worked out.

Acknowledgement

Supported by the Wellcome Trust. We thank Robert Knight, UCL, for technical support and members of the MBL BIE workshop for encouragement.

References

1. Jeng J-Y, Ceriani F, Olt J, Brown SDM, Holley MC, Bowl MR, Johnson SL, Marcotti W (2020) Pathophysiological changes in inner hair cell ribbon synapses in the ageing mammalian cochlea. J Physiol 598:4339–4355

2. Ashmore J (2008) Cochlear outer hair cell motility. Physiol Rev 88:173–210

3. Ashmore JF (1987) A fast motile response in guinea-pig outer hair cells: the cellular basis of the cochlear amplifier. J Physiol 388:323–347

4. Brownell WE, Bader CR, Bertrand D, de Ribaupierre Y (1985) Evoked mechanical responses of isolated cochlear outer hair cells. Science 227:194–196

5. Mammano F, Ashmore JF (1996) Differential expression of outer hair cell potassium currents in the isolated cochlea of the guinea-pig. J Physiol 496:639–646

6. Goutman JD, Pyott SJ (2016) Whole-cell patch-clamp recording of mouse and rat inner hair cells in the intact organ of Corti. Methods Mol Biol 1427:471–485

7. Mistrik P, Daudet N, Morandell K, Ashmore JF (2012) Mammalian prestin is a weak Cl(−)/HCO(3)(−) electrogenic antiporter. J Physiol 590:5597–5610

8. Albuquerque AAS, Rossato M, de Oliveira JAA, Hyppolito MA (2009) Understanding the anatomy of ears from guinea pigs and rats and its use in basic otologic research. Brazilian J Otorhinol 75:43–49

9. Housley GD, Ashmore JF (1992) Ionic currents of outer hair cells isolated from the guinea-pig cochlea. J Physiol 448:73–98

10. Frank G, Hemmert W, Gummer AW (1999) Limiting dynamics of high-frequency electromechanical transduction of outer hair cells. PNAS 96:4420–4425

11. Santos-Sacchi J, Iwasa KH, Tan W (2019) Outer hair cell electromotility is low-pass filtered relative to the molecular conformational changes that produce nonlinear capacitance. J Gen Physiol 151(12):1369–1385. https://doi.org/10.1085/jgp.201812280

12. Jean P, Anttonen T, Michanski S, de Diego AMG, Steyer AM, Neef A, Oestreicher D, Kroll J, Nardis C, Pangršič T et al (2020) Macromolecular and electrical coupling between inner hair cells in the rodent cochlea. Nat Commun 11:3208

Chapter 12

Endocochlear Potential Measures, Local Drug Application, and Perilymph Sampling in the Mouse Inner Ear

Kevin K. Ohlemiller, Jared J. Hartsock, and Alec N. Salt

Abstract

Functionally and anatomically, the laboratory mouse inner ear is comparable to other mammals, except that it is considerably smaller and operates over a higher acoustic frequency range. Other than miniaturization of methods that might be applied to the inner ears of guinea pigs, gerbils, or chinchillas, the major difference lies in fewer points of access, due both to small number of cochlear turns ($2\frac{1}{4}$) and reduced access to cochlear scalae in any single turn. These features do not particularly complicate auditory brainstem, compound action potential, or distortion product emission recording. Due to the close proximity to generators, these responses can be quite large in mice. Instead, they present challenges for endocochlear potential (EP) measurement and for manipulations and measurements of inner ear fluids. Without appropriate modifications, considerable technical challenges and potential pitfalls can render such measurements uninterpretable in mice. This chapter outlines experimental techniques for targeting inner ear fluid manipulations and fluid measures in mice. We specifically consider methods for EP measurement, perilymph sampling, and introduction of chemical agents into the middle or inner ear.

Key words Endocochlear potential, Auditory bulla, Intratympanic injection, Intralabyrinthine injection, Round window membrane, Posterior semicircular canal

Abbreviations

CA	Cyanoacrylate adhesive
EP	Endocochlear potential
IHCs	Inner hair cells
IT	Intratympanic
OHCs	Outer hair cells
PSCC	Posterior semicircular canal
RWM	Round window membrane
TM	Tympanic membrane

Andrew K. Groves (ed.), *Developmental, Physiological, and Functional Neurobiology of the Inner Ear*, Neuromethods, vol. 176, https://doi.org/10.1007/978-1-0716-2022-9_12,

1   Introduction

Laboratory mice (*Mus musculus*) have become the dominant experimental model for hearing science [1, 2]. This reflects their economy, short life span, and especially genetic standardization that facilitates the use of molecular tools to study both the physiology and pathophysiology of hearing. The vast majority of known deafness genes have been discovered and characterized using parallel studies of humans and mice. Instances where mice do not reproduce the phenotype of particular human mutations have typically reflected the use of different protein isoforms or expression by different subsets of cells in humans and mice (e.g., [3, 4]). As the promise of therapeutics against hearing loss has grown (e.g., [5–8]), local delivery of these to the mouse inner ear has accelerated. This requires mouse protocols specifying where to access inner ear fluids for injection of therapeutic agents, and where and how to sample inner ear fluids to verify the effective dose.

Manipulations and measurements of the mouse inner ear are made challenging by the very small dimensions and volumes involved for the mouse inner ear. Likewise, attempts to extrapolate findings from the mouse ear to their likely outcome in the human ear can be distorted by the large differences in volume involved. Table 1 compares some fluid volumes and lengths for mice, guinea pigs and humans. The values are derived from 3D reconstructions of thin sheet laser imaging datasets of fixed, intact ears. The total perilymph volume of the mouse is close to 1.0 μL, which is approximately 200-fold less than the perilymph volume of the human ear.

Table 1
Inner ear dimensions

	Human	Guinea pig	Mouse
Volume (μL, mm³)			
Inner ear perilymph	196.9	20.0	1.0
Cochlea endolymph	7.7	2.0	0.3
Scala tympani	40.5	6.0	0.3
Scala vestibuli + vestibule	84.9	10.8	0.5
Length (mm)			
Scala tympani	28.5	17.1	4.3

2   Endocochlear Potential Recording

The endocochlear potential (EP) is a positive DC potential recorded from the endolymphatic space of the cochlea. In mice, the normal EP is typically >100 mV, although this is impacted by basal–apical gradients, and can vary by inbred strain. Ion measurements made with glass microelectrodes (EP and ion-specific measures using ion-selective electrodes) are no more difficult in the mouse ear than in other mammals used experimentally, although the number of access points is reduced. As described here, they require a small fenestration of the bony otic capsule 20–30 μm in diameter through which a microelectrode is inserted. Given suitable resources (operating scope, micromanipulators) the small volume of the endolymphatic compartment is not limiting for the measurement.

2.1   Rationale

Sensitive hearing requires a normal middle ear and a full complement of cochlear outer hair cells (OHCs), along with their attending supporting cells in the organ of Corti. The electromechanical responses of OHCs in turn help provide the necessary mechanical input to inner hair cells (IHCs), the "coding" sensory cells, to paint a detailed picture of the incoming sound spectrum for processing by the auditory nerve [9, 10]. Both OHCs and IHCs require appropriate electrochemical gradients to drive sound-driven receptor currents. The electrical portion of the gradient includes the EP, which depends on a normally functioning stria vascularis and spiral ligament (collectively, the cochlear lateral wall), plus a continuous lining of ion-tight boundaries of endolymph [11]. In supporting EP generation, cells of the lateral wall must regulate a number of ions, including Na^+, K^+, Cl^-, H^+, and Ca^{2+} [12]. While K^+ represents the dominant ion in this process, the EP appears more sensitive to fluctuations in pH and Ca^{2+} [12, 13]. In any event, because of its reliance on multiple ions, a normal EP indicates that a host of metabolic processes are normal, and provides the single most informative metric regarding the microenvironment of hair cells.

The lining of the endolymphatic space is fragile and unforgiving of intrusions (e.g., [14]). Numerous technical difficulties, such as drift in DC potentials, preclude chronic EP recording even though long-term monitoring—and even modulation—of the EP have been the goal of some studies [15, 16]. For this reason, we present only an acute (terminal) EP recording procedure using a ventral approach.

2.2   Materials

The main equipment required includes the following.

2.2.1   Head Holder

For most procedures accessing the inner ear, it is essential to be able to hold the animal's head rigidly. A head holder suitable for making EP recordings is illustrated in Fig. 1. This holder provides

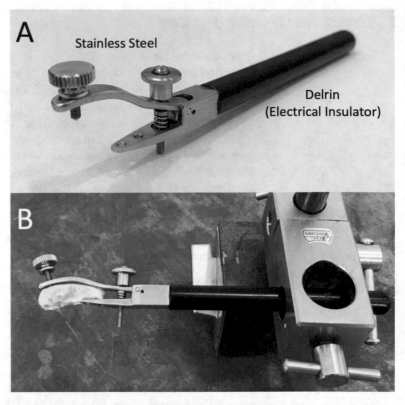

Fig. 1 (**a**) A custom-made mouse head holder. The lower rigid block has two holes through which the upper incisor teeth are inserted. The two locations are to accommodate different head sizes. The mobile upper bar is tightened to hold the head firmly in place with unimpeded access to both ears. The black shaft of the holder is made from Delrin, which is an electrical insulator. (**b**) The head holder, shown with a mouse skull in place, can be mounted to the table with a magnetic micromanipulator base

two-point support (upper jaw and vertex of the skull) offering stability that is more than sufficient to make EP recordings. Other variants are possible, such as the inclusion of ear bars to limit lateral movements, although these can also restrict access to the inner ear.

2.2.2   Electrometer

An intracellular amplifier with high input impedance suitable for use with glass microelectrodes is required, with the output routed to a digital display, voltmeter, or data logging system. The amplifier must be capable of making DC potential measurements (no AC-coupling, such as the "sound card" input on a laptop). The ability to offset the DC voltage to zero at the amplifier or output device is useful, but not essential.

2.2.3   Glass Microelectrodes

The active measurement electrode is an electrolyte-filled glass pipette, pulled to a fine tip in an electrode puller. The tip size formed by the puller is not critical, but the overall shape of the tip is important. If it is too blunt it will not insert easily, and insertion may be blocked by the bony fenestra. If it is too long and fine, it will bend upon contact with the tissues and will not insert easily. A tip length of about 8 mm works well. The tip of the pipette can either be broken (with fine forceps) to a tip diameter of 2–5 μm or beveled to that diameter on a beveling wheel (Narishige Model EG-40). The advantage of a beveled tip is that it is sharp (like a hypodermic needle) so it penetrates the tissues of the lateral wall more easily. It also allows visual verification of whether the electrode was inserted successfully without breaking. Filling the electrode with electrolyte is simplified when the electrode glass contains a filling fiber (filament glass) that runs down the inner wall of the tubing. This helps remove bubbles from the tip during the filling process. The pipette is typically filled with 150–500 mM KCl solution, to which electrical contact is made using an Ag/AgCl wire (detailed below). Alternatively, a variety of coupling devices are available that incorporate an Ag/AgCl contact (e.g., WPI model MEH3SF15).

The recording electrode-to-amplifier and reference electrode-to-amplifier junctions are critical for minimizing DC offsets. There are a variety of ways of providing stable fluid/metal junctions (such as using the calomel cells made for pH meters). For EP measurements, we have found that Ag/AgCl wires contacting solutions of fixed Cl^- concentration provide electrically stable junctions. Either commercially available Ag/AgCl pellets or Ag wire, cleaned and after current is passed to give an AgCl coating. This is readily performed by connecting the Ag wire (as the anode) to the positive pole of a 3 V battery, and some other wire as the cathode (connected to the negative side of the battery). With both wires immersed in an electrolyte solution containing Cl^-, the silver contact turns gray as AgCl is deposited on the surface. Voltage from an Ag/AgCl wire is stable when the Cl^- concentration of the solution it contacts is constant. If the wire is placed in direct contact with the animal, or near the back of a glass electrode that is "drying up" with time, the Cl^- concentration will change and the voltage generated by the wire will drift with time. In contrast, when the AgCl-coated wire is in contact with solution of stable Cl^- concentration, the voltage remains stable over time. For the active EP electrode, the wire should be inserted down to the shoulder of the glass pipette (where Cl^- concentration will be most stable) or an air-tight coupler with AgCl pellet utilized. For the ground electrode, a AgCl pellet connected to a saline bridge (a length of plastic tubing containing electrolyte) is used to maintain constant Cl^- at the pellet. It should be kept in mind that AgCl pellets are very fragile. They can crack at the stub on which they are mounted, or allow fluid to enter the casing and contact other metal components, causing instability. Having a supply of new ones available is recommended.

2.2.4   Audio Monitor

An audible baseline monitor (such as the model ABM previously available from World Precision Instruments, Sarasota) converts DC voltage to an audio signal. As the input voltage increases, an audible tone increases in frequency. This provides the operator with valuable feedback about voltage changes while inserting the electrode into endolymph, when attention is focused on the view through the operating microscope.

2.2.5   Micromanipulator/ Microdrive

A stereotaxic manipulator, either with magnetic base or bolted in position, is necessary to hold and position the electrode in the animal. A motorized electric or hydraulic microdrive is helpful but not required, as long as the drive allows fine resolution of movement (<5 μm per step) and does not bend or allow other uncontrolled movement when touched. Ideally, there should be some type of readout of microelectrode depth for verification of when the electrode has been returned to its original position.

2.2.6   Surgical Instruments

Fine surgical scissors and forceps are needed. We have found ocular retractors and small weights hung from bent wire to be useful for gaining access and clearing the view. The pick used to fenestrate the mouse cochlear capsule is a Bausch + Lomb 30 degree stapes pick (Storz N1705 80) which must be sharp and in good condition. The pick should not be used for other purposes. It is advisable to first thin the bone using the sharp edge of an "old" 30 degree pick, turned on its side.

2.2.7   Other Equipment

An adjustable "jack-stand" is required to hold the torso of the animal, ideally with a Plexiglas (or otherwise electrically insulated) bed. Since depressed body temperature will lower the EP, a heating pad is needed to maintain core temperature. A passive or DC current-based heater is preferred over one that relies on circulating water, as the variation in water flow may mechanically destabilize the preparation. While in some cases a ventilator system may aid good oxygen status, we have found a simple tracheostomy to suffice for this purpose. Care should be taken to prevent any fluids from entering the tracheostomy opening.

2.3   Detailed Methods

2.3.1   Initial Surgical Approach

Because EP recording is an invasive terminal procedure, the animal should be anesthetized fairly deeply. A number of anesthetics are adequate for this purpose. Light anesthesia risks jerking movements that destabilize the recording. Position the mouse ventral-side up with the upper incisor teeth inserted into the head holder and the tongue moved aside as necessary to ensure a clear airway. A midline incision is then made, as indicated in Fig. 2a, after which the underlying fibromuscular tissue is blunt dissected away from the midline to expose the trachea, followed by making a small opening (Fig. 2b). After the connective tissue over the trachea is retracted laterally, the auditory bulla will come into view, surrounded by

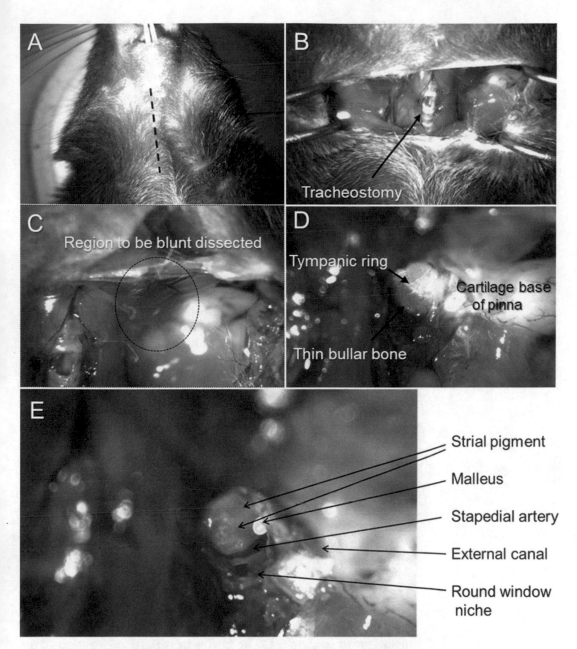

Fig. 2 Surgical approach to mouse inner ear for EP recording. (**a**) The location of the initial incision at the midline. (**b**) Ocular retractor reveals trachea, from which overlying muscles are blunt dissected and a single hole is made to facilitate breathing. (**c**) Location of musculature overlying the left auditory bulla, which is removed by a combination of blunt dissection, cutting at the posterior attachment, and retracting the anterior flap. (**d**) Exposed bulla, which is nearly transparent in a normal middle ear. Tympanic ring can be seen under the lateral half. (**e**) The bulla has been opened and a portion of the tympanic ring has been removed to expose the cochlear capsule, stapedial artery, and round window niche. In this highly pigmented strain (C57BL/6), the pigmented stripe of the underlying stria of both basal and apical turns can be seen

overlying muscle (Fig. 2c). This muscle can be addressed by a combination of blunt dissection, cutting at the posterior point of attachment on the bulla, and using a retractor or small clip to pull the muscle flap anteriorly. Exposure such as shown in Fig. 2d can be achieved with little bleeding as long as care is taken not to probe too deeply either medially (carotid artery) or posterior to the bulla (jugular vein). The bulla is then opened by drilling a small hole in the most translucent portion with a handheld needle, then removing small pieces with malleus nippers to fully expose the basal turn of the cochlea and stapedial artery (Fig. 2e). The thin bone is easy to remove. If the bulla is not translucent, or the bone or underlying mucosa appear vascularized, there may be middle ear pathology. Do not use a dental burr to open the bulla or remove large pieces at once, due to the damaging noise that could be transmitted to the cochlea. It is not necessary to remove the bony tympanic ring, although doing so improves the view (assuming no sound to be presented). Avoid touching the stapedial artery, as bleeding from the artery can be difficult to stop. Notably, this artery provides little of the cochlear blood supply, so if bleeding can be controlled using patient gentle pressure, the procedure may continue.

After the stapedial artery is exposed, cochlear capsule "targets" for fenestration will appear directly anterior (Fig. 2e). In pigmented mouse strains such as C57BL/6 or CBA/CaJ, the stria of both basal and second turns will often stand out as two medial–lateral-running dark stripes against the white of the capsule bone. For albino mouse strains (e.g., BALB/c) subtle contouring of the capsule and bony septa lines can be helpful. For initial training, it is highly recommended that a pigmented strain be used to learn the anatomic features. The majority of the exposed cochlear capsule borders scala media (providing limited access to other scalae), so that most approach errors are errors of electrode angle. In general, it is best to center the point selected for fenestration along the medial-lateral dimension of the capsule.

2.3.2 Fenestration of the Cochlear Capsule

The method for measuring the EP we adapt here was established by Teruzo Konishi [17]. Konishi may not have invented the method, but he perfected it and taught it to a considerable number of investigators in the field. The goal is to make a fenestration of the bone with minimal damage to the underlying tissues of the spiral ligament. The main elements of the technique are illustrated in Fig. 3, whereby the selected point for fenestration is thinned by shaving the surface. If too little bone shaved, unnecessary effort is needed to make the fenestra through the remaining thick bone. If, however, too much bone is shaved, the remaining bone cracks easily when contacted, and an overly large fenestration is created.

While shaving the surface, the anatomy should repeatedly be evaluated to adjust the site where the fenestration will be located. As the bone is thinned, place the pick gently on the bone at this site

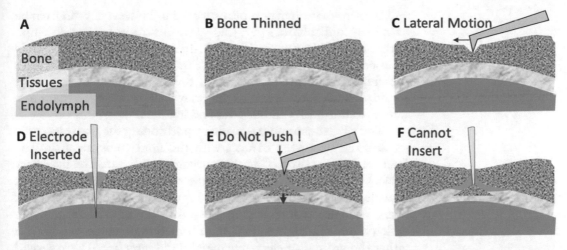

Fig. 3 Making a fenestration in the bone for endocochlear potential measurement. From the initial state (**a**), bone is first thinned by shaving with the sharpened side of an old pick (**b**). A sharp pick is then gently placed at site for fenestration and moved laterally away from the site to scrape bone away (**c**). The pick is returned gently to the same site and then moved in a different direction. Performed repeatedly in random radial directions, this creates a slowly enlarging conical depression in the bone. When fluid is seen, the EP measurement can be taken (**d**). It is important never to push on the pick. This can cause a cone-shaped fragment of bone to be pushed into the tissue/fluid space (**e**), damaging the underlying tissues and blocking entry of the glass electrode (**f**)

and move it *radially* away from the site. Next, gently reposition the pick at the same site and again move it away radially, this time in a different direction. Continue placing the pick on the same site and moving it outward radially from the site, moving bone (invisibly) with the tip of the pick. This causes the depression to become deeper with time without applying any pressure to the bone. The thinner the bone becomes, the more careful placement must be. Ultimately a point is reached when fluid emerges as the pick is placed. At this point, white edges of the fenestration or just a white dot of bone dust, about 20–30 μm diameter may be visible. During the fenestration process it is important not to force the pick, but to slowly let the tool do the work. It should come as a "surprise" when the appearance of fluid confirms the fenestration is complete. If the bone is thin and the pick is sharp, it may take just three to five repetitions for the fenestration to succeed. If many repeats of the pick have been performed without a fenestra appearing, it may be necessary to shave the bone further and start the process again. In older animals the bone is thicker and typically requires more shaving before starting the picking process.

Figure 3 also shows how *not* to make the fenestration. If the pick is pushed prematurely into the thinned bone, it can cause a cone-shaped bone fragment to be pushed into the cochlea (Fig. 3e). The operator will see fluid and through the microscope, and the fenestration may appear successful. However, it can then be extremely difficult to insert the glass electrode into endolymph without breaking the tip (Fig. 3f). It may sometimes be possible to insert the electrode at the edge of the fenestration, to the side of the fragment. Typically, however, the bone fragment will have already detached tissues from the bony wall or damaged them to a degree that the EP will be lower and less stable.

The EP measurement is made by positioning the glass electrode so that the tip makes contact with the fluid emerging from the fenestration. If a drop of fluid has accumulated, it may be necessary to wick it away before positioning the electrode. After fluid contact is made, zero the electrometer to designate this contact potential as "zero voltage" with respect to ground. If the initial voltage varies widely from animal to animal, then something may be wrong with either the ground or electrode metal/fluid interface. There should be minimal potential drift over time. If potential remains stable (<1 mV change over 10 min) then advance the electrode. Be aware that Ag/AgCl wire interfaces require time to stabilize. An electrode that is used immediately after fabrication may show a drift in voltage. If the electrode tip bends as the electrode advances it may be necessary to readjust the manipulator. Commonly, as the tip enters the fenestra, there is a "quiet period" when you know it has passed through the bony aperture but no voltage change is apparent. When the electrode encounters a voltage increase by digital readout or audio monitor, the electrode should briefly be moved faster to "pop it" through the tissues. As soon as EP reaches a high, stable value, stop advancing.

If there appears no voltage change as the electrode is advanced, the electrode tip may have missed the endolymphatic space, instead entering scala tympani or scala vestibuli. If scala media has not been breached, it may still be possible to adjust the angle of entry. If the recording is ultimately unsuccessful, it may be worthwhile for future attempts to dissect the preparation to determine how the target was missed.

After a successful EP measurement is made, withdraw the electrode to the "open circuit" reading and reposition it just in contact with the fluid at the fenestration site. Compare the voltage reading with the initial zero voltage reading before the measurement. If it is within a few milliVolts, this gives confidence that the measurement was valid. If the reading differs markedly from zero then either (1) The electrode drifted electrically, (2) The electrode tip was broken during insertion, or (3) The electrode was blunt, and inserted deeply so that a large perforation of the endolymphatic boundary was created, locally polarizing perilymph. Any of these conditions lowers the degree of trust in the validity of the EP measurement.

Repeated or Prolonged EP Measurements

Some protocols may call for repeat EP measurements in the same preparation. In general, even in the best of circumstances the EP will read a few mV lower than the initial reading. If the electrode is blunt or was broken or too-deeply inserted, repeat EP measures can be substantially lower and less meaningful. Repeated EP measurements are generally unreliable.

Monitoring EP changes over time during a procedure is particularly difficult in the mouse. If the head holder lacks ear bars to laterally stabilize the head, there will be a tendency for head position to drift with time, slowly displacing the recording electrode. Variability in the depth of anesthesia, or attempting to maintain anesthesia with an injectable anesthetic, can also mechanically disturb the recording electrode position. Beware also of thermostatically controlled heating pads that rely on circulating water, as the animals' head and body may shift between cycles. Finally, the overall surgical arrangement should not require the experimenter to support or stabilize their hands on the head holder or platform. While continuous EP recording (that is, over the course of a procedure) is possible in mice, the cost and quality of the equipment needed versus brief recordings may be prohibitive.

Choices and Significance of EP Recording Location

The mouse cochlea offers three primary recording sites for the EP: the lower basal turn as described above (~45 kHz frequency place), through the round window membrane (~60 kHz frequency place), and the lower apical turn (~6 kHz frequency place) [18]. Due in part to the short length of the mouse cochlea (~5.5 mm) [19] compared to the estimated 2.0 mm length constant of the EP in rodents [20], there is little frequency specificity to any mouse EP recording. For most research questions, recording from the lower base will suffice to assess strial function in any animal, as most mouse studies have done.

A few papers have described EP recording through the mouse round window membrane (RWM) on the anterior wall of the round window niche (Fig. 2e), which can be accessed using the surgical approach described above. In this approach, the electrode is inserted directly through the RWM, then through a very thin expanse of scala tympani, then through the organ of Corti adjacent to the osseous spiral lamina. Access to the round window requires more aggressive posterior exposure of the cochlea than for other recordings, and may be prohibitively difficult in some inbred strains. Because a rush of perilymph as the RWM is perforated will quickly alter the optics of this approach, and also because of the anatomic disruption the electrode typically causes, the round window approach is especially unsuitable for repeated entry or monitoring the EP over time. EP values reported in mice using this method vary from similar to those obtained from the lateral wall of the lower basal turn [21] to 10–15 mV lower in the same mouse strains [22, 23], and it is difficult to say whether this reflects a real spatial gradient or is an artifact of the approach. The round window approach has value for research goals other than EP recording, such as recording saccular and utricular endolymphatic potentials [22, 23]. However, it offers little advantage over the lower basal turn approach described here for EP assessment.

A slight adjustment of location (Fig. 2e) converts the lower basal turn EP recording into a lower apical turn EP recording. The room for error in electrode angle and depth is reduced in the apex, and the extreme thinness of the capsule at this location renders it more susceptible to damage. Typically, thinning of the bone is not needed, and even a light touch with a pick will penetrate the cochlear capsule. Normally the apical turn EP is about 10 mV lower than in the lower base (Fig. 4a). Work in guinea pigs [13] has correlated the EP gradient to endolymphatic Ca^{2+} levels. The spatial EP gradient may be smaller in albino animals. Pathological conditions such as aging and broadband noise exposure can both eliminate or even reverse this gradient (Fig. 4b, c). This may reflect preferential effects of most noise exposure protocols and aging on the stria of the basal turn.

2.4 EP Changes in Pathologic Conditions

The EP can be somewhat selectively and temporarily reduced by loop diuretics such as furosemide [26], and may also be affected by genetic mutations (e.g., [27–29]) and aging [15, 30]. Permanent EP reduction by noise exposure appears uncommon in mice, but may occur as a function of inbred strain, exposure level, and age at exposure [31, 32]. Nevertheless the universal assertion that a low EP must indicate pathology, and that a higher EP is always better, is false. Acute, mechanically induced organ of Corti displacements in guinea pigs can increase EP by over 10 mV [33], demonstrating there may be conditions when a high EP is "abnormal". In addition, since hair cells act as a sink to the currents sourced by the stria, ablation of hair cells or their stereociliary bundles by ototoxins or noise exposure could yield abnormally high EP values. A comparison of noise-exposed recombinant inbred mouse strains formed from C57BL/6 and BALB/c [34] identified strains with acutely increased EP and those with an acutely decreased EP. Thus the direction of acute EP change for a given exposure condition appears explicitly genetic, potentially reflecting genes and alleles that impact current-sinking by hair cells.

2.5 Troubleshooting EP Recording

A common problem disrupting EP recordings is one of "drift" or instability of the measured electrical potential. Avoidance of drift requires stable electrical interfaces between the metal wires of the amplifier and the aqueous environment of the animal. Correctly grounding the animal is essential for stable EP measurements. The "grounding electrode" (Ag/AgCl pellet and saline bridge or calomel cell) should be the only ground on the animal (Fig. 5a). If the

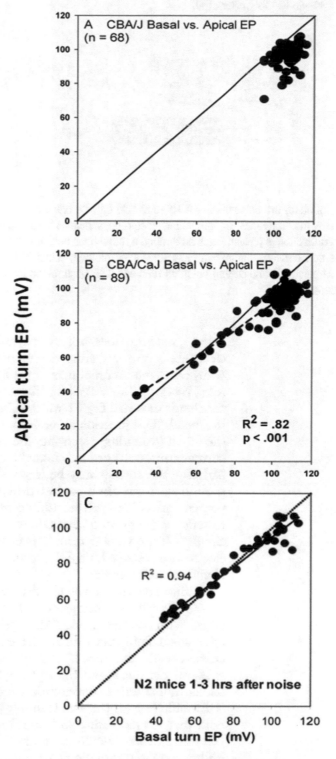

Fig. 4 EP measurements from both basal and apical cochlear turns under three conditions.(**a**) Healthy young CBA/J mice show that the EP in the basal turn averages ~10 mV higher than in the apex. (**b**) Aged CBA/CaJ mice, in which the EP declines with age, show reversal of the normal gradient. (**c**) Similar result for EP measured acutely after noise exposure in N2 backcross mice formed from C57BL/6 and CBA/J. (Adapted from [24, 25])

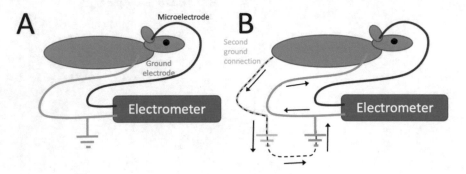

Fig. 5 Grounding the preparation. (**a**) There should only be one electrical pathway to ground on the animal, provided by the "reference" or "grounding" electrode, shown as the green line. (**b**) In the case where a second ground connection is present, such as through a metal head holder, IV line, or rectal probe (yellow line), large currents can circulate through the low resistance pathway (black dotted lines), causing instability of the reference electrode and drifting potential readings. Left uncorrected the reference electrode can be permanently damaged

voltage reading does not go "open circuit" when the ground electrode is removed from the animal, then something is wrong. If any other metal component of the head holder or animal support (e.g., rectal probe or IV line) effectively grounds the animal, then the circuit shown in Fig. 5b will develop. Moreover, if the voltage of the metal/fluid junction at the second ground differs from that of the initial grounding electrode, the voltage difference will drive a current around the ground circuit "loop". As this is a low-resistance pathway, the current may be large. Current passing through the ground electrode can cause it to drift, and with time, can damage it permanently. When voltage drift occurs, the first step in diagnosis is to remove the ground electrode to verify the reading goes "open circuit". If the voltage reading is little changed when the ground electrode is removed, then the second ground connection must be identified and removed.

Other sources of voltage drift can include old or damaged Ag/AgCl cells or inappropriately placed Ag/AgCl pellets or wires. With commercial Ag/AgCl pellet electrodes, instability can arise when fluid seeps down the edge of the pellet and makes contact with the metal wire underneath. Circulating current between the metal and the AgCl will cause the voltage to drift. Similarly, if chloride concentration at the location of the AgCl wire is unstable (e.g., if the wire is simply placed on a cut muscle, or in fluid that is evaporating with time) then voltage will drift. Silver wires and pellets should always be located in some type of fluid bridge containing solution with a stable chloride concentration.

3   Local Drug Application Methods

3.1   Rationale

Local drug application to the inner ear is typically targeted to perilymph since the endolymphatic space is miniscule and does not respond stably to breach. The preferred method will depend on the balance between the permissible degree of invasiveness and the required degree of control of drug levels. Intratympanic applications (IT, through the tympanic membrane) are relatively noninvasive and there is little risk of compromise to the inner ear, as long as the drug itself is not toxic. Because drug levels in perilymph will be governed by the balance between entry from the middle ear and the rate of loss to the vasculature and other compartments, control of perilymph drug levels cannot be precise. Most drugs injected into the middle ear are lost to the vasculature through the middle ear mucosa and potentially the lymphatic system. Mice possess an extremely active mucosa; In our hands, IT solutions that completely fill the middle ear may be entirely absent within 3 h, and ABR thresholds can recover within the same day. Loss of solution to the pharynx through the Eustachian tube may also contribute to the loss of injected solutions. In mice, moreover, this can lead to quick aspiration and death, and the inclusion of a "vent" hole is emphasized in our protocol. Middle ear retention may be improved by delivering agents in a gel, including Poloxamer formulations that solidify at normal body temperature [35]. These may at least temporarily resist flowing back through the injection hole in the tympanic membrane (TM) and may be less inclined to pass through the Eustachian tube. In our hands, however, 17–20% Poloxamer formulations remain in a gel state for less than 30 min in the mouse middle ear. It may be just as effective to simply keep animals on their sides (injected ear up) as they recover from anesthesia.

An effective IT injection fills the middle ear, being limited only by the 6.0–8.0 μL volume of the mouse bulla [36, 37]. As stated, most of this is lost, yet mice may also offer more effective routes into perilymph than do larger animal models. In the mouse, entry routes into perilymph include not just the RWM, but also the stapes, annular ligament, and even the thin, porous bone of the cochlear apex. In fact, in mice the RWM does not appear to be the major route of drug entry into the cochlea [38, 39]. As one consequence, basal–apical perilymph drug gradients can be much smaller in mice than in other models. Although the importance of the RWM as a drug entry route will depend on the size and polarity of the drug, as well as the existence of transporters that happen to transport the drug [38, 39], these other routes in mice are more "leaky" and less selective in a way that often works in the investigator's favor. For these reasons, it may be appropriate to reconsider protocols that call for invasive recovery surgeries to place drugs directly onto the RWM. These come with complications and regulatory headaches, and may hold no advantage since they do not take advantage of the multiple possible routes into the mouse inner ear.

3.1.1 *Limitations of Intratympanic Drug Application*

After intratympanic application, perilymph drug levels are highly dependent on middle ear retention time, which appears to be shorter in mice than in larger animals. For most drugs, this results in a rapid decline in perilymph concentration. In some cases, it may therefore be preferable to apply drug directly into perilymph. This is more invasive but allows more precise control of the perilymph concentration to be achieved. Any perforation of the otic capsule, however, results in perilymph release, driven by cerebrospinal fluid entering ST through the cochlear aqueduct. If the perilymph leakage cannot be adequately controlled in all animals, then reliable estimates of drug delivery may not be possible. Finally, just because a drug is applied to the perilymph of one ear does not guarantee that its effects will be local to that ear. Due to the small size of the mouse skull, mice may be particularly prone to the "Schreiner effect", whereby the contents of perilymph on one side of the head may be able to reach CSF via the cochlear aqueduct [40, 41]. From there, drugs may reach remote parts of the brain and the opposite cochlea. Although this phenomenon has yet to be explored systematically in mice, the two ears of any mouse subjected to intratympanic or intralabyrinthine injection should not be assumed to be independent.

3.2 Materials

3.2.1 *Intratympanic Injections*

In microsurgical terms, the dimensions of the mouse auditory bulla are forgiving, and one might be tempted to simply hold the mouse and inject through the TM by the unaided eye. The ossicles are easy to dis-articulate, however, and the experiment is compromised if the stapedial artery, which cannot readily be seen through the TM, is breached. We have found that consistent injections can be performed by placing the animals ventral-side-up in a head holder on a stable platform of the type used for EP recording. We inject by mating a 1 cc disposable Luer-lock syringe filled with the experimental compound to a #16 gauge syringe needle, which in turn is inserted into a short length of 1.5 mm ID polyethylene tubing. This size of tubing matches the OD of WPI 1B150F-4 glass filament microcapillary tubing, which is pulled on a forge and then broken to ~50 μm using forceps under a magnifier. When the microcapillary part of the assembly is affixed to a holder and placed on a manual micromanipulator/magnetic stand, it can be maneuvered through the TM under an operating scope. One advantage of this arrangement is that the entire assembly can be moved into place, then operated with one hand, freeing the other hand to reposition the pinna as needed. If the injected media is a gel that may solidify, the microcapillary tip can be placed in or near an ice bath between animals. We do not recommend using a syringe needle of any type for direct IT injections in mice. If the animal and microcapillary tube are mechanically stable, there is little danger of breaking off any part of the tip during the injections. Because only the tip of the microcapillary tube enters the middle ear space, the only danger of infection arises by contamination of the injection media, which should be sterilized by filtration. The variety of mouse strains we have worked with appear extremely resistant to middle ear infections, which are nearly nonexistent in our IT-injected animals. For easy visualization of the filling process, we add trace Evans blue to the media.

3.2.2 Intralabyrinthine Injections

Direct injections into perilymph require a syringe pump capable of injection rates below 1 μL/min, with the syringe connected to a glass or polyimide delivery pipette. As injection rates and volumes are typically very small, it is generally not suitable to use any form of flexible tubing between the syringe and the delivery pipette. Compliance of the tubing results in slow on and off time courses and makes flow rates subject to mechanical disturbance by any movement of the tubing. Instead, a manipulator-mounted syringe pump (e.g., WPI model UMP3 UltraMicroPump) with a plexiglass coupler (WPI MPH6S Microelectrode Holder) allows the pump to be directly connected to the injection pipette. The tip size necessary for injection varies in different applications but can range from 10 μm to 50 μm in diameter. The choice of syringe also varies in different applications, but should be kept as small as possible, with a volume of around 3× to 10× the total volume to be injected. Suitable gas-tight syringes with volumes of 50 μL and less are available from a number of manufacturers.

Also required is a fenestration pick (detailed earlier) and adhesives and sealants necessary to seal the injection pipette in place in the bony otic capsule, to control fluid leakage or to seal fenestration sites after the injection procedure. There are many brands and types of adhesive suitable.

3.3 Detailed Methods

3.3.1 Intratympanic Injections

Once an animal is lightly anesthetized and the injection assembly is in place, the entire IT procedure typically requires less than 10 min. For such a short procedure, no heating may be needed and the animals quickly recover. Because of the placement of the ossicles and the angle of the cochlea within the skull, it is not possible to inject compounds directly onto or through the RWM using an IT approach. Our strategy instead has been to select two widely separated points in the TM, one as a vent hole and the other for injection (Fig. 6). If the entry point is below the vent, this encourages the ME to fill from bottom to top, with the excess eventually emerging from the vent. Our typical entry point (Fig. 6b, c) offers a safe location for insertion of the microcapillary tube, which is inserted just through the TM. If the media is highly viscous or prone to gelling, it may be necessary to break back the

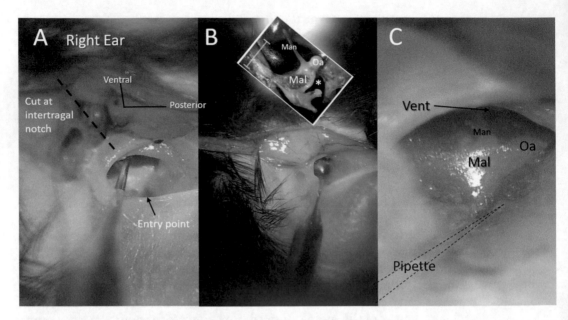

Fig. 6 Intratympanic injection into the right ear of a mouse. (**a**) View is optimized by bloodless cut at intratragal notch (dotted line). In this view, the same pipette that is used to make injection is first used to make a vent hole that is ventral (above) and remote from the injection site. (**b**) Injection pipette with Evans blue-labeled Poloxamer 407 is inserted at location indicated by inset (asterisk). (**c**) After fill, blue coloration appears behind the tympanic membrane. *Mal* malleus, *Oa* orbicular hypophysis, *Man* manubrium. (Inset in (**b**) modified with permission from [42]). Scale bar in inset 1 mm

microcapillary tube to a larger size, which may require a more ventral entry hole. The use of a vent hole reduces, but does not eliminate, the danger of media passing through the Eustachian tube and being aspirated. Since the quality of the ME fill depends on a good view, we recommend a slight, bloodless cut at the intratragal notch (Fig. 6a). Placement of additional small weights on hooks as retractors further improves the view. We do not attempt to measure the actual amount of media that is injected, but visual filling of the ME (coloration plus emergence from the vent hole) will mean the delivery of ~6.0–8.0 μL of media.

3.3.2 Intralabyrinthine Injections

The two most common sites for intralabyrinthine injections in mice are (1) through the round window membrane and (2) via the posterior semicircular canal (PSCC). Injections through the RWM of the mouse are problematic, due to the close proximity of the cochlear aqueduct. Perforating the RWM with an injection pipette results in immediate CSF efflux at the injection site. The resulting washout of drug has been quantified in guinea pigs [43] but not in mice, where the influence is likely to be larger due to smaller perilymph volumes. This is not to say that some agents (such as adenoviruses) can't be delivered by round window injections, especially in early postnatal specimens [44]. Nevertheless, the method is unlikely to be reliable and effective for smaller drugs in adult mice. Some groups have reported the use of RWM injections combined with canal fenestration for gene therapy in mice using a viral suspension [45]. Opening the canal relieves the backflow of CSF and passage of CSF through the perilymphatic spaces may help distribute the adenovirus. However, the high degree of washout would also make the technique unsuitable for use with small drug molecules.

Injections into the PSCC obviate most of the problems associated with RWM injections. Injection at this site, which is distant from the cochlear aqueduct, actually causes flow toward the aqueduct, allowing drug loading through almost the entire perilymphatic space. The surgical approach to the posterior canal is shown in Fig. 7. After a postauricular incision, soft tissues are cleared until the surface of the temporal bone is visible. Following the facial nerve provides a guide to the lateral portion of the temporal bone. The lateral and posterior semicircular canals are seen as small tube-shaped "bumps" joining at right angles at the edge of the temporal bone.

The procedure for sealing a drug injection pipette into the bony canal is illustrated in Fig. 8. We have found it to be extremely difficult to prevent perilymph leaks when adhesives are applied after the injection pipette is inserted, after the bone is fenestrated and wet. At this point, fluid channels between the bone and the

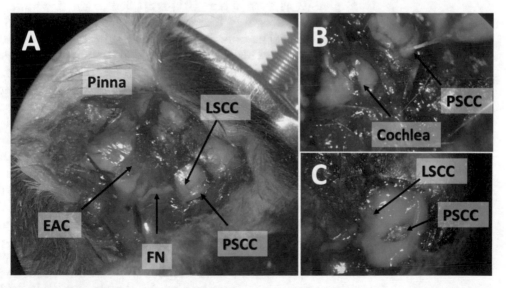

Fig. 7 Injection into the posterior semicircular canal of the mouse. (**a**) Structures revealed after the postauricular incision. The LSCC and PSCC are seen as a pair of "bumps" orientated at right angles on the surface of the temporal bone. (**b**) Specimen in which fluorescent solution was injected from a pipette sealed into the PSCC. The cochlea is seen to be filled with fluorescent solution. (**c**) The posterior canal prepared for perilymph sampling. The green color is a silicone adhesive that surrounds the site on the PSCC that will be fenestrated. Abbreviations: *EAC* external auditory canal, *FN* facial nerve, *PSCC* posterior semicircular canal, *LSCC* lateral semicircular canal

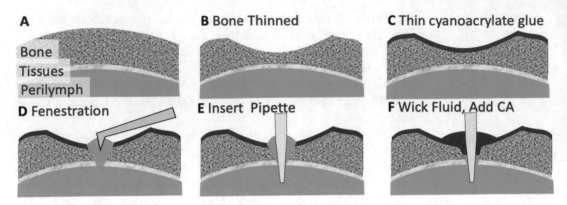

Fig. 8 Process for sealing a pipette into a perilymphatic fenestration anywhere on the otic capsule. From the initial state (**a**) the bone is first thinned with a flap knife or other sharp micro blade (**b**). Bone is then carefully dried, and a thin layer of fast-setting cyanoacrylate (CA) glue is applied (**c**). CA thickness is greatly exaggerated in the figure and should be as thin as possible. A fenestration is then made though the glue and bone (**d**), allowing the pipette to be inserted (**e**). Perilymph will be leaking at this time. A wick is used to remove the perilymph from the surface and fenestration and a drop of CA glue applied as wick is simultaneously removed (**f**). CA entering the fenestration sets immediately

adhesive continue to allow small rates of fluid loss. The loss may occur at very low rates, which makes it difficult to visualize, even with concentrated dye in the solution. Instead, we have found it more effective to make the fenestration after a layer of cyanoacrylate (CA) glue (3 M Permabond 101) is applied to the dry bone. CA glue is more hydrophobic than bone, allowing it dry more quickly and thoroughly. The bone is first thinned, if necessary, by shaving with a suitably sized blade, such as a flap knife or sharpened edge of a pick (Fig. 8b). The bone is dried and a droplet of thin, fast-setting CA is applied to the bone (Fig. 8c). A small polyethylene scraper, such as a 1–2 mm wide piece cut from tubing (PE 240 Becton Dickinson), can aid in spreading the droplet, leaving as thin a layer as possible on the surface of the bone. A fenestration through both the adhesive and bone is made with a sharp pick (Fig. 8c), after which the drug injection pipette is inserted (Fig. 8e). An injection pipette can be made in a similar fashion to the microelectrodes that were described in the previous section. Nonfiber glass is 1 mm outer diameter, pulled by a glass puller to a tip length of 8 mm. The tip diameter is broken to 10–50 μm in size, which helps to ensure a successful injection. Perilymph will continue to escape both before, during and after the pipette insertion, accumulating as a droplet at the fenestration site. An absorbent wick is used to remove the accumulating fluid. When the pipette is ready to be sealed, a new dry wick is touched to the pipette/fenestration site, while instantaneously applying a small droplet of CA. At the moment the CA is applied, the wick is removed. Ideally, the wick will transiently remove fluid from the conical fenestration at the instant the CA is applied and CA in the fenestration will set immediately (Fig. 8f).

3.4 Troubleshooting   If too much fluid is present at the fenestration site when CA is applied, the CA will float over the fluid and allow a fluid bubble to be formed underneath the CA. A small bubble is acceptable and can be covered with more CA to make it more rigid. Larger fluid bubbles, stretching and thinning the CA, need to be removed with a wick and resealed.

The calculated disruptive influence of perilymph leaks on drug distribution and time courses when drug solutions are directly injected into perilymph are illustrated in Fig. 9. Calculated spatial distribution (Left column) and time courses (Right column) are shown for a 0.1 μL/min injection of a solution with concentration 1000 (arbitrary units) into the PSCC of the mouse. If the pipette is completely sealed in place with no fluid leak, this injection rate would displace the 0.8 μL perilymph volume between the injection site and the cochlear aqueduct within about 10 min. The results, as expected, are that the basal part of ST (the region furthest from the injection along the perilymphatic space) becomes filled with drug solution within about 10 min. The result of a successful injection with fluorescein in the injection medium is shown in Fig. 7b. The perilymph of the cochlea is visible through the bone and is seen to contain fluorescein at a concentration similar to that in the injection pipette, seen sealed into the PSCC at the right.

If the pipette is not adequately sealed in place, however, a completely different outcome will result. The lower row in Fig. 9 shows the calculated results for the same injection when there is a small leak at the injection site. In this instance, the leak is set to 0.1 μL/min, but real leakage rates can be much higher; sometimes greater than 1 μL/min if preventive steps are not taken. With a leakage rate of 0.1 μL/min, drug concentrations in SV are 5–20× lower than the injected concentration, and almost no drug reaches scala tympani. Even more dramatic is the drug washout after the 30 min injection, which occurs in just 5–10 min in the presence of the leak. These calculations show that it is not sufficient simply to inject drug solutions into the ear. Accurate evaluation of drug effects requires the utmost diligence to deal with even the smallest leak to prevent drug washout from occurring. Also of importance is that the volumes and rates of leakage required to wash out drugs are extremely small. For this reason, visual observation of the injection site, even with dye in the solution, may be inadequate to confirm the absence of leakage.

4 Perilymph Sampling

4.1 Rationale   Despite the potential advantages of sampling endolymph for some research questions, endolymph volumes in mice are too small to sample or selectively manipulate without artifacts (Table 1). While the volume of perilymph in mice is also small, it can be accessed and

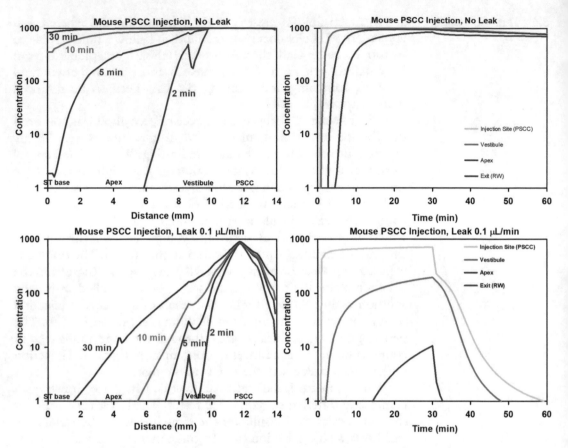

Fig. 9 *Upper row.* Calculated spatial drug distribution (left side) with a 30 min, 0.1 µL/min injection from a pipette sealed into the PSCC of a mouse. Time course at different perilymph locations is also shown (right side). With the pipette sealed, all perilymph can be loaded with drug within about 10 min, and drug is retained well in the ear after injection ceases. *Lower row.* Same calculation when the pipette is incompletely sealed, allowing a sustained fluid leak of just 0.1 µL/min. Perilymph concentrations in the cochlea are substantially lower, especially in scala tympani (0–4 mm on the plot) and drug is rapidly washed away after 30 min injection ceases

manipulated at multiple locations without compromising cochlear function. For sampling purposes, however, it very much matters where perilymph is accessed. Sampling from the RW results in fluid that is primarily CSF, not perilymph. As the cochlear aqueduct enters next to the RW, fluid taken from the ear is readily replaced by CSF entering at the aqueduct. Thus, pure perilymph samples cannot be collected at the RW of the mouse. Instead, sampling at a distance from the aqueduct using the posterior semicircular canal allows relatively pure perilymph to be collected. A fluid sample of 0.8–1.0 µL collected from the PSCC, represents most of the perilymph from one mouse inner ear, including the canals, vestibule, scala vestibuli, and scala tympani.

4.2 Materials

The picks and adhesives required for perilymph sampling are identical to those required for intralabyrinthine injections except that a two-part silicone (Kwik-Cast 2-part silicone adhesive (WPI, Sarasota)) is also required. Perilymph samples are collected into hand-held capillary tubes (VWR catalog # 53432-728 10 μL or Drummond 2-000-005 5 μL). Samples are diluted and stored in 0.5 mL polyethylene vials with screw top caps incorporating a silicone O-ring. (USA Scientific Cat# 1405-9304).

4.3 Detailed Methods

Accessing the posterior semicircular canal for sampling begins with the same surgical approach described above for intralabyrinthine injections. Fluid collection must be performed without contamination or loss, so our approach has been to construct a "cup" from silicone adhesive at the perforation site to isolate the emerging fluid. The appearance of the silicone cup over the PSCC during surgery is shown in Fig. 7c. The procedure to make and secure the silicone cup is illustrated in Fig. 10. Similar to drug injection, the bone is thinned (Fig. 10b) and coated with a thin layer of cyanoacrylate (Fig. 10c). The silicone is then applied over the CA glue (Fig. 10d). This sequence is necessary because silicone does not readily attach to bone without the CA and is easily detached by mechanical movements during sample collection. Applying the adhesives in order ensures the cup is firmly attached to the bone and is resistant to the moist surrounding tissues. When a fenestration is made through the adhesives, the emerging fluid forms a ball on the silicone surface, allowing it to be readily collected by

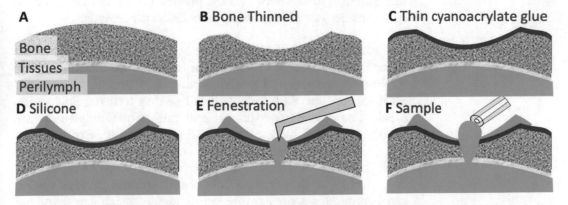

Fig. 10 Perilymph sampling from a fenestration anywhere on the otic capsule. From the initial state (**a**) the bone is first thinned with a flap knife or other sharp micro blade (**b**). The bone is then dried, and a thin layer of cyanoacrylate glue is applied to the bone (**c**), followed by a layer of silicone, kept thin in the middle and built up at the edges to form a cup shape (**d**). A fenestration is then made though both the adhesives and bone (**e**), allowing perilymph to collect in the silicone cup. A calibrated microcapillary tube is used to draw a precise volume of perilymph as it emerges (**f**)

capillarity into handheld microcapillary tubes (VWR catalog # 53432-728 10 μL or Drummond 2-000-001 5 μL micropipettes). The latter are demarcated in 1 μL increments. The exact sample volume is measured by measuring sample length under a dissecting microscope with an eyepiece reticule. Sample volume is calculated by comparing sample length with the capillary length to the calibrated marking. Samples are then expelled into a small volume (25–150 μL) of diluent according to the requirements of the measurement assay. The diluent may be aqueous for fluorescence analysis or organic (methanol or acetonitrile) for HPLC/mass spectrometry. Samples are expelled into the diluent using the tubing and capillary adaptor that is included with the micropipettes. The tubing is intended for expelling fluid samples by mouth, but instead we expel them by connecting the tubing to a 1 mL syringe, in which the plunger can be manipulated to drive the sample into the diluent. Diluent is then drawn into the capillary and expelled a few times to rinse the pipette to the maximum degree possible.

4.3.1   Sample Handling

It can be challenging to handle samples of less than 1 μL in volume. Even in the capillary, they will evaporate with time, preventing quantitative interpretation. We also avoid the technique of "spraying" the sample into the collection tube by applying a high pressure. The viscosity of water means that in fine capillaries a substantial portion of the sample may remain on the walls of the capillary tube, or may be lost as mist. We therefore prefer the technique of carefully expelling each sample into a fluid diluent of substantially larger volume. The rinsing of the pipette interior as the diluent is drawn back into the pipette a few times and expelled helps ensure that all the sample is included for analysis.

4.4   Troubleshooting

The musculature over the PSCC is highly vascularized, especially ventral and posterior to the point of fenestration. Bleeding can be minimized by patient blunt dissection and occasional pressure. Heat cautery should be avoided, as this may heat the fluid in the canal. The ease of fenestration and rate of perilymph flow are greatly impacted by the choice of mouse strain. Some strains (e.g., C57BL/6, BALB/c) feature both very thin bone and relatively large internal PSCC diameter, which promote rapid fluid flow. Other strains (e.g., CBA-related strains) possess thicker canal bone plus a smaller internal diameter, which can delay and slow sample flow. Although the PSCC houses both perilymphatic and endolymphatic compartments, the endolymphatic space collapses when the canal is fenestrated, and there is little risk of sample contamination [46].

5   Conclusions

We describe a small set of inner ear research procedures adapted to the laboratory mouse. These procedures—EP recording, local drug delivery, and perilymph sampling—enhance the value of mouse models for the study of stria vascularis pathology, inner ear fluid regulation, and testing of potential therapeutic agents against hearing loss. In mice, these methods leverage genetic standardization and a growing list of molecular tools, imparting unparalleled versatility for hearing research.

Acknowledgement

Thanks to Ruth Gill for assistance with some of the figures.

References

1. Bowl MR, Dawson SJ (2015) The mouse as a model for age-related hearing loss-a mini-review. Gerontology 61:149–157

2. Ohlemiller KK, Jones SM, Johnson KR (2016) Application of mouse models to research in hearing and balance. J Assoc Res Otolaryngol 17:1–31

3. Hosoya M, Fujioka M, Kobayashi R et al (2016) Overlapping expression of anion exchangers in the cochlea of a non-human primate suggests functional compensation. Neurosci Res 110:1–10

4. Hosoya M, Fujioka M, Ogawa K et al (2016) Distinct expression patterns of causative genes responsible for hereditary progressive hearing loss in non-human primate Cochlea. Sci Rep 6:1–12

5. Crowson MG, Hertzano R, Tucci D (2017) Emerging therapies for sensorineural hearing loss. Otol Neurotol 38:792–803

6. Wang J, Puel JL (2018) Toward cochlear therapies. Physiol Rev 98:2477–2522

7. Bielefeld EC, Kobel MJ (2019) Advances and challenges in pharmaceutical therapies to prevent and repair cochlear injuries from noise. Front Cell Neurosci 13:285

8. Szeto B, Chiang H, Valentini C, Yu M, Kysar JW, Lalwani AK (2020) Inner ear delivery: challenges and opportunities. Laryngoscope Invest Otolaryngol 5:122–131

9. Dallos P, Zheng J, Cheatham MA (2006) Prestin and the cochlear amplifier. J Physiol 576:37–42

10. Carney LH (2012) Peripheral anatomy and physiology. In: Tremblay KL, Burkard RF (eds) Translational perspectives in auditory neuroscience: normal aspects of hearing. Plural, San Diego, pp 91–111

11. Jahnke K (1975) The fine structure of freeze-fractured intercellular junctions in the guinea pig inner ear. Acta Otolaryngol 336:1–40

12. Wangemann P (2006) Supporting sensory transduction: cochlear fluid homeostasis and the endocochlear potential. J Physiol 576(1):11–21

13. Gill S, Salt AN (1997) Quantitative differences in endolymphatic calcium and endocochlear potential between pigmented and albino guinea pigs. Hearing Res 113:191–197

14. Suzuki J, Hashimoto K, Xiao R et al (2017) Cochlear gene therapy with ancestral AAV in adult mice: complete transduction of inner hair cells without cochlear dysfunction. Sci Rep 7:1–12

15. Schmiedt RA (2010) Chapter 2: the physiology of cochlear presbycusis. In: Gordon-Salant S, Frisina RD, Popper AN, Fay RR (eds) Springer handbook of auditory research, The aging auditory system, vol 34. Springer, New York, pp 9–38

16. Mercier PP, Lysaght AC, Bandyopadhyay S et al (2012) Energy extraction from the biologic battery in the inner ear. Nat Biotechnol 30:1240–1243. https://doi.org/10.1038/nbt.2394

17. Tl K (1979) Some observations on negative endocochlear potential during anoxia. Acta Otolaryngol 87:506–516

18. Hirose K, Liberman MC (2003) Lateral wall histopathology and endocochlear potential in the noise-damaged mouse cochlea. J Assoc Res Otolaryngol 4:339–352

19. Saunders JC, Garfinkle TJ (1983) Peripheral anatomy and physiology I. In: Willott JF (ed) The auditory psychobiology of the mouse. Charles C. Thomas, Springfield, IL, pp 131–168

20. Johnstone BM, Johnstone JR, Pugsley ID (1966) Membrane resistance in endolymphatic walls of the first turn of the guinea-pig cochlea. J Acoust Soc Am 40:1398–1404

21. Schrott A, Melichar I, Popelar J et al (1990) Deterioration of hearing function in mice with neural crest defect. Hearing Res 46:1–7

22. Liu H, Zhao L (2015) Recording potentials from scala media, saccule and utricle in mice. J Otol 10:87–91

23. Li Y, Liu H, Zhao X et al (2020) Endolymphatic potential measured from developing and adult mouse inner ear. Front Cell Neurosci 14:431

24. Ohlemiller KK, Dahl AR, Gagnon PM (2010) Divergent aging characteristics in CBA/J and CBA/CaJ mouse cochleae. J Assoc Res Otolaryngol 11:605–623

25. Ohlemiller KK, Rosen AD, Gagnon PM (2010) A major effect QTL on chromosome 18 for noise injury to the mouse cochlear lateral wall. Hearing Res 260:47–53

26. Sewell W (1984) The effects of furosemide on the endocochlear potential and auditory nerve fiber tuning curves in cats. Hearing Res 14:305–314

27. Steel KP, Barkway C (1989) Another role for melanocytes: their importance for normal stria vascularis development in the mammalian inner ear. Development 107:453–463

28. Ingham NJ, Carlisle F, Pearson S et al (2016) S1PR2 variants associated with auditory function in humans and endocochlear potential decline in mouse. Sci Rep 6:1–13

29. Nishio A, Ito T, Cheng H, Fitzgerald TS, Wangemann P, Griffith AJ (2016) Slc26a4 expression prevents fluctuation of hearing in a mouse model of large vestibular aqueduct syndrome. Neuroscience 329:74–82

30. Ohlemiller KK (2009) Mechanisms and genes in human strial presbycusis from animal models. Brain Res 1277:70–83

31. Ohlemiller KK, Rosen AR, Rellinger EA et al (2011) Different cellular and genetic basis of noise-related endocochlear potential reduction in CBA/J and BALB/cJ mice. J Assoc Res Otolaryngol 12:45–58

32. Ohlemiller KK, Kaur T, Warchol ME et al (2018) The endocochlear potential as an indicator of reticular lamina integrity after noise exposure in mice. Hearing Res 361:138–151

33. Salt AN, Brown DJ, Hartsock JJ et al (2009) Displacements of the organ of Corti by gel injections into the cochlear apex. Hearing Res 250:63–75

34. Ohlemiller KK, Kiener AL, Gagnon PM (2016) QTL mapping of endocochlear potential differences between C57BL/6J and BALB/cJ mice. J Assoc Res Otolaryngol 17:173–194

35. Salt AN, Hartsock J, Plontke S et al (2011) Distribution of dexamethasone and preservation of inner ear function following intratympanic delivery of a gel-based formulation. Audiol Neurotol 16:323–335

36. Saunders JC, Crumling MA (2001) The outer and middle ear. In: Willott JF (ed) Handbook of mouse auditory research. CRC Press, New York, pp 99–115

37. Richter CA, Amin S, Linden J et al (2010) Defects in middle ear cavitation cause conductive hearing loss in the Tcof1 mutant mouse. Hum Mol Genet 19:1551–1560

38. Salt AN, Plontke SK (2009) Principles of local drug delivery to the inner ear. Audiol Neurotol 14:350–360

39. Salt AN, Hirose K (2018) Communication pathways to and from the inner ear and their contributions to drug delivery. Hearing Res 362:25–37

40. Schreiner L (1999) Translation: recent experimental and clinical findings retarding an interlabyrinthine connection. Laryngo-Rhino-Otologie 78:387–393

41. Stöver T, Yagi M, Raphael Y (2000) Transduction of the contralateral ear after adenovirus-mediated cochlear gene transfer. Gene Ther 7:377–383

42. Mason MJ (2013) Of mice, moles and guinea pigs: functional morphology of the middle ear in living mammals. Hearing Res 301:4–18

43. Plontke SK, Hartsock JJ, Gill RM et al (2016) Intracochlear drug injections through the round window membrane: measures to improve drug retention. Audiol Neurotol 21:72–79

44. György B, Sage C, Indzhykulian AA et al (2017) Rescue of hearing by gene delivery to inner-ear hair cells using exosome-associated AAV. Mol Ther 25:379–391. https://doi.org/10.1016/j.ymthe.2016.12.010

45. Yoshimura H, Shibata SB, Ranum PT et al (2018) Enhanced viral-mediated cochlear gene delivery in adult mice by combining canal fenestration with round window membrane inoculation. Sci Rep 14:2980. https://doi.org/10.1038/s41598-018-21233-z

46. Hirose K, Hartsock JJ, Johnson S et al (2014) Systemic lipopolysaccharide compromises the blood-labyrinth barrier and increases entry of serum fluorescein into the perilymph. J Assoc Res Otolaryngol 15:707–719

Part IV

The Central Auditory Pathway

Chapter 13

In Vivo Whole-Cell Recording in the Gerbil Cochlear Nucleus

Hsin-Wei Lu and Philip X. Joris

Abstract

This chapter describes how to use patch-clamp electrodes to perform blind in vivo whole-cell recording in the cochlear nucleus of anesthetized gerbils. This technique offers low-noise and stable recording of the intracellular membrane responses to sound stimuli, which is important for understanding the mechanisms underlying diverse sound-encoding schemes in the nucleus. With biocytin inside the pipette, it is also possible to label the recorded cell and link its anatomical features with physiology. Although in vivo whole-cell recording in rodents has been developed for more than a decade and has been abundantly used in cortical areas, it has seen little use in the auditory brainstem. This chapter discusses the techniques and challenges specific to recording in the gerbil cochlear nucleus.

Key words Cochlear nucleus, Sharp electrode, Whole-cell patch clamp, In vivo recording, Single-cell recording, Single-cell labeling

1   Introduction

The cochlear nucleus is the first processing center in the central ascending auditory pathway. It is the obligatory and only target of the auditory nerve, and thoroughly transforms the acoustic information carried by the latter structure. It is divided into three regions: the dorsal cochlear nucleus (DCN), anteroventral cochlear nucleus (AVCN), and posteroventral cochlear nucleus (PVCN). Each region contains several types of projection and interneurons with distinct anatomical and physiological properties [1]. Anatomically, the neurons in the DCN are organized in a three-layer structure: the molecular layer, principal cell layer, and deep layer. The principal layer contains principal neurons (fusiform cells), whereas the other two layers contain mostly glycinergic/GABAergic interneurons. The AVCN and PVCN do not have such layered organization, and different types of neurons are more intermingled. One exception is the octopus cell area (OCA) of the PVCN, which largely contains octopus cells and is immunohistochemically devoid of glycinergic activity [2, 3]. Starting in the 1970s, in vivo

Andrew K. Groves (ed.), *Developmental, Physiological, and Functional Neurobiology of the Inner Ear*, Neuromethods, vol. 176,
https://doi.org/10.1007/978-1-0716-2022-9_13,

physiologists found increasingly effective ways to study and classify the diversity of single-cell responses using simple acoustic stimuli, and to relate these classifications to cell types with distinct morphology and projection patterns [4]. Particularly for AVCN and PVCN, the temporal characteristics of the response to brief (~ 25 ms) tone bursts at the characteristic frequency (CF) of neurons has proven to be distinctive for morphologically different cell types. For example, bushy cells in the AVCN show temporal responses similar to that of the auditory nerve, except with enhanced synchronization to low-frequency tones [5, 6]. In contrast, octopus cells in the PVCN only fire an onset spike to high frequency tones [7–9], while T-stellate cells in PVCN/AVCN show regular spiking at intervals unrelated to the stimulus period [10, 11]. Although initial structure–function relationships were hypothesized based on extracellular recordings localized to specific CN regions (e.g., [7, 12–14]), recording and labeling of neurons with sharp microelectrodes had a prominent role in confirming these relationships. These recordings either targeted cell bodies in the CN [8, 10, 15, 16] or their axons [17–20]. An important insight resulting from this body of work is that, through its diversity of physiological responses and projection targets, the CN creates parallel sound processing channels to encode different sound features in the temporal and/or spectral domain (for reviews *see* refs. 4, 21, 22).

How do different cell types in the CN carry out these transformations? Anatomical studies of the circuit wiring diagram, combined with knowledge of the cellular/synaptic properties of each cell type, have resulted in numerous mechanistic proposals and models to explain these diverse computations [22–27]. However, direct measurement of these transformations, which requires in vivo intracellular recordings of transitions between sub- and suprathreshold potentials during sound stimulation, is very limited. While it is possible to infer excitatory synaptic efficacy during sound-evoked spiking from in vivo loose patch juxtacellular recordings, this method is more applicable to giant synapses such as the Endbulb of Held or Calyx of Held where presynaptic spikes and postsynaptic EPSPs/spikes can be recorded with a juxtacellular pipette [28–31]. Moreover, this method is not able to track the intracellular membrane potential changes from the postsynaptic cell. Thus, for most neurons intracellular recording remains the most informative approach to measure synaptic / cellular mechanisms underlying sound processing in vivo. Although previous studies have made efforts to obtain such recordings by intracellular sharp microelectrode penetrations [8, 10, 19, 32–35], the short-lived nature of these recordings typically only provides brief glimpses of the membrane dynamics in response to sounds and does not always yield consistent results.

At present, the best way to obtain qualitative, stable intracellular recordings in vivo is with whole-cell recording via a patch pipette. Due to the tight seal formed between the patch electrode and cell membrane and the latter's tolerance for deformation, in vivo whole-cell recording offers great stability despite periodic pulsatile motions from heartbeats and breathing or other movements [36, 37]. In vivo whole-cell recording in mammals was developed and first applied in the visual cortex in cats [38, 39], but the technique for rodents did not gain popularity until Margrie et al. published detailed guidelines on blind-patching in vivo in rats and mice [37]. Since then there have been many advanced or modified in vivo patch-clamp related techniques developed for rodents such as visually guided patching [40–42], awake/behavioral recordings [43, 44], automated recordings [45, 46], whole-cell recordings combined with single-cell initiated viral tracing [47] or drug delivery [37, 48] and others [49–51]. It is worth noting that the technique has been much more commonly used in superficial and easily reached structures (e.g., cerebral cortex) compared to deeper structures (e.g., brainstem). Indeed, several technical aspects such as tip clogging, capacitive coupling, and frequent electrode changes make it less conducive for recordings in deep brain regions. For auditory brainstem recordings, additional difficulties include that cell bodies tend to be small; are often enveloped by heavily myelinated bundles; and that many neurons have very fast membrane characteristics: all factors that impose challenges for obtaining high-quality recordings. Nevertheless, successful in vivo whole-cell recordings of the mammalian auditory brainstem have been achieved and have provided surprising insights that were outside the realm of other techniques [48, 52–56].

This chapter presents the procedures of blind in vivo whole-cell recording in the cochlear nucleus of anesthetized gerbils. We have successfully implemented this technique to obtain stable intracellular recordings from several neuron types. The basic steps are similar to descriptions in the literature [37, 50]: (1) apply positive high pressure to the pipette as it advances to the target area; (2) lower the positive pressure and search for signs of contacting a cell, revealed by pulsatile changes in the tip resistance; (3) once the electrode is close enough to a cell, release the positive pressure and form a gigaohm seal; (4) apply brief suction to break into whole-cell configuration and start recording. We use a posterior-fossa approach and aspirate parts of the cerebellum to expose the cochlear nucleus from its dorsal side. The invasive nature of the surgery and deep location of the target structure introduces challenges to obtaining successful, stable recordings. We address these issues in the Subheadings 3 and 4.

2   Materials

2.1   Animal

We use adult gerbils of both sexes weighing between 35–75 g (~ P35 to P70). In younger (< P40) gerbils, the softer and thinner bone facilitates the craniotomy. Also, perhaps due to less myelination, obtaining successful whole-cell recordings is more likely in young animals. However, it is still possible to obtain whole-cell recordings at around P70 or older.

2.2   Anesthetics

1. Ketamine + xylazine mixture (for induction): mix the following in volume ratios and store them in syringes at 4 °C. Dosage: 0.05 ml mixture/20 g gerbil (i.e., 1 mg ketamine/10 g gerbil; 0.15 mg xylazine/10 g gerbil). In our experience the mixture lasts for at least 2 weeks at 4 °C.

   (a)   Ketamine (100 mg/ml): 0.4×.

   (b)   Xylazine-M (2% or 20 mg/ml): 0.3×.

   (c)   Saline (0.9% NaCl): 0.3×.

2. Diluted Ketamine (for maintenance): Dilute Ketamine with saline to 0.4×. Dosage: 1/3 to 1/4 of the induction dose.

3. Diazepam (Valium; for maintenance): No dilution. Stored in syringes at 4 °C. Note that diazepam is water-insoluble, hence do not mix diazepam and ketamine in the same syringe. Dosage: 10 µg/10 g gerbil.

4. Atropine (atropine methyl nitrate; to reduce mucus secretion): 10 µg/ml with distilled water. Dosage: 0.2 µg/10 g gerbil.

2.3   Surgical Tools

1. Dissection scissors.

2. Forceps (standard and Dumont #5, FST instruments).

3. No. 11 Blade Scalpel.

4. Pearson Rongeur (FST instruments 16015-17).

5. Spatula (FST instruments 10091-12).

6. Custom-made head bar.

7. UV-cured dental cement and adhesives (CLEARFIL SE Bond; Kuraray).

8. Glass disposable Pasteur pipettes (VWR): for aspiration of cerebral spinal fluid (CSF), blood, and cerebellar tissue. Before use, heat the tip of each pipette with a Bunsen/alcohol burner so that it can be bent/lengthened at an optimal angle/length for aspiration.

9. Vacuum pump (e.g., BVC Vacuubrand) to connect with the Pasteur pipette for aspiration.

10. Homeostatic heating pad and temperature controller (TC-1000, Cwe Inc.).

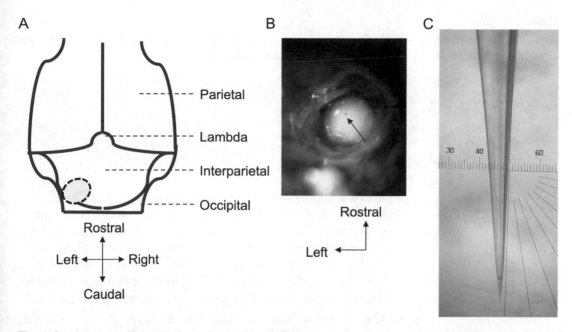

Fig. 1 Craniotomy and pipette shape for in vivo whole-cell recording in the gerbil cochlear nucleus. **a** and **b** share the same axes. (**a**) A schematic drawing of the location of craniotomy (dashed circle) in **b** on the gerbil skull. (**b**) A photo of the cochlear nucleus viewed through the craniotomy in the interparietal bone. The red dot (arrow) in the cochlear nucleus is made by a penetration of the patch pipette in (**c**). (**c**) A photo of the recording electrode that had successfully recorded from an octopus cell in the gerbil cochlear nucleus in (**b**). The smallest division equals 20 µm

11. Gelfoam (Pfizer) and Bone wax (VWR) to stop bleeding.

12. Electrocautery. A brief zap can effectively stop bleeding in bone or soft tissue. We use "glowing wire" type cauteries (handheld: WPI, or with line power: Geiger TCU), or sometimes a simple handheld, battery-operated soldering iron (Weller).

2.4 Electrophysiology

1. Recording electrode: Borosilicate glass pipettes (1B120F-4, World Precision Instruments). We typically obtain whole-cell recordings using a tip resistance of 5–7 MΩ. From our experience the taper should be longer (*see* Fig. 1 for an example) if the target area is deeper (e.g., > 500 µm) to minimize tissue damage during repeated penetrations. An example of pulling parameters for the Sutter puller (**item 3**) is shown in Table 1.

2. Internal solution (in mM): 115 Potassium gluconate; 4.42 KCl; 10 sodium phosphocreatine; 10 HEPES; 0.5 EGTA; 4 Mg-ATP; 0.3 Na-GTP; 0.2–0.3% (wt/vol) biocytin (Invitrogen) with pH adjusted to 7.30 with KOH and osmolality to 290–300 mOsm with sucrose.

3. Pipette puller (e.g., Model P-1000, Sutter Instrument).

Table 1
An example of a three-step pulling program for in vivo patch clamp electrodes. The tip resistance is around 5–7 MΩ when loaded with K-gluconate internal solution

Heat	Pull	Velocity	Time	Pressure
508	0	65	250	500
508	0	30	250	
503	0	24	250	

4. Micromanipulators: We mount a single-axis hydraulic microd-rive (Trent-Wells Inc.) on a manual stereotaxic manipulator (Narishige SM-11). The Narishige manipulator is compact and offers a high degree of freedom for adjusting the angles of the electrode with respect to the cochlear nucleus. Also, it has a long working distance. We find these features helpful for setting up the penetration angle and location of the pipette.

5. Custom-made frame to hold head bar and ear bars including acoustic assemblies with probe microphone and speakers enclosed in metal canisters.

6. Patch-clamp amplifier (Dagan BVC-700A) and acquisition interface (TDT RX6 and/or HEKA InstruTECH ITC-18).

7. Microforge (Narishige MF2): for fire-polishing the pipette tip.

8. Recording software (using IGOR Pro or Matlab). We use DataPro (https://github.com/JaneliaSciComp/DataPro) for patch clamp-related stimuli generation and data acquisition. We also use custom-written software in Matlab for sound stimuli generation and electrophysiology data acquisition.

2.5 Acoustic System

1. Double-walled sound booth (IAC, Niederkrüchten, Germany).

2. Digital system for sound generation and acoustic calibration, for example, TDT System 3 (Tucker Davis Technologies).

3. Speakers (Radio Shack Supertweeter or Etymotic ER1 or ER2).

3   Methods

3.1   Surgery

3.1.1   Anesthesia: Induction and Maintenance

Induction of general anesthesia is achieved by an intraperitoneal (i.p.) injection of a ketamine and xylazine mixture. An i.p. injection of atropine 30 minutes prior to induction helps to prevent mucus accumulation in the airways during experiment, but is not necessary. To maintain anesthesia, an intramuscular (i.m.) or i.p. injection of diluted ketamine followed by i.m. injection of diazepam is performed when the animal shows pinch reflexes. Occasionally, if the animal still shows signs of reflex, a small dose (~1/3 to 1/4 of the induction dose) of ketamine and xylazine mixture is administered i.p.. *See* Subheading 2.2 for dosage of drugs used here.

3.1.2 Mounting the Head Bar

Mount the head bar on the frontal and parietal bone. Make sure the head bar and the dental cement do not block the interparietal bone just rostral to the nuchal ridge, where we will perform the craniotomy.

1. Use scissors to remove the hair on the scalp and make an incision to expose the frontal to the interparietal part of the skull.

2. Use spatulas to scrape off the connective tissue and periost on the parietal and frontal bones. Make sure the surface is clean and dry. If bleeding occurs, use Gelfoam to stop it.

3. Apply dental adhesives ("primer" first, "bond" second) to the frontal part of the skull. Air dry and UV-cure the dental adhesives. Then, place a layer of dental cement on top of the dental adhesives and glue the head bar on top. We find that applying dental adhesives first strengthens the bond between dental cement and the skull.

3.1.3 Craniotomy and Exposure of the Cochlear Nucleus

1. Craniotomy: We use scalpels and rongeurs rather than (electric) drills to make the craniotomy, to avoid noise-induced elevation of hearing thresholds. The goal is to make a 4-mm-diameter round or a 4-mm-by-4-mm square craniotomy on the lateral part of the interparietal bone (*see* Fig. 1). Use the scalpel to gently make incisions on the outline of the square on the interparietal bone, and subsequently use forceps to carefully lift the bone flap up. It usually requires at least three to five repetitive scalpel strokes on one edge to penetrate the skull. Apply bonewax or Gelfoam on the site of the incision in case of bleeding. Once the bone flap is removed, use fine forceps to retract the dura over the edge of the craniotomy, which helps to stop bleeding from the bone edges.

2. Cerebellar aspiration: connect a glass Pasteur pipette to a diaphragm vacuum pump and aspirate the lateral aspects of the cerebellum (mainly Crus II and parts of the Crus I and PML) until the cochlear nucleus is exposed. Use saline-soaked pieces of cotton or paper near the pipette tip while aspirating to stem bleeding. *See* **Note 1** for more tips to prevent serious bleeding. It is easy to identify the cochlear nucleus as it is a clean oval-shaped structure without readily apparent blood vessels running on its surface (Fig. 1). Just before exposing the cochlear nucleus, one will usually encounter some parts of the

choroid plexus [57]. During the removal of the choroid plexus, be careful with the handling of the aspiration pipette as the cochlear nucleus lies just beneath the plexus. A significant advantage of the dorsal approach to the CN is that much of its dorsal surface is part of the lateral recess of the IV ventricle (Luschka aperture) and is thus ependymal. This surface is easy to penetrate with an electrode and easy to keep blood-free.

3. Refine the craniotomy: Due to the depth of the cochlear nucleus (> 8 mm below the surface of the interparietal bone), it is possible that the initial craniotomy made in **step 1** is not wide enough to allow a pipette to freely move into or out of the cochlear nucleus. Thus it may be necessary to refine the craniotomy after exposing the cochlear nucleus. Mount a dummy electrode on the manipulator and see if any parts of the interparietal/occipital bone hinder its path into the cochlear nucleus. If so, use a rongeur to chip the bones away (usually some parts of the occipital bone, *see* Fig. 1). *See* **Note 2** if edema occurs.

4. After making sure the electrode can freely move inside the aspirated region, the experimenter can proceed to acoustic system calibration and electrophysiological recording.

3.1.4 Electrophysiology

1. Pulling pipettes: We find that a pipette with a relatively long taper (Fig. 1) is essential to obtain whole-cell recordings in the cochlear nucleus. Usually several pipette penetrations are required before obtaining a whole-cell recording, and a pipette with a long taper minimizes the damage to the penetrated tissue. An example three-step pulling program is shown in Table 1. The tip resistance is around 5–7 MΩ. We use a microforge to fire-polish the pipette tip. *See* **Note 3** for the importance of an optimal pipette.

2. Load the pipette with internal solution and mount it on the head stage.

3. Wrap the grounding electrode with a thin layer of gauze, soak it with saline, and place it underneath the skin of the animal.

4. Advancing the electrode to the target area: Before the pipette touches the brain surface, apply high pressure (4–5 psi) through the tubing system that connects to the pipette holder. Switch the patch clamp amplifier to voltage-clamp and apply a voltage pulse (e.g., a 3-ms step pulse at 2.5 mV repeated every 50 ms) to continuously monitor the pipette resistance. The pipette resistance now should be infinitely large as the circuit is open (pipette not touching the CSF/brain yet). Then, use the remote control on the micromanipulator to advance the pipette until it touches the surface of the brain (judged either by visual cues or a sudden DC-shift and decrease in the pipette

resistance). Zero the position of the pipette on the micromanipulator, make a note of the pipette resistance, and keep advancing the pipette at high speed (>100 μm/s) until it reaches ~100 μm above the target depth (e.g., for PVCN the target depth is around 600 μm below the surface). Lower the pressure to about 0.3–0.4 psi. Note that during this high-speed penetration the pipette resistance may increase, presumably due to penetration of cells or blood vessels, however the high pressure at the tip should prevent tissue debris from sticking to it and hence the pipette resistance should return to its initial value quickly. If the pipette resistance has stayed at >0.2 MΩ for more than a minute from its original value (e.g., from 5.5 to 5.8 MΩ) [45], the tip is probably contaminated and will not form a GΩ seal. The user should retract the pipette, replace it with a new one and repeat this step again.

5. Searching for a neuron: When using other recording techniques (e.g., sharp or metal microelectrodes), the searching for neurons is usually based on the continuous presentation of "search sounds" while advancing the electrode to identify the presence of neuron(s) near the electrode tip. The search strategy with whole-cell recordings is completely different: since spike activity is only very rarely observed during the advancement of the electrode, this approach is entirely dependent on continuously monitoring the pipette resistance. If the pipette resistance stays within 0.2 MΩ from its original value, advance the pipette in 1–2 μm steps and pay attention to the changes in the pipette resistance. If the pipette resistance suddenly increases and oscillates at the rate of heartbeat, keep advancing the electrode for two to three more steps and see if the pipette resistance increases further. If so, immediately proceed to **step 5** to form a GΩ seal. If not, apply a brief positive pressure (~1 psi for 0.5 s) to clean the tip of the pipette (i.e., the pipette resistance should return to its initial value) and repeat this step again.

6. Forming a GΩ seal: quickly release the pipette pressure to 0 and set the holding potential to −70 mV. The pipette resistance should rapidly increase to >1 GΩ. If not, a brief negative pressure (~0.15 to 0.5 psi) may help to form the seal. If no GΩ seal is observed, the pipette tip is likely contaminated and no longer suitable for whole-cell recording. Retract the pipette and start from **step 2** with a fresh pipette.

7. Whole-cell recording: Once a GΩ seal is formed, apply brief suction to the pipette to enter the whole-cell configuration. If the cell is a neuron, the user should almost immediately observe abundant synaptic currents in the oscilloscope. Switch the patch amplifier to current clamp with zero current injection, and subsequently perform capacitance neutralization and

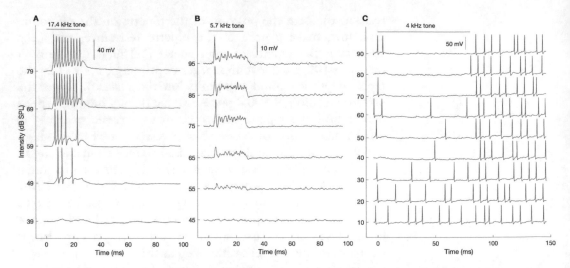

Fig. 2 Whole-cell responses from three cells to tones at different levels. Note the differences in the action potential amplitudes among these cells. (**a**) A vertical cell in DCN shows a monotonic increase in rate with intensity. (**b**) An octopus cell in PVCN shows a small onset spike to high frequency tones. (**c**) A fusiform cell in DCN shows inhibition as tone intensity increases

bridge balance. The user should be able to see action potentials in response either to sound stimulation or current injection. However, it is also possible that there is no spiking response at all and that there is a linear voltage-current relation to a series of current steps. In this case a glia cell is probably recorded (*see* **Note 4** for glia features). The user should then retract and replace the pipette, then repeat the procedures again starting from **step 2**.

8. After obtaining the whole-cell configuration with a neuron, the user can now apply acoustic stimuli and measure the intracellular responses. Figure 2 shows several intracellular responses from different cell types to pure tone stimulation. Despite the pulsations from heartbeat and breathing, the recording should be quite stable. In our hands it usually lasts at least 30 min, and at most around 2 h. Hence there is ample time for the user to present different stimulation protocols. However, the user should keep an eye on the pipette capacitance and reperform capacitance neutralization in between acquisition of datasets, as the CSF is likely to accumulate in the aspirated part of the brain. Due to the depth of the cochlear nucleus (> 8 mm from the surface of the skull), such accumulation can increase the pipette capacitance by a large amount and may heavily filter the kinetics of the membrane potential waveform. *See* **Note 5** for ways to prevent CSF accumulation.

3.1.5   Perfusion

1. After recording, switch the patch clamp amplifier to voltage clamp mode and set holding voltage at −70 mV. Slowly (~3 μm/s) retract the pipette while monitoring the pipette resistance. During this process the membrane should reseal, causing the pipette resistance to increase and eventually forming an outside-out patch [47]. Set holding voltage to 0 mV and apply brief suction to the pipette to remove the membrane at the tip. The pipette resistance should now return to its initial value and it should now be safe to retract the pipette without damaging the cell. Increase the withdrawal speed (> 50 μm/s) to remove the pipette from the brain.

2. After pipette withdrawal, wait for another 20 min to let biocytin diffuse throughout the cell. Prepare the perfusion system while waiting. *See* **Note 6** for reasons why terminating the experiment at this point. Overdose the animal with pentobarbital and transcardially perfuse the animal with saline followed by 4% paraformaldehyde in 0.1 M phosphate buffer. Remove the brain and set the brain in the fixative overnight. Use a vibratome to section the brain at 70 μm and perform DAB staining (ABC kit, VECTASTAIN). Our experience is that ~50% of recorded neurons can be successfully labeled. An example of a labeled cell is shown in Fig. 3.

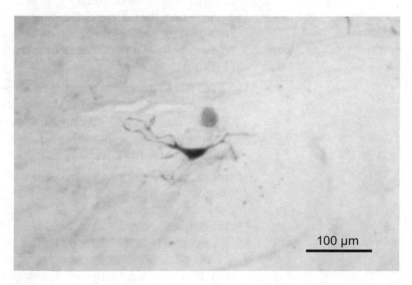

Fig. 3 An octopus cell labeled by biocytin through an in vivo whole-cell recording pipette

4   Notes

1. Bleeding is inevitable during aspiration of cerebellar folia; however, it typically quickly subsides as long as major blood vessels are not damaged. Avoid large blood vessels near the pia on the side of the occipital bone while aspirating the cerebellum. Damaging those blood vessels often causes massive bleeding and forms a hard-to-remove clot that blocks the view of the cochlear nucleus [45].

2. In case of edema: sometimes the cerebellum in the aspirated region will develop edema and block the view of the cochlear nucleus. This often indicates that the depth of anesthesia is decreasing. Apply a maintenance dose of anesthetics and use the glass aspirator to remove the edematous tissue. Also, remake the anesthetics stock, which can degrade over time.

3. A clean pipette tip is key to successful recording. This is often judged by whether the pipette is clogged (i.e., an increase in pipette resistance) during penetration. As observed by others [45], we find that if the increase is >0.2 MΩ the pipette often will not form a GΩ seal. Hence when such increase occurs the user should replace the pipette and try again. We also find that a pipette with lower resistance (<6.4 MΩ) tends not to clog, presumably due to a wider tip diameter. However, when the resistance is <5.0 MΩ the pipette tip is probably too wide and in our hands cannot form a GΩ seal even though there is no sign of clogging. Hence, the user should spend time fiddling with pipette pulling programs to find optimal parameters that render a low pipette resistance and a relatively wide tip (1–2 μm).

4. There is a high chance to form GΩ seals with nonspiking cells. Presumably these are glia, endothelial cells, or membrane debris. Interestingly, we find that these nonspiking cells tend to correlate with a low-value GΩ seal, that is, around 1 GΩ. With neurons, the value is usually higher than 2 GΩ.

5. In case of CSF accumulation: CSF is likely to accumulate in the craniotomy, which will increase the pipette capacitance and slow the apparent kinetics of the membrane potential. This can be detrimental to the recording quality if the target cell has extremely fast EPSPs or action potentials (time constant <1 ms), such as in bushy or octopus cells. We find that placing two to three long strips of Kimwipes paper or tissue paper along the edge of the craniotomy can remove CFS accumulation efficiently. In contrast to cortical experiments, where many studies place agarose around the electrode to stabilize pulsations, we find this approach drastically increases the pipette capacitance, presumably due to the >8 mm depth of the

craniotomy. We thus do not recommend the use of agar for cochlear nucleus experiments, unless the experimentalist has very fine surgical skills to trim the agarose to a thin layer inside the cranial window.

6. We terminate the experiment once a successful whole-cell recording has been obtained, for several reasons. We put a premium on morphological examination of the recorded cell: assignment of a labeled neuron to the associated recording becomes problematic if multiple cells were recorded in the cochlear nucleus. Also, there is the possibility to damage a previously labeled neuron when performing a subsequent new penetration. Finally, we want to minimize time between successful recording and perfusion, because there is always a risk of losing the animal before perfusion.

5   Conclusion

This chapter presents the procedures and challenges of performing in vivo whole-cell recordings in the cochlear nucleus in anesthetized gerbils. We find that the success rate depends on the depth of the target neuron: we estimate a >30% success rate (i.e., one successful recording per three animals) for DCN principal neurons but <15% for principal neurons in the PVCN, such as the octopus cells. We have not tried to record from cells in the AVCN, but if one uses the posterior fossa approach as presented in this chapter we would expect the success rate to be even lower as the depth would be more than 1000 μm from the surface and the myelination would be denser. A lateral transbulla approach through the niche of the flocculus may be more suitable: due to the enlarged middle ear bulla in the gerbil, one can easily access the cochlear nucleus via opening of the anterior chamber of the bulla with limited cerebellectomy (flocculus) [28, 58]. However, the lateral approach constrains the angle of the electrode and leaves a rather small operational space in the lateral bulla, thus access to the cochlear nucleus is more restrictive compared to the dorsal approach presented in this chapter.

Despite the low success rate, the tight seal with the membrane during in vivo whole-cell recordings makes them much more stable (lasting >30 min) compared to intracellular recordings made by sharp microelectrodes (in our hands often <5 min) which are inherently leaky due to the impalement of the cell membrane with the electrode. The tight seal, together with the lower access resistance from whole-cell recordings also provides a better signal-to-noise ratio compared with sharp recordings. This allows the user not only to record high quality sound-evoked responses but also to characterize intrinsic membrane properties by current injection

[56]. An important potential advantage of the in vivo whole-cell technique is that it not only allows investigation of the membrane mechanisms that transform synaptic inputs to spiking output in response to physiologically relevant stimuli, but also that it allows one to manipulate single-cell physiology or pharmacology via the patch pipette or with a piggyback pipette [37, 48]. Recent advances also show that it is possible to transfect single cells with plasmids to subsequently transfect the neuron with modified rabies virus, thereby obtaining both in vivo intracellular physiology and the presynaptic wiring diagram for a single cell [47, 59]. Hence, mastering the technique opens doors for functional microcircuit mapping at the single-cell level in vivo.

References

1. Cant NB, Benson CG (2003) Parallel auditory pathways: projection patterns of the different neuronal populations in the dorsal and ventral cochlear nuclei. Brain Res Bull 60(5–6):457–474. https://doi.org/10.1016/S0361-9230(03)00050-9

2. Wickesberg RE, Whitlon D, Oertel D (1994) In vitro modulation of somatic glycine-like immunoreactivity in presumed glycinergic neurons. J Comp Neurol 339(3):311–327. https://doi.org/10.1002/cne.903390302

3. Wickesberg RE, Whitlon D, Oertel D (1991) Tuberculoventral neurons project to the multipolar cell area but not to the octopus cell area of the posteroventral cochlear nucleus. J Comp Neurol 313(3):457–468. https://doi.org/10.1002/cne.903130306

4. Yin TCT, Smith PH, Joris PX (2019) Neural mechanisms of binaural processing in the auditory brainstem. Compr Physiol 9(4):1503–1575. https://doi.org/10.1002/cphy.c180036

5. Joris PX, Smith PH, Yin TC (1994) Enhancement of neural synchronization in the anteroventral cochlear nucleus. II. Responses in the tuning curve tail. J Neurophysiol 71(3):1037–1051

6. Joris PX, Carney LH, Smith PH, Yin TC (1994) Enhancement of neural synchronization in the anteroventral cochlear nucleus. I. Responses to tones at the characteristic frequency. J Neurophysiol 71(3):1022–1036

7. Godfrey DA, Kiang NYS, Norris BE (1975) Single unit activity in the posteroventral cochlear nucleus of the cat. J Comp Neurol 162(2):247–268. https://doi.org/10.1002/cne.901620206

8. Rhode WS, Oertel D, Smith PH (1983) Physiological response properties of cells labeled intracellularly with horseradish peroxidase in cat ventral cochlear nucleus. J Comp Neurol 213(4):448–463. https://doi.org/10.1002/cne.902130408

9. Lu H-W, Smith PH, Joris PX (2018) Submillisecond monaural coincidence detection by octopus cells. Acta Acust United Acust 104(5):852–855. https://doi.org/10.3813/AAA.919238

10. Smith PH, Rhode WS (1989) Structural and functional properties distinguish two types of multipolar cells in the ventral cochlear nucleus. J Comp Neurol 282(4):595–616. https://doi.org/10.1002/cne.902820410

11. Rhode WS, Smith PH (1986) Encoding timing and intensity in the ventral cochlear nucleus of the cat. J Neurophysiol 56(2):261–286

12. Godfrey DA, Kiang NYS, Norris BE (1975) Single unit activity in the dorsal cochlear nucleus of the cat. J Comp Neurol 162(2):269–284. https://doi.org/10.1002/cne.901620207

13. Pfeiffer RR (1966) Anteroventral cochlear nucleus:wave forms of extracellularly recorded spike potentials. Science 154(3749):667–668. https://doi.org/10.1126/science.154.3749.667

14. Pfeiffer RR (1966) Classification of response patterns of spike discharges for units in the cochlear nucleus: tone-burst stimulation. Exp Brain Res 1(3):220–235. https://doi.org/10.1007/BF00234343

15. Feng JJ, Kuwada S, Ostapoff E-M, Batra R, Morest DK (1994) A physiological and structural study of neuron types in the cochlear nucleus. I. Intracellular responses to acoustic stimulation and current injection. J Comp Neurol 346(1):1–18. https://doi.org/10.1002/cne.903460102

16. Ostapoff E-M, Feng JJ, Morest DK (1994) A physiological and structural study of neuron types in the cochlear nucleus. II. Neuron types and their structural correlation with response properties. J Comp Neurol 346(1):19–42. https://doi.org/10.1002/cne.903460103

17. Smith PH, Joris PX, Yin TCT (1993) Projections of physiologically characterized spherical bushy cell axons from the cochlear nucleus of the cat: evidence for delay lines to the medial superior olive. J Comp Neurol 331(2):245–260. https://doi.org/10.1002/cne.903310208

18. Smith PH, Massie A, Joris PX (2005) Acoustic stria: anatomy of physiologically characterized cells and their axonal projection patterns. J Comp Neurol 482(4):349–371. https://doi.org/10.1002/cne.20407

19. Friauf E, Ostwald J (1988) Divergent projections of physiologically characterized rat ventral cochlear nucleus neurons as shown by intra-axonal injection of horseradish peroxidase. Exp Brain Res 73(2):263–284. https://doi.org/10.1007/BF00248219

20. Spirou GA, Brownell WE, Zidanic M (1990) Recordings from cat trapezoid body and HRP labeling of globular bushy cell axons. J Neurophysiol 63(5):1169–1190. https://doi.org/10.1152/jn.1990.63.5.1169

21. Joris PX, Schreiner CE, Rees A (2004) Neural processing of amplitude-modulated sounds. Physiol Rev 84(2):541–577. https://doi.org/10.1152/physrev.00029.2003

22. Oertel D, Young ED (2004) What's a cerebellar circuit doing in the auditory system? Trends Neurosci 27(2):104–110. https://doi.org/10.1016/j.tins.2003.12.001

23. Oertel D, Wright S, Cao X-J, Ferragamo M, Bal R (2011) The multiple functions of T stellate/multipolar/chopper cells in the ventral cochlear nucleus. Hear Res 276(1–2):61–69. https://doi.org/10.1016/j.heares.2010.10.018

24. Joris PX, Smith PH (2008) The volley theory and the spherical cell puzzle. Neuroscience 154(1):65–76. https://doi.org/10.1016/j.neuroscience.2008.03.002

25. Oertel D (1999) The role of timing in the brain stem auditory nuclei of vertebrates. Annu Rev Physiol 61(1):497–519. https://doi.org/10.1146/annurev.physiol.61.1.497

26. Golding NL, Oertel D (2012) Synaptic integration in dendrites: exceptional need for speed. J Physiol 590(22):5563–5569. https://doi.org/10.1113/jphysiol.2012.229328

27. Manis PB, Campagnola L (2018) A biophysical modelling platform of the cochlear nucleus and other auditory circuits: from channels to networks. Hear Res 360:76–91. https://doi.org/10.1016/j.heares.2017.12.017

28. Kuenzel T, Borst JGG, van der Heijden M (2011) Factors controlling the input-output relationship of spherical bushy cells in the gerbil cochlear nucleus. J Neurosci 31(11):4260–4273. https://doi.org/10.1523/JNEUROSCI.5433-10.2011

29. Lorteije JAM, Rusu SI, Kushmerick C, Borst JGG (2009) Reliability and precision of the Mouse Calyx of Held Synapse. J Neurosci 29(44):13770–13784. https://doi.org/10.1523/JNEUROSCI.3285-09.2009

30. Mc Laughlin M, van der Heijden M, Joris PX (2008) How secure is in vivo synaptic transmission at the Calyx of Held? J Neurosci 28(41):10206–10219. https://doi.org/10.1523/JNEUROSCI.2735-08.2008

31. Keine C, Rübsamen R, Englitz B (2017) Signal integration at spherical bushy cells enhances representation of temporal structure but limits its range. elife 6:e29639. https://doi.org/10.7554/eLife.29639

32. Britt R, Starr A (1976) Synaptic events and discharge patterns of cochlear nucleus cells. I. Steady-frequency tone bursts. J Neurophysiol 39(1):162–178

33. Britt R, Starr A (1976) Synaptic events and discharge patterns of cochlear nucleus cells. II. Frequency-modulated tones. J Neurophysiol 39(1):179–194

34. Romand R (1978) Survey of intracellular recording in the cochlear nucleus of the cat. Brain Res 148(1):43–65. https://doi.org/10.1016/0006-8993(78)90377-3

35. Rouiller EM, Ryugo DK (1984) Intracellular marking of physiologically characterized cells in the ventral cochlear nucleus of the cat. J Comp Neurol 225(2):167–186. https://doi.org/10.1002/cne.902250203

36. Petersen CCH (2017) Whole-cell recording of neuronal membrane potential during behavior. Neuron 95(6):1266–1281. https://doi.org/10.1016/j.neuron.2017.06.049

37. Margrie TW, Brecht M, Sakmann B (2002) In vivo, low-resistance, whole-cell recordings from neurons in the anaesthetized and awake mammalian brain. Pflüg Arch 444(4):491–498. https://doi.org/10.1007/s00424-002-0831-z

38. Jagadeesh B, Wheat HS, Ferster D (1993) Linearity of summation of synaptic potentials underlying direction selectivity in simple cells of the cat visual cortex. Science

262(5141):1901–1904. https://doi.org/10.1126/science.8266083

39. Pei X, Volgushev M, Vidyasagar TR, Creutzfeldt OD (1991) Whole cell recording and conductance measurements in cat visual cortex in-vivo. Neuroreport 2(8):485–488. https://doi.org/10.1097/00001756-199108000-00019

40. Kitamura K, Judkewitz B, Kano M, Denk W, Häusser M (2008) Targeted patch-clamp recordings and single-cell electroporation of unlabeled neurons in vivo. Nat Methods 5(1):61–67. https://doi.org/10.1038/nmeth1150

41. Komai S, Denk W, Osten P, Brecht M, Margrie TW (2006) Two-photon targeted patching (TPTP) in vivo. Nat Protoc 1(2):647–652. https://doi.org/10.1038/nprot.2006.100

42. Margrie TW et al (2003) Targeted whole-cell recordings in the mammalian brain in vivo. Neuron 39(6):911–918. https://doi.org/10.1016/j.neuron.2003.08.012

43. Lee AK, Epsztein J, Brecht M (2009) Head-anchored whole-cell recordings in freely moving rats. Nat Protoc 4(3):385–392. https://doi.org/10.1038/nprot.2009.5

44. Lee D, Shtengel G, Osborne JE, Lee AK (2014) Anesthetized- and awake-patched whole-cell recordings in freely moving rats using UV-cured collar-based electrode stabilization. Nat Protoc 9(12):2784–2795. https://doi.org/10.1038/nprot.2014.190

45. Kodandaramaiah SB, Franzesi GT, Chow BY, Boyden ES, Forest CR (2012) Automated whole-cell patch clamp electrophysiology of neurons in vivo. Nat Methods 9(6):585–587. https://doi.org/10.1038/nmeth.1993

46. Li L et al (2017) A robot for high yield electrophysiology and morphology of single neurons in vivo. Nat Commun 8:15604. https://doi.org/10.1038/ncomms15604

47. Rancz EA, Franks KM, Schwarz MK, Pichler B, Schaefer AT, Margrie TW (2011) Transfection via whole-cell recording in vivo: bridging single-cell physiology, genetics and connectomics. Nat Neurosci 14(4):527–532. https://doi.org/10.1038/nn.2765

48. Franken TP, Roberts MT, Wei L, Golding NL, Joris PX (2015) In vivo coincidence detection in mammalian sound localization generates phase delays. Nat Neurosci 18(3):444–452. https://doi.org/10.1038/nn.3948

49. Schramm AE, Marinazzo D, Gener T, Graham LJ (2014) The touch and zap method for in vivo whole-cell patch recording of intrinsic and visual responses of cortical neurons and glial cells. PLoS One 9(5). https://doi.org/10.1371/journal.pone.0097310

50. DeWeese MR (2007) Whole-cell recording in vivo. Curr Protoc Neurosci 38(1):6.22.1–6.22.15. https://doi.org/10.1002/0471142301.ns0622s38

51. Zhu Z, Wang Y, Xu X-Z, Li C-Y (2002) A simple and effective method for obtaining stable in vivo whole-cell recordings from visual cortical neurons. Cereb Cortex 12(6):585–589. https://doi.org/10.1093/cercor/12.6.585

52. Franken TP, Bremen P, Joris PX (2014) Coincidence detection in the medial superior olive: mechanistic implications of an analysis of input spiking patterns. Front Neural Circuits 8:42. https://doi.org/10.3389/fncir.2014.00042

53. van der Heijden M, Lorteije JAM, Plauška A, Roberts MT, Golding NL, Borst JGG (2013) Directional hearing by linear summation of binaural inputs at the medial superior olive. Neuron 78(5):936–948. https://doi.org/10.1016/j.neuron.2013.04.028

54. Zhou M, Li Y-T, Yuan W, Tao HW, Zhang LI (2015) Synaptic mechanisms for generating temporal diversity of auditory representation in the dorsal cochlear nucleus. J Neurophysiol 113(5):1358–1368. https://doi.org/10.1152/jn.00573.2014

55. Zhou M, Tao HW, Zhang LI (2012) Generation of intensity selectivity by differential synaptic tuning: fast-saturating excitation but slow-saturating inhibition. J Neurosci 32(50):18068–18078. https://doi.org/10.1523/JNEUROSCI.3647-12.2012

56. Franken TP, Joris PX, Smith PH (2018) Principal cells of the brainstem's interaural sound level detector are temporal differentiators rather than integrators. eLife 7. https://doi.org/10.7554/eLife.33854

57. Perin P, Voigt FF, Bethge P, Helmchen F, Pizzala R (2019) iDISCO+ for the study of neuroimmune architecture of the rat auditory brainstem. Front Neuroanat 13:15. https://doi.org/10.3389/fnana.2019.00015

58. Frisina RD, Chamberlain SC, Brachman ML, Smith RL (1982) Anatomy and physiology of the gerbil cochlear nucleus: an improved surgical approach for microelectrode studies. Hear Res 6(3):259–275. https://doi.org/10.1016/0378-5955(82)90059-4

59. Vélez-Fort M et al (2014) The stimulus selectivity and connectivity of layer six principal cells reveals cortical microcircuits underlying visual processing. Neuron 83(6):1431–1443. https://doi.org/10.1016/j.neuron.2014.08.001

Chapter 14

Measurement of Human Cochlear and Auditory Nerve Potentials

Eric Verschooten and Philip X. Joris

Abstract

Understanding the properties of the peripheral auditory system in humans is important in explaining auditory perception, in diagnosing auditory pathologies, and in reconstructing normal hearing. Auditory peripheral properties cannot be investigated in humans at the level of detail possible in experimental animals and is therefore largely modeled on detailed knowledge obtained from such animals. However, the auditory periphery shows significant differences between species, and small-bodied experimental animals such as rodents are not necessarily good models for humans. Sound-evoked mass-potentials provide a bridge between investigations in human and experimental animals. These potentials are synchronized responses of cochlear and neural origin to sounds, and can be noninvasively recorded at different locations.

To assess cochlear and neural properties in humans, we developed a transtympanic procedure to obtain evoked mass-potential recordings using a minimally invasive procedure via the middle ear of awake subjects. In parallel, several stimulus and analysis paradigms were developed to target specific peripheral auditory properties such as frequency selectivity, phase-locking, and efferent effects. We describe the practical aspects of obtaining transtympanic recordings. It is our conviction that such recordings have great potential to address several outstanding issues in both normal and impaired hearing, such as the role of cochlear efferent activation and the prevalence of synaptopathy.

Key words Electrophysiology, Mass potentials, Cochlear microphonics, Neurophonic, ECochG, Transtympanic, Cochlea, Auditory nerve

---

1 Introduction

Recording of responses to sound of individual neurons of the auditory nerve is the electrophysiological method of choice to investigate the output of the cochlea and its input to the central nervous system. This recording method has only been used in anesthetized laboratory animals due to its invasiveness: it requires surgery to obtain access to the auditory nerve and only axons of spiral ganglion cells are accessible without opening the cochlea. The closest alternative electrophysiological technique to explore properties of the auditory periphery is the recording of sound-evoked

Andrew K. Groves (ed.), *Developmental, Physiological, and Functional Neurobiology of the Inner Ear*, Neuromethods, vol. 176, https://doi.org/10.1007/978-1-0716-2022-9_14,

mass-potentials, generated through the synchronized activity of receptor cells and auditory nerve fibers. In clinical settings this technique is known as electrocochleography (ECochG) [1, 2] and was first applied during surgical procedures early in the previous century [3–5]. Over time, this technique evolved into a clinical diagnostic tool in patients, but with varying degrees of interest. More details on the history and evolution of ECochG can be found in [6].

Peripheral sound-evoked mass potentials containing components from the cochlea and auditory nerve can be recorded at several anatomical locations. Two areas are most commonly targeted: extratympanic (ET) recordings are obtained at the external auditory canal or tympanic membrane (TM), while transtympanic (TT) recordings are obtained at the middle ear, most notably from the cochlear promontory and niche of the round window. TT recordings require an electrode beyond the barrier of the TM and are therefore more invasive than ET recordings, for which a ball or gel electrode on the skin of the external ear structures or the external surface of the TM suffices. However, the location of TT recordings is closer to the sources of interest, resulting in a 2–20 times greater response, depending on stimulus type (tones, clicks, chirps, etc.), level, frequency and the component of interest (e.g., [7, 8]). The closer the recording location to the cochlea, the greater the response, with the best location for obtaining good signal magnitudes and reproducibility being the niche of the round window [9]. A third approach during which auditory peripheral mass potentials can be recorded is from the auditory nerve in neurosurgical settings, for example, for removal of tumors in the pontocerebellar angle or surgery for neurovascular conflicts (e.g., [10]).

In humans, the signal-to-noise ratio of peripheral mass potentials measured extra- or transtympanically is quite small compared to species routinely used for animal experiments, thus by default much more averaging is required to suppress non-stimulus related background noise to an acceptable level. Since the noise reduction is proportional to the square root of the number of repetitions, the ET method requires drastically more measurement time ($\times 100$ or more) to obtain the same signal-to-noise ratio as obtained with TT recordings. Alternatively, one can increase the signal-to-noise ratio by resorting to stimuli that produce more synchronous responses across receptor cells or neurons, using stimuli such as clicks and chirps. While this strategy may be adequate for certain goals (e.g., clinical or diagnostic), it reduces the potential of the recordings to provide physiological insight because it hampers dissection of the generators contributing to the potentials, for example in terms of cochlear frequency region or of receptor cells versus neurons.

Traditionally, three main components are distinguished in stimulus-evoked mass-responses recorded near the cochlea: the *summating potential* (SP) which is thought to reflect a DC response from the receptor cells, an AC response component which is an ongoing signal that mimics the acoustic input signal and is referred to as the *cochlear microphonic* (CM), and a transient response called the *compound action potential* (CAP) which is generated by the auditory nerve. The SP, in particular its relation to the CAP, is of clinical interest (i.e., in Ménière's disease, [6, 11–13]), but its generation is ill-understood [14–17]. The CM is usually regarded as a purely receptor-generated potential (dominated by outer hair cells) so that investigators have usually turned to the CAP to evaluate the auditory nerve and by extension their excitatory source: the inner hair cells. The CAP to acoustical tone pips is used to extract objective audiograms in animal experiments and has been used toward that purpose in patients as well. However, because it depends on synchronicity among nerve fibers induced by stimulus onset, the CAP is inherently limited in its usefulness to examine sustained neural responses to spectrally pure stimulus components and is particularly problematic to probe low frequencies.

The CM has recently regained attention because it does not always contain only the mass receptor potential, as is generally assumed: at lower frequencies it also contains a neural contribution [18–24]. This component is a reflection of synchronized neural phase-locking in the auditory nerve and is also referred to as the auditory nerve neurophonic or simply neurophonic [25]. Given the limitations of the CAP mentioned above (transient synchronicity, poor at low frequencies), its measurement provides a welcome complement to assess neural properties. It provides the opportunity to probe low-frequency neural responses and the phenomenon of phase-locking, which is of obvious importance in a "low-frequency" species such as humans especially given the important role that has been attributed to temporal coding in normal and impaired hearing ([26, 27] but *see* ref. 28). The simplest way to extract a nerve contribution is to quantify the second harmonic component originating from the non-linear neural rectification process in the synaptic transmission between inner hair cell and auditory nerve fiber. The justification for this is the assumption that the receptor potential, produced primarily by the outer hair cells [29], is nearly linear at low and moderate intensities. Remarkable linearity is indeed observed in several species, such as in guinea pig, cat, monkey and human, but still the degree of linearity varies among species and has to be verified. Our own experience is that the harmonic distortion of the mass receptor potential (second Harmonic/first Harmonic) is very small in cat but is much larger in gerbil; in human, we observed a small harmonic distortion of the receptor potential, with a second Harmonic that is ~7 times smaller compared to that of the neurophonic. Besides the issue of linearity of the receptor potential, another disadvantage of using the second harmonic component to gauge neural responses is that it ignores the largest neural contribution, which is at the fundamental stimulus frequency. This component is strongly present at low frequencies and can easily exceed the contribution of the receptors particularly at low intensities [30], but it is heavily entangled with the receptor potentials and is often not recognized as a separate generator (e.g., [8]). We developed and verified a method in animal models to extract the fundamental neural component of acoustically evoked mass-potentials [24], based on a method of [31] for the assessment of CM contamination in the frequency following response. Our method allowed us to measure the upper limit of neural phase-locking using mass-potentials from the round window [32]. To quantify the neural response, the method makes use not of harmonic distortion, but of another property found at the level of the auditory nerve but not in receptor cells: adaptation to a forward masker [33].

The observation that the CM contains both receptor and neuron contributions makes the term "cochlear microphonic" somewhat of a misnomer or at least ambiguous as to its underlying generators. We use the term in a historical, operational sense to refer to the potential that is recorded in the immediate vicinity of the cochlea, not to refer to the potential generated by the receptor cells. A misunderstanding that is frequently encountered is that simple reversal of stimulus polarity is a sufficient basis for a clean separation between neural and "cochlear" (i.e., receptor) generators. Indeed, averaging of responses to, for example, tone pips that alternate in polarity removes large AC response components that are locked to the stimulus waveform. However, it not only removes (linear) receptor potentials, but also the largest component of the neural contribution. Only at frequencies beyond the upper limit of phase-locking (several kHz, depending on the species), does this procedure result in a good separation of neural and receptor contributions. At lower frequencies, it discards not only receptor potentials but also the large and arguably most interesting neural contributions. Simply put, at frequencies below a few kHz, both the receptor and neural potentials "follow" the stimulus waveform, and subtler differences between these two generators (harmonic distortion, adaptation) must be exploited to disambiguate them.

Based on procedures and acoustic paradigms previously developed in animals [24, 32, 34], we developed a TT-procedure to assess different fundamental properties of the human peripheral auditory system, such as sharpness of neural tuning and the limit of neural phase-locking [35]. In addition, we also assessed the medial olivocochlear efferent system, for an ipsilateral [36] and contralateral acoustic elicitor [37]. These methods open opportunities for fundamental or clinical study of auditory pathologies in patients. For instance, the simultaneous assessment of CM, neurophonic and CAP could be useful in the diagnosis of auditory neuropathy spectrum disorder.

In developing the TT method and procedure, there were a number of concerns we had to address, including the safety of the subject, accurate acoustical reproduction, and mechanical and electrical reliability. In the past, there was a concern about the invasive nature of the TT method, but the method has been found to be generally safe and well-tolerated by patients [38]. In this chapter we describe our TT-procedure with a focus on the practical aspects of the recordings. The rationale of the stimuli and analyses are briefly addressed but can be found more fully in our published work [35, 36]. Likewise, we do not discuss medical considerations and only briefly mention various factors affecting the quality and duration of the recordings.

2   Material and Methods

Our procedure, protocol and materials for the recording of mass-potentials via the human middle ear are described below. The TT-procedure and custom tools were developed, extensively tested, and practiced on more than 20 fresh human cadavers in the university hospital before applying it to awake subjects. Our recordings were done with approval of our institution's ethics committee (EC Research UZ/KU Leuven). Before the start of the procedure subjects gave written informed consent, in accordance with the Declaration of Helsinki.

2.1   Subject Screening and Preparation

Before the procedure, the subject's hearing is screened, including an inquiry for hearing problems, a pure tone audiogram (thresholds <20 dB nHL, 125 Hz to 8 kHz), and tympanometry to assess middle ear function. An otomicroscopic examination of the ear canal and tympanic membrane is performed by an otolaryngologist to examine the anatomy of the ear canal and the ability to clearly visualize the tympanic membrane, as well as to assess any signs toward middle or external ear disease. In our studies of efferent effects, at both ears the threshold for eliciting middle ear contraction by ipsilateral and contralateral acoustic stimulation was assessed with tympanometry (ZODIAC 901) using a 1 kHz tone and also for broadband white noise. Subjects who passed initial screening were requested to avoid exposure to loud sounds in the days preceding the transtympanic measurement session.

2.2   Custom Earmold

For acoustic control and reproducibility throughout the procedure and to preserve low-frequency performance during the recordings, a custom silicone earmold ① (large orange object Figs. 1 and 2; Dentsply, Aquasil Ultra XLV regular) was made for every subject. We usually combined this step with the preceding step (audiological and otological exam). The earmold was constructed such that it contained two casted openings (diameter 2 mm) ②③ used for

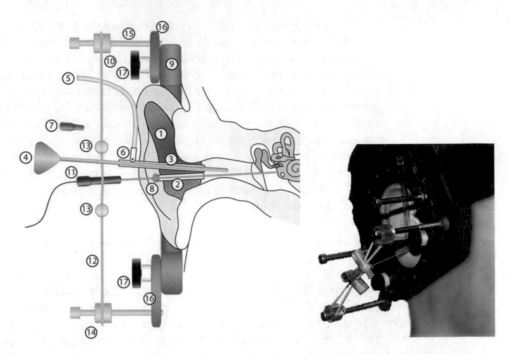

Fig. 1 Schematic representation (left) and photo (right) of the trans-tympanic arrangement around the ear

Fig. 2 Picture of a custom earmold with view on the two openings on the ear canal side. Note the presence of a tube (cyan-colored) that is used during casting and needle placement

visualization, acoustic stimulation, calibration, and the placement of the needle electrode. The position of the openings, and the angle between them, is critical to ensure good visualization and successful placement of the needle. To ensure this, a simple tube configuration (Fig. 3) was developed and used during casting. Two straw-like plastic tubes with outer diameter of 2 mm were glued together at a predetermined angle with a piece of foam at the front (e.g., Oto-block). The foam secures the tubes in the ear canal, protects the ear canal from the tubes, and prevents silicone deposition on the tympanic membrane during casting. The correct placement of the

Fig. 3 Picture of the disposable casting tool used to cast the openings in the earmold

tube configuration involves the use of a camera-equipped small rigid endoscope ④ (R. WOLF, 8654.402 25 degree PANOVIEW; ILO electronic GmbH, XE50-eco X-TFT-USB), inserted through the upper tube ③ and enabling visualization of the ear canal and eardrum.

During casting, the person lies on his/her side with the disposable casting tool (Fig. 3) facing upward, whereby the tube for the actual placement of the needle electrode is locked by a custom needle holder ⑩ which is mounted on an adjustable frame ⑨. The settings of the positioning of the adjustable frame and needle holder are noted down or photographed and reused later in the procedure before placing the needle electrode. After the earmold has cured, the whole assembly is carefully removed and the disposable casting tool is removed from the earmold. The part of the earmold protruding into the ear canal is trimmed to allow more freedom of movement later when inserting the needle electrode.

Protocol:

1. Mounting of the adjustable frame ⑨ on the subject's head with a Velcro strap, with pinna through opening.

2. Insertion of the tube configuration (Fig. 3) in the ear canal. The Otoblock foam is inserted to a depth that is completely in the ear canal but at a safe distance from the eardrum.

3. Adjustment of the positioning of the tube configuration using endoscope. The eardrum must be visible through both tubes (future openings ②③).

4. Adjust the swivel arms ⑯ attached to the frame ⑨, on which the metal guides ⑮ with the needle holder ⑩ are mounted, such that one of the two plastic tubes of the disposable casting tool is locked by needle holder and thereby supports this assembly.

5. Casting of silicone earmold. The silicone has to cover the inside of the pinna and the orifice of the ear canal.

6. During curing confirm the position of the tubes again with endoscope.

7. Remove earmold ① and trim ear canal part (Fig. 2).

2.3   In Situ Acoustic Calibration

Once a satisfactory earmold is casted and before the placement of the needle electrode, the complete acoustic system is calibrated in situ using a tube earphone-speaker (Etymotic Research, ER1 or ER2) and a probe tube microphone (Etymotic Research, ER-7C) close to the TM (inserted via ③). Before reinserting the earmold, it is recommended to apply Vaseline on the earmold where it touches the pinna to ensure that it is airtight. Sound from the tube earphone-speaker ⑤ is delivered through a side entry of a custom plastic T-piece ⑥ inserted into opening ③ of the earmold. The T-piece allows passage of the endoscope ④ to visualize the TM even when the speaker-tube is already connected. This is especially important later in the procedure to minimize the number of manipulations once the needle electrode is inserted. With the endoscope, placement of the probe tube microphone relative to the eardrum can be visually observed. During calibration, and also later during the recordings, the earmold is made airtight except for a tiny opening in the plastic T-piece to prevent static pressure buildup. After removal of the endoscope, the T-piece is closed with a custom silicone plug ⑦ (Dentsply, Aquasil Ultra XLV regular) and the opening of tube ② containing the silicone microphone tube is sealed with ear impression compound ⑧ (Audalin Bulk; Microsonic). Note that for safe handling of the endoscope, a spacer (not shown in the figure) is placed at the end of the shaft to limit the reach of the endoscope to stay well clear of the TM.

Protocol:

1. While the subject is seated, two skin electrodes (not shown) are applied: a standard ECG electrode at the ipsilateral mastoid and an earring pinch to the earlobe. Both electrodes are coated with a conductive gel (GVB-geliMED, Neurgel) to ensure a stable prolonged electrical contact, and before attaching the electrodes the skin is first treated with an mild abrasive gel (ADIN STRUMENTS, MLA1093) to improve skin conductivity. Thereafter the electric wires are connected to the electrodes and fixed with medical tape.

2. The frame ⑨ is mounted again on the head and the person is asked to take a supine position on a bed.

3. The earmold ① is inserted again with a little Vaseline on the inside to make it airtight. The T-piece ⑥ and a dummy plastic tube ② are inserted in the openings of the earmold and the speaker tube ⑤ is connected to the side input of the T-piece.

4. The silicone tube of the microphone is inserted in the tube ② and placed close to the tympanic membrane under visual guidance via the endoscope ④.

5. After placing the microphone tube and after removal of the endoscope, the T-piece is closed with a silicone plug ⑦ and the tube ② is sealed with soft earmold impression material ⑧ (Audaline).

6. The acoustic system is calibrated.

7. After calibration, the tubes are cleared and the positions of the swivel arms ⑯ and needle holder ⑩ on the frame are preset again to prepare insertion. The positions are verified with a dummy needle electrode "⑪" under endoscopic vision.

8. The entire earmold assembly is carefully removed.

2.4 Local Anesthetics

For subject's comfort, before insertion of the needle electrode, the tympanic membrane and ear canal were locally anesthetized with Bonain's solution (1:1:1 volume mixture of cocaine hydrochloride, phenol and menthol). The solution was applied for 30 min and then aspirated. The aspiration must be done thoroughly to avoid anesthetics entering the middle ear through the subsequent small perforation caused by insertion of the needle electrode. The brief sliding of the needle through the tympanic membrane (in the next **step 5** in the procedure) is audible by the subject. Subjects usually have a brief and vague sensation of touch during puncturing of the tympanic membrane, which then quickly disappears.

Protocol:

1. Under visual control with an otological microscope (ZEISS, OPMI Pico), Bonain's solution is applied in the ear canal and on the eardrum.

2. After 30 min, the solution is aspirated.

2.5 Transtympanic Electrode Placement

A short sterile plastic tube (length < 1 cm, diameter 2 mm) is placed in opening ② of the earmold. In this opening an ENT surgeon inserts a sterile needle electrode ⑪ (TECA, sterile monopolar disposable, 75 mm × 26G, 902-DMG75-TP). Thereafter, the needle electrode is advanced and placed in the region of the posterior inferior quadrant (third quadrant) of the tympanic membrane, toward the cochlear promontory or the niche of the round window. These underlying structures are sometimes visible through the TM. Once the needle touches the promontory or niche, the needle is fixed with a rubber band held by sliders ⑭ on the metal guides ⑮ attached to the frame via the swivel arms. The sliders are moved to put the needle under slight tension. After the recording session the needle electrode is gently pulled back and the ear mold removed. The session is concluded with an otomicroscopic examination by an ENT-surgeon. The eardrum perforation is limited to the size of the needle electrode (0.46 mm) and recovers fast. Following the TT-procedure, it is recommended to avoid external pressure and water in the ear (e.g., showering) for 10 days. During this period, the ENT-surgeon was available to address any concerns of the subject related to the recording session.

During the actual recording, subjects were asked to remain supine in a comfortable position. Slight and careful adjustments in position by the subjects were possible, as the entire assembly was lightweight and safely strapped to the head of the subject. An investigator was with the subject inside the sound booth at all times, to facilitate communication with the investigators outside the sound booth.

Protocol:

1. The person with the stereotactic frame is asked to take a supine position on a bed.

2. The skin electrodes, the mastoid electrode, and sterile needle electrode are connected to a battery-operated differential amplifier (which is at this point switched off).

3. The tube for calibration is replaced by a new but sterile tube ②. The earmold ① is then reinserted and the T-piece ⑥ connected (with speaker tube already attached).

4. The needle holder ⑩ is put in place on the frame ⑨.

5. The sterile needle electrode ⑪ is inserted through the open elastic bands ⑫ (without touching it), such that the front of the needle electrode is resting in the sterile tube ②. The holder of the needle electrode ⑪ (purple plastic carrier at backside) is loosely held between the elastic bands; the sliding shutters ⑬ on the elastic band are in the "open" position, that is, away from the electrode needle. This allows at the same time support and some degree of freedom to the needle electrode. Note that at this stage subjects and experimentalists, while in the sound booth, have to be grounded to booth and amplifier with an antistatic wrist strap to prevent electrostatic discharge when inserting the needle electrode.

6. The needle electrode is advanced by the ENT surgeon toward the TM, under visual guidance of the endoscope operated by an assistant. While moving the needle electrode forward in the direction of the posterior inferior quadrant of the TM, the sliders ⑭ on the metal guides ⑮ will also slide forward until the needle electrode touches the promontory or niche. At that point, the needle is held in place by the ENT surgeon.

7. Next, the shutters ⑬ are moved toward the needle electrode and gently closed. The sliders ⑭ are then gently pushed forward over the metal guides ⑮ so as to obtain slight tension on the needle by the spring action of the elastic bands. The ENT surgeon releases the needle which is now retained securely by the needle holder under slight tension due to friction between the plastic sliders and metal guides.

8. The endoscope is removed, the two open tubes are closed and sealed again by the silicone plug and soft impression material, respectively. The amplifier is switched on and the measurement session can start.

9. After the session the needle is gently retracted (painless) and the ear is inspected by the ENT surgeon with a surgical microscope.

2.6   Adjustable Frame

In order to maintain the position of the needle electrode ⑪ relative to the unrestricted head, a lightweight custom head frame ⑨ is used (Fig. 1). The frame consists of a ring centered on the external ear and is fastened around the subject's head with a Velcro strap. Our assembly was custom made from Polyetherimide (PEI) which is easy to machine and clean.

On this ring two swivel arms ⑯ with metal guides ⑮ (partially threaded M6 socket cap screws) are mounted on the frame with thumbscrews ⑰. To these screws, a needle holder ⑩ is attached, consisting of two elastic bands ⑫, two plastic sliders ⑭ and two custom plastic shutters ⑬ to constrict the elastic bands and squeeze the plastic head of the needle electrode ⑪. The needle holder provides stable support for the needle electrode and can move freely with the sliders over the metal guides. The combination of slide/guides with a certain pre-tension on the elastic bands gives the sliders the right amount of friction on the guides so that the needle electrode can be brought under light tension. This ensures good and reliable mechanical and electrical contact between the needle electrode and the promontory or niche throughout the recording.

2.7   Experimental Apparatus: Measurement Booth

As in most auditory, neurophysiological investigations it is desirable to perform the experiments in a double-walled soundproofed and faradized booth (e.g., Industrial Acoustics Company, Niederkrüchten, Germany) to exclude or minimize electrical and acoustical interference and allow full acoustic control. Of course, the awake subject and accompanying investigator are themselves potential sources of uncontrolled sound. It is advisable to have continuous visual and/or auditory means of communication with the investigators outside the booth, and to set up time slots when the investigator inside the booth queries the subject regarding his or her comfort.

2.8   Experimental Apparatus: Acoustical Stimulation

In our experiments, stimuli were generated with custom software and a digital sound system (Tucker-Davis Technologies, system 2, sample rate: 125 kHz/channel) consisting of a digital-to-analog converter (PD1), a digitally controlled analog attenuator (PA5), a headphone driver (HB7) and an electrically shielded earphone-speaker (Etymotic Research, ER-1 or ER-2).

2.9   Experimental Apparatus: Electrophysiological Recordings

Acoustically-evoked mass-potentials are recorded with a low-noise differential preamplifier (Stanford Research Systems, SR560). The signal input is connected to the transtympanic needle electrode; the reference input is connected to an earlobe clamp coated with conductive gel; and the ground is connected to a standard disposable surface electrode placed at the mastoid, also coated with conductive gel. All contacts are made on the side ipsilateral to the recording. For electrical safety, the battery-operated preamplifier is galvanically isolated (A-M systems, Analog stimulus isolator Model 2200) from the mains-powered equipment. Note that this type of optical isolator introduces additional system noise so that more averaging is required to improve the signal-to-noise ratio. Before the signal is recorded (TDT, RX8, ~100 kHz/channel, max. SNR 96 dB), stored and analyzed (The MathWorks, Matlab), the signal is further amplified (DAGAN, BVC-700A) to a total gain of ×100k and band-pass filtered (for frequency tuning: 30 Hz to 30 kHz, for neural phase-locking 0.03 Hz to 30 kHz; cut-off slopes 12 dB/octave). During the sessions the most relevant signals are visualized on an oscilloscope (LeCroy, WaveSurfer 24Xs) and monitored with an amplified loudspeaker. The differential preamplifier and the galvanic separator were inside the sound booth, the computer and the rest of the equipment were stationed outside.

2.10 Stimulus Paradigms and Data Processing

To increase the signal-to-noise ratio of the acoustically-evoked mass potentials, responses are averaged across multiple repetitions (between 128 and 1024). The polarity of the stimulus is alternated to allow splitting of the response into polarity-dependent and -independent components. This gives two sets of recordings that are each averaged (e.g., Fig. 5Ba) and which can then be added or subtracted to provide the responses of interest. Addition (Fig. 5Bb) provides the stimulus polarity-independent responses, such as the CAP (with summating potential), the even harmonics of the neurophonic and also those of the CM if it were not linear. Subtraction (Fig. 5Bc) provides the fundamental and other odd harmonics of the neurophonic and the CM. An illustration of summed, polarity-alternated and averaged responses in human to 6 kHz short tones (10 ms; with 5 ms on/off gating) is shown in Fig. 4 for increasing intensities (black to red: background activity, 55, 60, ... 80 dB SPL). A CAP is noticeable in the different traces, which increases and decreases in amplitude and peak-delay with stimulus intensity, respectively. Note that the CM is completely cancelled by the summation.

To measure the neurophonic we developed a paradigm to temporally separate the heavily entangled neurophonic and CM. The paradigm is based on neural adaptation, that is, the decline in activity after stimulus onset and recovery after stimulus cessation, which can be modeled with several time constants [31, 39–41] and which is not present in the receptor potential. The basis of the masking paradigm is a probe, usually a pure tone, which is preceded by a higher intensity masker, usually a tone or noise. The preceding masker adapts the response of neurons such that the response to the probe is temporarily reduced or suspended,

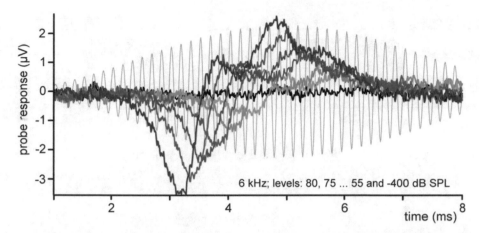

Fig. 4 Averaged evoked summed responses to polarity alternating tones of 6 kHz for different intensities after stimulus onset. The most prominent feature is the CAP. The amplitude and timings of the peaks (N1, P1, etc.) are dependent on the stimulation level. The original stimulus shape (not compensated for acoustic delay of 1 ms) is shown in the background in gray. Additional stimulus specifications: 10 ms, 5 ms on/off gating

depending on the relative masker vs. probe levels. Note that the receptor potential is not affected by the preceding masker. An example of the representation of the stimulus (only positive polarity shown) with its averaged response are illustrated in Fig. 5A. More information about the stimulus and how the response is obtained can be found in the caption. When the masked probe response (Fig. 5Bd) is subtracted from the non-masked probe response (Fig. 5Bc) the result consists solely of neural contributions (Fig. 5Be). Note that after a while the neuron readapts to the state before the masker so that the extracted neural response exists only temporarily (order of magnitude 10 ms). This short period is usually sufficient to obtain a measurement of the amplitude of the neurophonic. Vice versa, this paradigm also allows measurement of the CM uncontaminated by the neurophonic [31, 42]. The measurement indication between traces Bd and Be in Fig. 5 illustrates the onset delay between CM and neurophonic and thus represents the time span in 5Bd free of evoked neural contributions.

3 Conclusion

In this chapter, we describe a TT-procedure with protocol and material for recording peripheral acoustically evoked mass-potentials in awake subjects. The method is designed to allow safe and stable measurements for a long period of time with minimum of discomfort for the subject. In practice, the duration of effective recording varied rather widely between subjects (2–6 h), and this was also true regarding the quality of the recordings. Tense subjects typically provided for sessions limited in duration and quality,

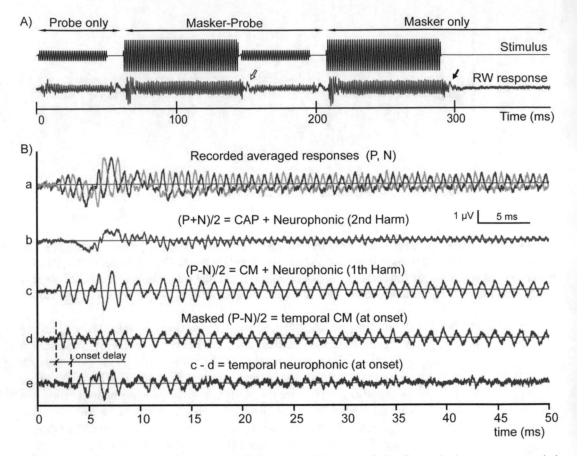

Fig. 5 Stimulus to measure neurophonic and CM together with averaged stimulus evoked responses recorded in human to long tones of 800 Hz. (**A**) Schematic representation of half the stimulus representation (upper black trace; positive polarity) with the average response (lower cyan trace). The stimulus consists of three segments, the first segment contains only the probe (50 ms), the second segment contains the masker (83.71 ms) followed by a probe, and the last segment contains only the masker. The purpose of the masker-only segment is to remove (subtract from the "mask-probe" segment) a possible lingering masker offset response, as illustrated by the two arrows. Moreover, the masker-only segment contains an additional stimulus-free period (~80 ms) to recover from the masker, so this is before the beginning of the other half stimulus representation with opposite polarity. The masker-probe interval was 1 ms, and the interval between different stimuli was at least 10 ms. The probe and masker were gated with a 1 ms raised cosine. The masker level is 10 dB above the probe level. (**B**) Example of averaged processed signals recorded with the paradigm in **A**. (a) A pair of raw unmasked probe responses for opposite stimulus polarity (from the probe-only segment in panel **A**). (b) Summation of responses in a, with CAP and second harmonic, here mainly from neural origin. (c) Subtraction of responses in (a). (d) As in c but now the tone is preceded by the masker (from the masker-probe segment minus masker-only segment in panel **A**). (e) The result after subtraction of the masked (d) from the non-masked (c) summed response. The calculated response shows the neurophonic (~fundamental component) as a decaying component due to the recovery of adaptation from the masker. Note that the neurophonic has an onset delay compared to CM, consistent with their different origin. Stimulus parameters: probe level 75 dB SPL, masker level 85 dB SPL, frequency of both tones is 800 Hz. Note that the responses in **B** are from another subject than that in **A**

mainly because of myogenic noise. In some subjects, middle ear muscle artifacts interfered with the recording. The TT recordings enabled us to assess important basic properties of the human auditory peripheral system including frequency selectivity, phase-locking, and efferent effects, and to compare these with data from experimental animals. We believe the TT approach is underused and has much promise to advance our understanding of hearing impairment, for example to address the controversies surrounding auditory neuropathy [43]. In this chapter, we only addressed the recording and not the medical aspects of the TT-procedure, which clearly require the involvement of a qualified ENT surgeon for guidance and performance of critical aspects of the procedure. The main drawback of the method is that the placement of the needle electrode is blind, that is, without clear visualization of middle ear structures beyond the TM. One approach to reduce the "blind" aspect of the needle electrode placement is to limit the procedure to patients with perforation of the TM.

References

1. Harrison RV, Aran JM (1982) Electrocochleographic measures of frequency selectivity in human deafness. Br J Audiol 16:179–188

2. Eggermont JJ (1977) Compound action potential tuning curves in normal and pathological human ears. J Acoust Soc Am 62:1247–1251

3. Fromm B, Nylén CO, Zotterman Y (1935) Studies in the mechanism of the Wever and Bray effect. Acta Otolaryngol 22:477–486. https://doi.org/10.3109/00016483509118125

4. Lempert J, WEVER EG, LAWRENCE M (1947) The cochleogram and its clinical application: a preliminary report. Arch Otolaryngol 45:61–67. https://doi.org/10.1001/archotol.1947.00690010068005

5. Perlman HB, Case TJ (1941) Electrical phenomena of the cochlea in man. Arch Otolaryngol 34:710–718. https://doi.org/10.1001/archotol.1941.00660040766003

6. Eggermont JJ (2017) Ups and downs in 75 years of electrocochleography. Front Syst Neurosci 11. https://doi.org/10.3389/fnsys.2017.00002

7. Mori N, Saeki K, Matsunaga T, Asai H (1982) Comparison between AP and SP parameters in trans- and extratympanic electrocochleography. Audiology 21:228–241

8. Noguchi Y, Nishida H, Komatsuzaki A (1999) A comparison of extratympanic versus transtympanic recordings in electrocochleography. Audiology 38:135–140. https://doi.org/10.3109/00206099909073015

9. Roland PS, Yellin MW, Meyerhoff WL, Frank T (1995) Simultaneous comparison between transtympanic and extratympanic electrocochleography. Am J Otol 16:444–450

10. Møller AR, Jannetta P, Bennett M, Møller MB (1981) Intracranially recorded responses from the human auditory nerve: new insights into the origin of brain stem evoked potentials (BSEPs). Electroencephalogr Clin Neurophysiol 52:18–27. https://doi.org/10.1016/0013-4694(81)90184-X

11. Hornibrook J, Kalin C, Lin E et al (2012) Transtympanic electrocochleography for the diagnosis of Ménière's disease. Int J Otolaryngol. https://www.hindawi.com/journals/ijoto/2012/852714/. Accessed 10 Feb 2021

12. Iseli C, Gibson W (2010) A comparison of three methods of using transtympanic electrocochleography for the diagnosis of Meniere's disease: click summating potential measurements, tone burst summating potential amplitude measurements, and biasing of the summating potential using a low frequency tone. Acta Otolaryngol 130:95–101. https://doi.org/10.3109/00016480902858899

13. Schmidt PH, Eggermont JJ, Odenthal DW (1974) Study of MenièRe's disease by electrocochleography. Acta Otolaryngol 77:75–84. https://doi.org/10.1080/16512251.1974.11675748

14. Dallos P, Schoeny ZG, Cheatham MA (1972) Cochlear summating potentials. Descriptive aspects. Acta Otolaryngol Suppl 302:1–46

15. Davis H, Deatherage BH, Eldredge DH, Smith CA (1958) Summating potentials of the Cochlea. Am J Physiol Legacy Content 195:251–261. https://doi.org/10.1152/ajplegacy.1958.195.2.251

16. Pappa AK, Hutson KA, Scott WC et al (2019) Hair cell and neural contributions to the cochlear summating potential. J Neurophysiol 121:2163–2180. https://doi.org/10.1152/jn.00006.2019

17. Whitfield IC, Ross HF (1965) Cochlear-Microphonic and summating potentials and the outputs of individual hair-cell generators. J Acoust Soc Am 38:126–131. https://doi.org/10.1121/1.1909586

18. Choudhury B, Fitzpatrick DC, Buchman CA et al (2012) Intraoperative round window recordings to acoustic stimuli from cochlear implant patients. Otol Neurotol 33:1507–1515. https://doi.org/10.1097/MAO.0b013e31826dbc80

19. He W, Porsov E, Kemp D et al (2012) The group delay and suppression pattern of the cochlear microphonic potential recorded at the round window. PLoS One 7:e34356. https://doi.org/10.1371/journal.pone.0034356

20. Henry KR (1995) Auditory nerve neurophonic recorded from the round window of the Mongolian gerbil. Hear Res 90:176–184

21. Kamerer AM, Chertoff ME (2019) An analytic approach to identifying the sources of the low-frequency round window cochlear response. Hear Res 375:53–65. https://doi.org/10.1016/j.heares.2019.02.001

22. Lichtenhan JT, Cooper NP, Guinan JJ Jr (2012) A new auditory threshold estimation technique for low frequencies: proof of concept. Ear Hear. https://doi.org/10.1097/AUD.0b013e31825f9bd3

23. Patuzzi RB, Yates GK, Johnstone BM (1989) The origin of the low-frequency microphonic in the first cochlear turn of guinea-pig. Hear Res 39:177–188. https://doi.org/10.1016/0378-5955(89)90089-0

24. Verschooten E, Joris PX (2014) Estimation of neural phase locking from stimulus-evoked potentials. J Assoc Res Otolaryngol 15:767–787. https://doi.org/10.1007/s10162-014-0465-9

25. Snyder RL, Schreiner CE (1984) The auditory neurophonic: basic properties. Hear Res 15:261–280

26. Moore BC (2008) The role of temporal fine structure processing in pitch perception, masking, and speech perception for normal-hearing and hearing-impaired people. J Assoc Res Otolaryngol 9:399–406. https://doi.org/10.1007/s10162-008-0143-x

27. Zeng F-G, Kong Y-Y, Michalewski HJ, Starr A (2005) Perceptual consequences of disrupted auditory nerve activity. J Neurophysiol 93:3050–3063. https://doi.org/10.1152/jn.00985.2004

28. Verschooten E, Shamma S, Oxenham AJ et al (2019) The upper frequency limit for the use of phase locking to code temporal fine structure in humans: a compilation of viewpoints. Hear Res 377:109–121. https://doi.org/10.1016/j.heares.2019.03.011

29. Dallos P, Cheatham MA (1976) Production of cochlear potentials by inner and outer hair cells. J Acoust Soc Am 60:510–512. https://doi.org/10.1121/1.381086

30. Verschooten E (2013) Assessment of fundamental cochlear limits of frequency resolution and phase-locking in humans and animal models. (Document No LIRIAS1778220) [Doctoral dissertation, KU Leuven] Leuven Institutional Repository and Information Archiving System

31. Chimento TC, Schreiner CE (1990) Selectively eliminating cochlear microphonic contamination from the frequency-following response. Electroencephalogr Clin Neurophysiol 75:88–96. https://doi.org/10.1016/0013-4694(90)90156-E

32. Verschooten E, Robles L, Joris PX (2015) Assessment of the limits of neural phase-locking using mass potentials. J Neurosci 35:2255–2268. https://doi.org/10.1523/JNEUROSCI.2979-14.2015

33. Snyder RL, Schreiner CE (1985) Forward masking of the auditory nerve neurophonic (ANN) and the frequency following response (FFR). Hear Res 20:45–62

34. Verschooten E, Robles L, Kovačić D, Joris PX (2012) Auditory nerve frequency tuning measured with forward-masked compound action potentials. J Assoc Res Otolaryngol 13:799–817. https://doi.org/10.1007/s10162-012-0346-z

35. Verschooten E, Desloovere C, Joris PX (2018) High-resolution frequency tuning but not temporal coding in the human cochlea. PLoS Biol 16:e2005164. https://doi.org/10.1371/journal.pbio.2005164

36. Verschooten E, Strickland EA, Verhaert N, Joris PX (2017) Assessment of ipsilateral

efferent effects in human via ECochG. Front Neurosci 11:331. https://doi.org/10.3389/fnins.2017.00331

37. Verschooten E, Strickland E, Verhaert N, Joris P (2016) Effect of contralateral stimulation on low frequency hearing in human. In: 2016 midwinter meeting. Association for Research in Otolaryngology, San Diego, CA, p 197

38. Ng M, Srireddy S, Horlbeck DM, Niparko JK (2001) Safety and Patient Experience With Transtympanic Electrocochleography. The Laryngoscope 111:792–795. https://doi.org/10.1097/00005537-200105000-00007

39. Harris DM, Dallos P (1979) Forward masking of auditory nerve fiber responses. J Neurophysiol 42:1083–1107

40. Smith RL, Brachman ML (1982) Adaptation in auditory-nerve fibers: a revised model. Biol Cybern 44:107–120

41. Westerman LA, Smith RL (1984) Rapid and short-term adaptation in auditory nerve responses. Hear Res 15:249–260

42. Henry KR (1997) Auditory nerve neurophonic tuning curves produced by masking of round window responses. Hear Res 104:167–176. https://doi.org/10.1016/S0378-5955(96)00195-5

43. Bramhall N, Beach EF, Epp B et al (2019) The search for noise-induced cochlear synaptopathy in humans: mission impossible? Hear Res 377:88–103. https://doi.org/10.1016/j.heares.2019.02.016

Chapter 15

Strategies for Identification of Medial Olivocochlear Neurons for Patch-Clamp Studies of Synaptic Function Using Electrical Stimulation and Optogenetics

Kirupa Suthakar, Lester Torres Cadenas, and Catherine Weisz

Abstract

To understand how action potential patterns are generated in a single neuron, analysis of intrinsic electrical properties and the integration of synaptic activity are typically investigated using whole-cell patch-clamp electrophysiology experiments. However, standard patch-clamp recording techniques are hindered when target cells cannot be positively identified. The olivocochlear efferent circuitry of the auditory system, specifically medial olivocochlear (MOC) neurons, play critical roles in shaping responses to acoustic stimuli including gain control and detection of salient sounds, and through these mechanisms are also implicated in protection from noise damage and selective attention. MOC neurons are notoriously difficult to target for typical visually guided patch-clamping because their cell bodies are diffusely located in the superior olivary complex (SOC) among other cell types, and are intercalated amongst myelin dense trapezoid body fiber tracts, limiting the use of brightfield optics to identify morphology of unlabeled neurons. Therefore, studies that allow careful analysis of synaptic inputs and intrinsic electrical properties of MOC neurons are lacking. Here, we describe two techniques to identify MOC neurons for patch-clamp recordings in brain slices. We then describe experiments to analyze MOC neuron synaptic inputs at the single-cell level using electrical and optogenetic stimulation of presynaptic neurons.

Key words Medial olivocochlear neurons, Retrograde label, Patch-clamp electrophysiology, Optogenetic stimulation, Brainstem slices, Auditory

1 Introduction

The mammalian auditory system contains fast and precise neurons that allow the high-fidelity responses required for normal encoding of sound stimuli. All incoming auditory information at the sensory inner hair cells (IHCs) is distributed in a divergent manner via spiral ganglion neurons (SGNs) to neurons of the cochlear nucleus (CN), whose outputs divaricate to form multiple parallel pathways to

---

Kirupa Suthakar and Lester Torres Cadenas contributed equally with all other contributors.

Andrew K. Groves (ed.), *Developmental, Physiological, and Functional Neurobiology of the Inner Ear*, Neuromethods, vol. 176,
https://doi.org/10.1007/978-1-0716-2022-9_15,

handle processing of different features of sound such as pitch, level and timing. In addition to the complex ascending system, there is also considerable descending circuitry that feeds back to all stages of auditory processing [1, 2]. The final stage of the descending circuitry are the olivocochlear (OC) neurons consisting of medial olivocochlear (MOC) and lateral olivocochlear (LOC) subdivisions that innervate the organ of Corti in the cochlea [3–7]. MOC neurons are located in the ventral nucleus of the trapezoid body (VNTB) and send axons to the electromotile cochlear outer hair cells (OHC) [8–10], Fig. 1a. The net effect of MOC activation is to inhibit OHC, which reduces cochlear amplification enabling cochlear gain control in response to changing intensity levels [11–16], detection of salient sounds above background noise [17, 18], protection of the cochlea from acoustic trauma [19–25], and auditory attention [26–29].

The interrogation of neuronal circuits is ideally performed in an intact animal with normal sensory inputs and brain function. However, the use of awake in vivo electrophysiological techniques to measure synaptic activity is complicated by the ventral location of the SOC, access to which necessitates invasive surgical procedures. Further, the deep anesthesia required differentially alters auditory neuron activation patterns [30–35]. Thus, in vitro experiments using brain slices and cultured cells are still requisite in sensory neuroscience research for higher throughput analysis of synaptic activity and intrinsic electrical properties of single neurons. Patch-clamping is often guided by visualization of a neuron under specialized brightfield optics to allow precise contact between the patch pipette and the cell membrane and allows careful selection of the neuron or neuronal structure of interest [36–38]. While many neurons are readily identifiable, and consequently, easily targeted for visually guided patch-clamping because of their unique size or morphology, location, or presence in a homogeneous population of cells, this chapter is focused primarily on neurons for which traditional identification techniques are inadequate.

Until recently, investigation of the synaptic inputs to MOC neurons has been hampered by difficulties in accessing the neurons in the VNTB in unstained, live brainstem slice preparations typically used for patch-clamp electrophysiology because the few hundred MOC neurons are diffusely located among other types of neurons. Previous work has managed some recordings from MOC neurons, but relied on the more labor intensive task of surgical injection of neuronal tracers into live rats followed by a 2 day incubation with post-op care [39], or recordings from randomly selected neurons of the VNTB with post hoc neuronal identification based on dendritic morphology and groupings of current-clamp after-hyperpolarizations [40]. A higher-throughput method is required to visually identify MOC neurons in brain slices (Fig. 1b) to thoroughly investigate the synaptic circuitry that drives MOC function.

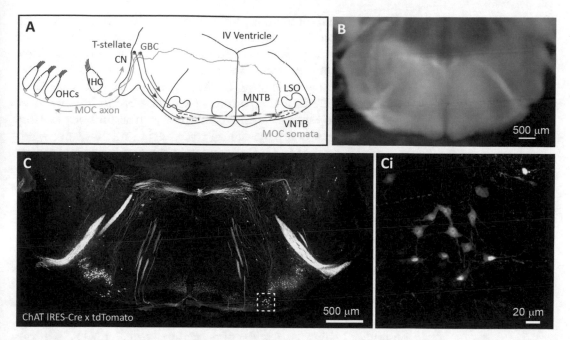

Fig. 1 MOC efferent circuitry. (**a**) As sound waves enter the cochlea, IHCs transduce this information into electrical signals which are conveyed to the CN by SGNs (light green). T-stellate neurons in the CN (sometimes referred to as planar multipolar cells; dark blue), project via the ventral acoustic stria (VAS aka trapezoid body) to innervate MOC neurons in the VNTB (light blue). Globular bushy cell (GBC) axons take the same route through the VAS to innervate neurons of the medial nucleus of the trapezoid body (MNTB, dark green). MOC axons project dorsally and turn sharply traveling via the olivocochlear bundle to either cross the midline and to form axon terminals on OHCs from the same ear (light blue) or the opposite ear (not shown). The recently characterized direct inhibitory input from MNTB to MOC neurons is indicated in magenta. (**b**) Reliable, positive identification of MOC somata in the VNTB using the ChAT-IRES-Cre × tdTomato transgenic mouse line. In this low-magnification micrograph, tdTomato labeling is observed in cells immunopositive for ChAT (modified from [44]). Bilaterally, MOC neurons are visible in VNTB, along with LOC neurons in the lateral superior olive (LSO). (**c**) Low-magnification image of representative coronal brain slice from which recordings were made with higher magnification image of VNTB and morphology of MOC somata shown in **ci**. Note the prominent facial nerves, which have a characteristic appearance in both brightfield and fluorescently tagged ChAT tissue such as that in panel "**b**." Abbreviations: *IHC* inner hair cell, *CN* cochlea nucleus, *MOC* medial olivocochlear efferent, *VNTB* ventral nucleus of the trapezoid body, *VAS* ventral acoustic stria, *MNTB* medial nucleus of the trapezoid body, *OHC* outer hair cell, *ChAT* choline acetyl transferase

Genetic technologies to enable precise identification or stimulation of target neurons are widespread; however, their experimental use requires careful validation. While multiple techniques exist to introduce genes of interest such as fluorescent proteins for cell localization, including the CRISPR-Cas9 system [41], and virus-mediated delivery [42], the Cre-Lox system is routinely used to induce transgenic mice through simple breeding schemes. The Cre-Lox system enables cell-type specific expression of a multitude of genes engineered with Cre-dependent sequences such as genes for fluorescent markers (e.g., tdTomato) for visual identification of cells based on their genetic profile. A plethora of mouse strains with Cre expression driven by different cell-type specific promoters are now commercially available, and can be easily paired with multiple Cre-dependent reporter strains generating fluorophores visible using different wavelengths of light [43] to localize cells. Best practices dictate that the pattern of Cre expression in any mouse line should be validated using techniques such as antibody labeling or conventional neuroanatomical tract tracing methods using retrograde tracer dyes [44]. Unfortunately, the Cre-lox system cannot be used for the combined expression of both a fluorescent reporter and other Cre-dependent tools such as optogenetic proteins, described below, and additionally, the large variety of commercially available lines are not yet available in other species commonly used for auditory research (e.g., rats, ferrets, gerbils).

We describe the use of commercially available mice expressing Cre recombinase driven by the choline acetyltransferase (ChAT) promoter crossed to a fluorescent reporter mouse line (Ai14 tdTomato, Fig. 1 c, ci). Then, because some experiments on MOC neurons may utilize the Cre-Lox system for other purposes briefly described below, we also detail procedures to label MOC neurons via retrograde transport of a lipophilic neuronal tracer applied to the cochlea. This alternative method for identification of MOC neurons has been used to label neuron subpopulations in other brainstem preparations, such as with labeling of CN globular bushy cell (GBC) axons and their Calyx of Held terminations in the MNTB [45] and was used by our group to characterize the ChAT-IRES-Cre × tdTomato line [44] for patch-clamp recordings from the fluorescent MOC neurons in brainstem slices to demonstrate inhibitory synaptic inputs from neurons of the MNTB, which are powerful enough to suppress action potentials in the MOC neurons [44]. This acute tracing procedure can be used for positive identification of other types of neurons that have a defined axon projection to a location that can be targeted for lipophilic dye application. In addition, use of a fluorescent tracer instead of a genetic label allows use of a different Cre line or other genetic tool for combinatorial experiments including positive identification of MOC neurons paired with Cre-driven expression of optogenetic proteins, genetically encoded calcium indicators, or any number of Cre-based tools available [46].

Either the genetic label of MOC neurons using the ChAT-IRES-Cre × tdTomato mouse line [44, 47] or retrograde label of MOC neurons from the cochlea can be used to perform patch-clamp recordings from identified MOC neurons to investigate their electrical properties and synaptic inputs. Here we describe experiments to investigate the single synaptic inputs to MOC neurons, using first a nonspecific electrical activation of presynaptic axons, then a cell-type specific optogenetic activation of presynaptic neurons using the recently discovered inhibitory synaptic inputs from the MNTB [44] as an example. Electrical stimulation of axons to generate action potentials and subsequent neurotransmitter release onto postsynaptic cells during patch-clamp recordings is a technique that has been used in many studies of MNTB synapses onto auditory neurons [48–52] and is a simple way to activate circuits in a high throughput, low cost, yet powerful manner. We then present an additional method for cell-type specific activation of presynaptic cells using light-activated optogenetic stimulation [53] of MNTB neurons expressing ChR2(H134R) (a channelrhodopsin variant) in a commercially available transgenic mouse line with ChR2 expression driven by the vesicular GABA transporter (VGAT) promoter in the VGAT-ChR2-EYFP mouse line [54], for expression in GABAergic and glycinergic neurons. Because the VGAT-ChR2-EYFP mouse line is not Cre dependent, it can be used with either of the methods described here to identify postsynaptic MOC neurons.

In addition to the synaptic stimulation methods described here, recordings from identified MOC neurons can be paired with techniques including, but not limited to, pharmacological manipulation of neurotransmitter receptors, selective stimulation of presynaptic axons, studies of intrinsic electrical properties, optogenetic stimulation of subpopulations of presynaptic neurons, neurotransmitter uncaging, and calcium and voltage imaging. The list of techniques that are amenable to brain slice preparations that support the temporal and spatial precision enabled by in vitro techniques continues to grow, and with it the improved opportunity to understand the underlying auditory circuitry.

2 Materials

2.1 Materials for Retrograde Label of Axons and Brain Slice Preparation

Artificial CSF (aCSF): Add in the following order (in mM) to 18.2 MΩ purified water: 124 NaCl (S271-1, Fisher Scientific), 1.3 MgSO$_4$ (M65-500, Fisher Scientific) 5 KCl (P217-500, Fisher Scientific), 1.25 KH$_2$PO$_4$ (P288-100, Fisher Scientific), 26 NaHCO$_3$ (S5761, Sigma-Aldrich), 10 dextrose (S5-500, Fisher Scientific), then gently bubble with 95% O$_2$/5% CO$_2$ for 5 min through either small diameter tubing or a pumice-tipped glass pipette (39533-12C, VWR). Add 1.2 or 2 CaCl$_2$ depending on the experiment (C7902, Sigma-Aldrich). Bring the final volume to 1 L, and pH if necessary to 7.4 with 1 M NaOH. A "slicing solution," used to block glutamate-evoked excitotoxicity during slicing, is prepared by dissolving 1 mM kynurenic acid (K3375-5G, Sigma-Aldrich) in a 250 mL aliquot of aCSF by sonicating for 10 min in a water bath. Chill the slicing solution on ice with continuous gentle bubbling to maintain pH.

<table>
<tr><td>

2.1.1 Materials for Retrograde Fluorescent Tracer Application

</td><td>

Surgical tools: 10 cm scissors with 4 cm blades for euthanasia (RS-6872, Roboz Surgical Instrument Co.), angled dissecting scissors (14082-09, Fine Science Tools), #4 forceps (501978, World Precision Instruments (WPI)), #5 forceps (500342, WPI), spatula with one tapered end and one rounded end (14374, Fisher Scientific), super glue, small scissors, single-edged carbon steel razor blades (12-640, Fisher Scientific).

</td></tr>
</table>

Additional materials: borosilicate glass pipette pulled to an ~1–5 μm tip (usually an unused patch-clamp pipette from a previous day's experiments), weighing paper (09-898-12A, Fisher Scientific), dextran fluorescein (D3306, Fisher Scientific), plastic petri dish 35 mm × 10 mm (FB0875711YZ, Fisher Scientific), and glass petri dish 100 mm diameter × 20 mm (08-747D, Fisher Scientific).

Dissecting microscope with continuous zoom from $1\times$ to $3\times$ (SM-1BN, Amscope) attached with a $0.5\times$ c-mount post (641809, Harvard Apparatus) with 150 W Dual Goose-neck Fiber Optic Illuminator (HL150-AY, Amscope).

2.1.2 Equipment for Brain Slice Preparation

Vibratome: V1200S Leica with ceramic (501631, WPI) or carbon steel blade (12-640, Fisher Scientific).

Interface chamber for slice incubation, chamber design by Dr. Lawrence Katz: Slices are stored in a custom humidified interface chamber (Fig. 2a, [50]) at 32 °C for 1 h, allowed to cool to room temperature (RT) for electrophysiological recordings and used within 4 h. Interface disks are fabricated by carefully gluing interface paper (1220823, Thomas Scientific) to one side of a custom acrylic disk that fits inside one of the chamber's four wells (Fig. 2a). The addition of venting holes around the circumference using a histological probe will aid in releasing trapped air bubbles (Fig. 2a). A chamber well is filled with aCSF, and an interface disk secured on top forming a semipermeable membrane atop which brain slices can be incubated permeated with aCSF while the chamber is being continuously bubbled with 95% O_2/5% CO_2. A coiled high resistance nickel chromium wire (155402E, Fisher Scientific) inside PTFE heat-resistant tubing (EW-06417-31, Cole Parmer) is glued to the inside base of the chamber, and attached to a power supply (GPS-3030D, Newark element14), and a thermistor (5SRTC-TT-J-30-72, Omega) attached to a temperature controller (CSC32J, Omega) enable precise control of chamber temperature during incubation. Depending upon experimenter preference, slices can alternatively be incubated floating in gently bubbled aCSF [55].

2.2 Materials for Patch-Clamp Recordings

Microscope: Nikon FN-1 upright microscope with CFI Plan Flour $4\times$/0.13 numerical aperture (NA) WD 17.2 mm and Nikon NIR Apo $40\times$/0.80 NA water-immersion objectives, and visible and infrared differential interference contrast (DIC) optics. Fluorescent cells were illuminated with an Excelitas X-Cite lamp with GFP

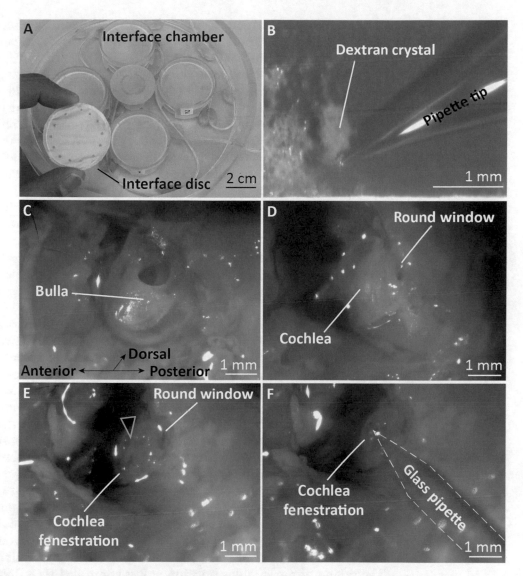

Fig. 2 Procedure for application of dextran crystals to the cochlea. (**a**) Custom-built interface chamber and interface discs used for incubation of brain slices. (**b**) High-magnification image of a dextran crystal approximately 0.5 mm in diameter on the end of a glass pipette used to transfer crystals from weighing paper to the cochlea. (**c**) The angle of approach is illustrated in this image of the left tympanic bulla viewed from the ventrolateral aspect following removal of the external auditory meatus, tympanic membrane, and middle ear ossicles. (**d**) Same field of view illustrated in "**c**" with bulla removed enabling clear visualization of the lateral face of the underlying cochlea. In this image, the basal turn of the cochlea is in focus, with the apical portion out of focus, but still visible anteriorly, to the left of the image. The round window niche is visible at the posterior most aspect of the cochlea, and serves as a convenient landmark in subsequent panels. (**e**) A fenestration to the otic capsule allows access to the modiolus of the cochlea. In this image, the middle turn of the cochlear is identifiable as indicated with the black arrowhead. (**f**) Placement of dextran crystals on the modiolus using a glass pipette. The orientation and field of view shown in panels **c–f** are comparable; however, minor transpositions were unavoidable during dissection steps and imaging

(ex/em 466/495 nm) and dsRed (ex/em 554/573 nm) filter sets. The images were collected with a QIClick, Mono 12 bit, camera and viewed using NIS-Elements software (Nikon).

Micromanipulator and recording chamber: Sensapex micromanipulator, RC-26G bath chamber (640236, Warner Instruments), slice hold-down (SHD-42/10 WI 64-1421, Warner Instruments), PM-1 in-line solution heater and temperature controller (641526, Warner Instruments), custom gravity fed perfusion system (60 mL syringe suspended about 1 foot above microscope stage, with tubing for flow into and out of chamber), peristaltic pump for fluid suction from chamber (PeriStar Pro, WPI).

Amplifier and digitizer: MultiClamp 700B amplifier and Digi-Data 1440A digitizer controlled by Clampex 10.6 or 11.3 software (Molecular Devices). The headstage was grounded using a 2.0 mm Ag/AgCl electrode (64-1310 (E-206), Warner Instruments) soldered to the headstage grounding wire and immersed in the bath solution using a magnetic clamp (MAG-7, 64-1554, Harvard apparatus).

Stimulus isolation unit and accompanying hardware: Electrical stimulation from the ISO-FLEX stimulus isolation unit (AMP Instruments) is applied to the tissue via a wire with an intact banana plug on one end and on the other end the wire is cut and exposed then soldered onto an ~3–4 cm length of silver wire (64-1320 (AG-15W), Warner Instruments), inserted into a 1.5 mm outer diameter glass pipette (BF150-110-10, Sutter Instruments) that has been pulled to a tip diameter of ~5 μm using a horizontal pipette puller (P1000, Sutter Instruments) or broken with a Kimwipe and half filled with aCSF. The ground for the stimulus isolation unit is a banana plug wire, cut on one end and connected to a 2.0 mm pellet ground Ag/AgCl electrode (64-1310 (E-206) and immersed into the bath solution. The stimulating pipette is clamped onto a ceramic probe rod (1404-0000E, Siskiyou) attached to a Sensapex or Sutter micromanipulator for positioning. The timing of the stimulus isolation unit is controlled via a BNC connection to the digital outputs on the digitizer.

Internal solutions and electrode glass: aCSF recording/bath solution as described above, no kynurenic acid. Internal solution contained the following (in mM): 56 CsCl, 44 CsOH, 49 D-gluconic acid, 1 $MgCl_2$, 0.1 $CaCl_2$, 10 HEPES, 1 EGTA, 0.3-Na-GTP, 2 Mg-ATP, 3 Na_2-phosphocreatine, 5 QX-314 to block voltage-gated sodium channels, and 0.25% biocytin, with 0.01 Alexa Fluor 488 hydrazide (A10436, Fisher Scientific). pH was adjusted to 7.2 with CsOH. 1 mL aliquots of internal solution are frozen, thawed immediately before use, loaded into a 1 cc syringe, then pushed through a Nalgene 4 mm diameter cellulose acetate 0.2 μm pore size syringe filter (09-740-34A, Fisher Scientific) attached to an electrode filler (CMF20G, World Precision Instruments) into recording pipettes pulled from 1.5 mm outer diameter borosilicate glass (BF150-110-10, Sutter Instruments) to tip resistances of 3–6 M using a horizontal pipette puller (P1000, Sutter Instruments).

2.3 Materials for Optogenetic Stimulation of Presynaptic MNTB Neurons

Imaging system for localization of fluorophore coupled to optogenetic protein and laser photostimulation: Custom VIVO multiphoton imaging system (3I: Intelligent Imaging Innovations) with a Chameleon Ultra infrared laser (Coherent) tuned to 810 nm for EYFP imaging, solid-state 488 nm laser targeted by the imaging galvos for ChR2 photostimulation controlled using SlideBook software v5.5. The timing of the 488 nm laser activation and duration was transmitted to the patch-clamp software via BNC connection to the digitizer. Many alternative light sources for optogenetic stimulation are also available (*see* **Note 1**).

Patch-clamp equipment: Zeiss Examiner D1 upright microscope with $20\times$ 1.0 N/A water dipping objective, Dodt contrast module for brightfield illumination for patch-clamp experiments, and epifluorescent light source (Excelitas) with red and green filter sets. The patch electrode position was controlled with PatchStar micromanipulators (Scientifica). Patch-clamp recordings were performed using a Multiclamp 700B amplifier with Digidata 1440A digitizer and pClamp v10.3 (Molecular Devices).

Solutions: aCSF as described above. Internal solution for current-clamp recordings from MNTB principal cells included (in mM): 56 KCl, 54 K-gluconate, 1 $MgCl_2$, 1 $CaCl_2$, 10 HEPES, 11 EGTA, 0.3 Na-GTP, 2 Mg-ATP, 5 Na_2-phosphocreatine. pH was adjusted to 7.2 with KOH.

Mouse line expressing an optogenetic protein driven by the appropriate promoter for expression in the cell type of interest, here MNTB principal neurons (VGAT-ChR2(H134R)-EYFP [54]; Jackson Labs stock number 014548).

3 Methods

3.1 Acute Application of Dextran Fluorescein Crystals to the Cochlea and Brainstem Slice Preparation

Patch-clamp recordings from identified MOC neurons can be performed using the recently characterized ChAT-IRES-Cre mouse line crossed with Ai14 tdTomato Cre reporter mouse line in which somata can be positively identified using epifluorescent imaging in brain slices yielding reliable MOC recordings [44]. The slicing procedure for MOC neurons has been described in detail elsewhere [56], with 300 μm brain slices used here. When the use of this transgenic strain is not plausible due to the utilization of a different Cre driver line to drive expression of Cre-dependent genes for molecules such as optogenetic proteins or calcium indicators, or for experiments in species other than mouse, it is beneficial to use the retrograde tracer method described here to label MOC neurons.

MOC somata are labeled via acute application of a fluorescent retrograde neuronal tracer (dextran fluorescein, 3000 MW, but *see* **Note 2**) to the cochlear modiolus. Upon application, the tracer is nonselectively taken up by axons, and transported back to efferent neuron somata demonstrating that tdTomato positive neurons in the VNTB have axons that project to the cochlea and are auditory efferents.

Mice of either sex are euthanized by CO_2 asphyxiation and decapitated with scissors according to approved NIDCD Animal Care and Use Committee guidelines. A partial dissection is performed under a dissecting microscope, making an incision along the saggital suture of the skull, and peeling back the overlying tissue and severing the external auditory meatus to expose the temporal bone. The tympanic membrane, ossicles and bulla are removed, and the cochlear modiolus exposed by chipping away at the otic capsule with #4 forceps to create a fenestration in the lateral wall of the cochlea. To ensure focal application of crystals to the modiolus, accumulated fluids (e.g. aCSF, perilymph and endolymph) at the site are carefully removed with finely rolled absorbent paper points prior to application of tracer crystals.

A small amount of fluorescein dextran crystals (a few mg, save unused crystals) are collected on weighing paper and a glass pipette (tip diameter ~ 1–1.5 μm) is used to carefully transfer crystals to the modiolus (Fig. 2b–f). The amount of crystal attached to the pipette will be variable due to static; crystals around ~0.25–0.5 mm in size yield consistent patterns of labeling.

The entire preparation (brain with attached cochlea) is incubated at room temperature for 30 min. The brain remains completely submerged in bubbling aCSF and the cochlea positioned above the fluid line to prevent dissolution of tracer crystals out of the cochlea and into the bath. Following incubation, the brain is dissected from the skull and transverse brain slices containing SOC were prepared as has been previously described [56] (but *see* **Note 3**).

3.2 Patch-Clamp Recordings from Fluorescent MOC Neurons in Brainstem Slices with Electrical Stimulation of Presynaptic Axons

Patch-clamp recordings were made from MOC neurons in brain slices from either ChAT-IRES-Cre × tdTomato mice, from a wild type (C57Bl/6) animal with fluorescein dextran retrograde tracer acutely applied to the cochlea (*see* Subheading 3.1), or from both genetic and tracer labeled neurons (Fig. 3a–aii). Slices are typically 300–400 μm thick and used within 4 h of preparation. Recordings can be performed at room temperature (RT, 22 ± 1 °C) or physiological temperature (35 ± 1 °C), depending on the experiment. Here, we used RT recordings. The fluorescence from the retrograde tracer crystals can be dim and bleach easily (Fig. 3ai), so use minimal fluorescent illumination to locate cells. Perform giga-ohm seal recordings from an MOC neuron, switching from fluorescent to DIC imaging as needed to locate the cell and form a seal, as previously described [56].

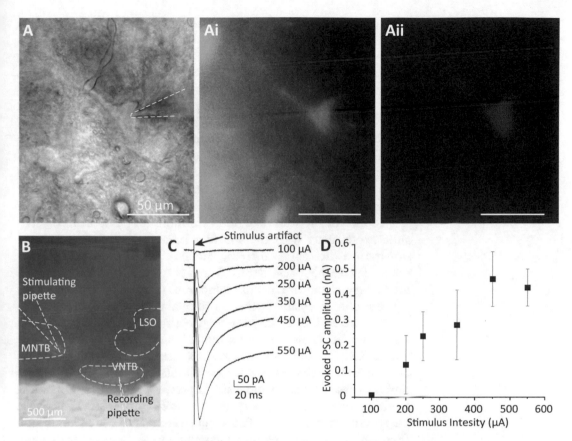

Fig. 3 Recording technique and evoked PSCs. (**a**) Example MOC cell during patch-clamp recordings visualized with differential interference contrast (DIC) with recording electrode outlined with dashed white lines, using a GFP filter to identify the retrogradely labeled cells (**ai**) and a DsRed filter to identify tdTomato positive cells (**aii**). (**b**) Brightfield image illustrating the position of the recording electrode positioned in the VNTB, stimulating electrode positioned in the MNTB, and dashed lines highlighting pertinent anatomical landmarks. (**c**) Example PSCs recorded from an MOC neuron evoked from different degrees of electrical stimulation of presynaptic axons. Each sample trace is an average of ten traces with an arrow indicating the stimulus artifact. The electrical stimulus intensity applied is indicated to the right of each trace. (**d**) Plot of ePSC amplitude by electrical stimulus intensity from the data shown in (**c**). Data presented as mean ± standard deviation

Presynaptic axons can be electrically stimulated to force action potentials and subsequent calcium-dependent release of neurotransmitter-containing vesicles, allowing detailed characterization of synaptic inputs to MOC neurons. Here, the stimulating electrode is placed in the MNTB axon bundle that projects throughout the SOC, including to the MOC neurons. Electrical stimulation is a nonspecific activation, so will also induce action potentials in axons from other cells traversing the region.

Prior to performing whole-cell recordings from an identified MOC neuron, position the stimulating electrode (connected to the stimulus isolation unit, *see* Subheading 2.2) at the ventrolateral edge of the MNTB by using the micromanipulator to gently push the tip of the stimulating electrode completely into the surface of the slice. Perform a patch-clamp recording from an MOC neuron, identified either by genetic label or retrograde tracer from the cochlea, in the voltage-clamp configuration with the membrane potential set to −60 mV and with appropriate capacitance and series resistance compensation [55–57]. Run a voltage-clamp protocol that includes a digital output command via a BNC cable to the stimulus isolation unit. Typical stimulus duration is 0.25–1 ms and single pulses can be applied once per trace with an inter-stimulus interval of 10–60 s, or in paired pulses or trains, as the experiment requires.

To electrically stimulate neurotransmitter release from presynaptic axons, start with the stimulus isolation unit set to 0, then slowly turn the dial to increase the amplitude of the stimulation until a fast (usually less than 0.5 ms) electrical artifact is visible in the voltage-clamp or current-clamp trace (Fig. 3c). Slowly increase the stimulus intensity further until evoked postsynaptic currents (ePSCs) are visible, which in these experiments (at RT), occur with a latency of ~2.6 ms from the stimulation artifact until the onset of the ePSCs. Carefully adjust the stimulation intensity until ePSCs occur in 50% or fewer of the single stimulations, "minimal stimulation", to evoke release from a single presynaptic axon. Stimulate at minimal stimulation intensity for at least 20 pulses. Slowly increase the stimulation intensity. ePSCs will typically increase in amplitude (Fig. 3c, d), sometimes incrementally and sometimes with large jumps depending on the number of presynaptic axons available, the number of synapses from a single presynaptic axon onto an MOC neuron, and the synaptic strength of individual synapses. Continue to increase the stimulation intensity until the amplitude of the ePSC plateaus (Fig. 3c, d. This is the "maximal stimulation intensity" at which all presynaptic axons proximal to the stimulation electrode in the slice are activated. Analyses including the "convergence ratio" [44, 49] or K-means clustering [44, 49, 58] are used to determine the number of presynaptic axons converging on an MOC. Additional synaptic function experiments can be performed including pharmacological isolation of inputs, tests of short-term plasticity, and a variety of other experiments.

3.3 Optimizing Optogenetic Stimulation of MNTB Neurons

Optogenetics allows cell-type specific activation of neurons to more precisely study the function of identified synapses compared to nonspecific electrical stimulation of axons. The commercially available VGAT-ChR2-EYFP mouse line (see **Note 4**) expresses the light-activated cation channel channelrhodopsin (ChR2 H134R variant) in glycinergic and GABAergic neurons, including MNTB principal neurons, with coupling of EYFP to allow fluorescent localization to confirm the appropriate cell-type specific targeting of the ChR2. 300 μm slices were cut through the brainstem as

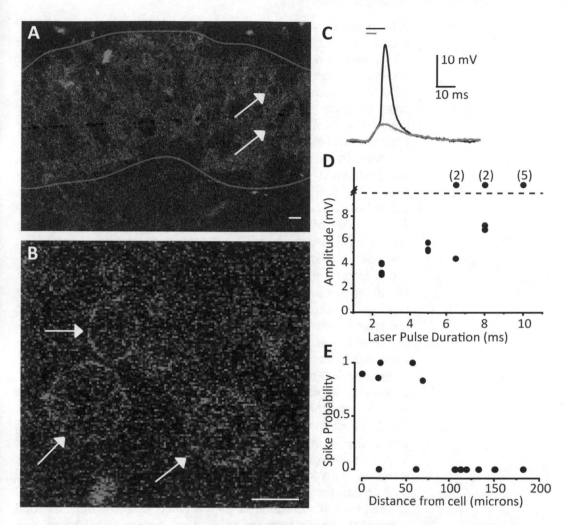

Fig. 4 Optogenetic stimulation of presynaptic MNTB neurons. (**a**) Two-photon image of the MNTB (outlined in gray) from an unfixed coronal brain slice from a VGAT-ChR2-EYFP mouse. Arrows indicate example MNTB principal cells with membrane labeling of ChR2-EYFP. Scale bar = 10 μm. (**b**) Higher magnification image of ChR2-EYFP localization to MNTB principal cell membranes from a different brain slice. Images are low resolution because exposure has been optimized to minimize photoactivation of neurons. (**c**) I-clamp traces from an MNTB neuron from a P3 VGAT-ChR2-EYFP mouse brain slice with 488 nm laser point stimulation. Duration of stimulation indicated by lines at the top of the traces. Gray indicates the response to laser stimulation directly on the cell being recorded from with a duration of 4.2 ms, resulting in subthreshold activation. Black indicates the response to laser stimulation at the same location and power with a duration of 8.4 ms, resulting in suprathreshold neuronal activation. (**d**) Relationship between laser pulse duration and response amplitude of an MNTB neuron expressing VGAT-ChR2-EYFP, same cell as "**c**." Response amplitudes increase with pulse duration resulting in an increased probability of action potentials. Points above dotted lines indicate action potentials, number of responses exhibiting action potentials indicated in parentheses. (**e**) Relationship between the probability of recording action potentials in a neuron with distance from the neuron. Data from Weisz and Kandler, unpublished

described above from a young (here P3) VGAT-ChR2-EYFP mouse and the MNTB was imaged under two-photon excitation with a laser wavelength of 810 nm at a resolution of 256×256. For live-cell experiments, imaging to localize the ChR2-EYFP is typically performed at a low resolution to minimize laser damage to cells, resulting in grainy, yet effective, images. ChR2-EYFP is detected in MNTB principal neurons identified by their unusually large round shape (Fig. 4a, b; Weisz and Kandler, unpublished).

Spatially and temporally precise channelrhodopsin activation was performed with point illumination via a solid state 488 nm laser (50 mW, typically used at 30–100% power; Weisz and Kandler, unpublished; also *see* **Note 1**). Patch-clamp recordings were performed from MNTB neurons expressing EYFP in current-clamp similar to methods described above for recordings from MOC neurons, in 300 μm coronal brainstem slices from VGAT-ChR2-EYPF mice. The laser stimulation timing was manually controlled via SlideBook software, and the illumination timing and duration was input to the patch-clamp software via a BNC cable connecting the laser to the digitizer. First, to determine the minimum light exposure required to activate neurons reliably, systematically modulate the laser power and illumination duration, which will vary from experiment to experiment depending on animal age, ChR2 expression, depth of the neuron within the slice, and other factors. For the example neuron shown, tetraethylammonium chloride (TEA, 10 mM) was included in the aCSF to block voltage-gated potassium channels and enhance optogenetic activation. The amplitude of optically evoked depolarization increased with laser duration from 2 to ~6 ms, reaching suprathreshold activation in 2/3 stimulations with a duration longer than 6 ms, and in all stimulations at durations of 10 ms (Fig. 4c, d). While the focal point of the laser is expected to be ~1 μm, focusing the illumination up to ~80 μm away from the soma could evoke reliable suprathreshold activation (Fig. 4e) due to either light scattering within the tissue or spread above and below the focal plane, or, more likely given the large activation radius, activation of ChR2 expressed in dendrites. Therefore, even though a single MNTB neuron can be targeted for optogenetic activation, surrounding neurons may also be excited. This technique can be used for specific activation of MNTB neurons while recording from any postsynaptic targets including MOC, LSO, medial superior olive (MSO), or superior peri-olivary (SPON) neurons.

4   Notes

1. Activation of ChR2-positive MNTB neurons was detailed here with a single-photon laser with a 488 nm wavelength, but a 474 nm laser is closer to the peak of ChR2 activation. Multiphoton activation with an infra-red laser can be used for more

focal activation but is far costlier. Focal illumination can also be achieved with a halogen lamp based [59] or LED based [44] system delivering light through an optical fiber, with the wavelength of light controlled by either optical filters or wavelength specific LEDs. Broad illumination via microscope epifluorescence is a simple way to evoke optogenctic activation, and can also be transmitted through a digital mirror device (DMD) to provide patterned illumination.

2. The selection of a neuronal tracer ultimately depends on the other fluorophores present in the tissue of interest. Here, we used fluorescein dextran to colocalize retrograde labeled cells with tdTomato positive somata labeling in the same slice; however, the use of the brighter rhodamine dextran may be preferred in non-transgenic tissue, or paired with a different Cre reporter mouse line.

3. In our experience, some blocking angles are optimized for cell visualization and accessibility, while others are more suitable for retaining complete circuits. A 300 μm slice with a traditional coronal slicing angle contains clusters of MOC neurons to improve yield of recordings, while the ~20° angle suggested in [56] is optimized for maximal retention of MNTB-MOC circuitry.

4. The Cre-Lox system can be used to introduce a growing number of molecular tools for characterization of neuron function, including expression of genes for calcium or voltage indicators, optogenetic proteins (e.g., channelrhodopsin), receptors for pharmacological cell death or activation/inactivation, or genetic knockout or knockin, among others. Unfortunately, Cre lines targeting different cell types cannot be combined in the same animal, requiring the use of other techniques such as the Flp or tetracycline systems, or injection of viral constructs for location-specific expression [46]. The use of the VGAT-ChR2-EYFP mouse here circumvents the need for an additional Cre line along with ChAT-IRES-Cre for MOC neuron targeting, but in the case that another Cre line is required for experiments, or a different channelrhodopsin variant available in viral construct or floxed transgenic mouse line is preferred, the retrograde tracer method of identifying MOC neurons can be a useful alternative.

5   Conclusions

Positive identification of MOC neurons for higher throughput patch-clamp recordings opens up the cells to a variety of cell- and circuit-specific manipulations including numerous imaging and photostimulation techniques. Here we presented two methods for identifying the cells, with commentary on when each technique may be most useful. In addition, we present two methods for stimulating presynaptic axons, namely, nonspecific but powerful, economical, and easy electrical stimulation of axons, as well as the costlier but more targeted optogenetic activation of presynaptic cells.

References

1. Ryugo DK, Fay R, Popper A (2011) Auditory and vestibular efferents. Springer Science Business Media, New York, NY

2. Warr W (1992) Organization of olivocochlear efferent systems in mammals. In: Webster DB, Popper A, Fay R (eds) The mammalian auditory pathways: neuroanatomy, Springer handbook of auditory research. Springer-Verlag, New York, NY, pp 410–448

3. Rasmussen GL (1960) Efferent fibers of the cochlear nerve and cochlear nucleus. In: Rasmussen GL, Windle W (eds) Neural mechanisms of the auditory and vestibular systems. C.C. Thomas, Springfield, IL, pp 105–115

4. Rasmussen GL (1946) The olivary peduncle and other fiber projections of the superior olivary complex. J Comp Neurol 84:141–219

5. Warr WB, Guinan JJ (1979) Efferent innervation of the organ of corti: two separate systems. Brain Res 173:152–155

6. Guinan JJ, Warr WB, Norris BE (1983) Differential olivocochlear projections from lateral versus medial zones of the superior olivary complex. J Comp Neurol 221:358–370

7. Warr WB (1975) Olivocochlear and vestibular efferent neurons of the feline brain stem: their location, morphology and number determined by retrograde axonal transport and acetylcholinesterase histochemistry. J Comp Neurol 161:159–181

8. Brown MC (1987) Morphology of labeled efferent fibers in the guinea pig cochlea. J Comp Neurol 260:605–618

9. Brown MC, Liberman MC, Benson TE, Ryugo DK (1988) Brainstem branches from olivocochlear axons in cats and rodents. J Comp Neurol 278:591–603

10. Brown MC (1993) Fiber pathways and branching patterns of biocytin-labeled olivocochlear neurons in the mouse brainstem. J Comp Neurol 337:600–613

11. Galambos R (1956) Suppression of auditory nerve activity by stimulation of efferent fibers to cochlea. J Neurophysiol 19:424–437

12. Desmedt J (1962) Auditory-evoked potentials from cochlea to cortex as influenced by activation of the efferent olivo-cochlear bundle. J Acoust Soc Am 34:1478–1496

13. Wiederhold ML, Peake WT (1966) Efferent inhibition of auditory-nerve responses: dependence on acoustic-stimulus parameters. J Acoust Soc Am 40:1427–1430

14. Wiederhold ML, Kiang NY (1970) Effects of electric stimulation of the crossed olivocochlear bundle on single auditory-nerve fibers in the cat. J Acoust Soc Am 48:950–965

15. Geisler CD (1974) Model of crossed olivocochlear bundle effects. J Acoust Soc Am 56:1910–1912

16. Guinan JJ, Gifford ML (1988) Effects of electrical stimulation of efferent olivocochlear neurons on cat auditory-nerve fibers. III. Tuning curves and thresholds at CF. Hear Res 37:29–45

17. Winslow RL, Sachs MB (1987) Effect of electrical stimulation of the crossed olivocochlear bundle on auditory nerve response to tones in noise. J Neurophysiol 57:1002–1021

18. Kawase T, Delgutte B, Liberman MC (1993) Antimasking effects of the olivocochlear reflex. II. Enhancement of auditory-nerve response to masked tones. J Neurophysiol 70:2533–2549

19. Rajan R (1995) Frequency and loss dependence of the protective effects of the olivocochlear pathways in cats. J Neurophysiol 74:598–615

20. Rajan R (1988) Effect of electrical stimulation of the crossed olivocochlear bundle on temporary threshold shifts in auditory sensitivity. I. Dependence on electrical stimulation parameters. J Neurophysiol 60:549–568

21. Reiter ER, Liberman MC (1995) Efferent-mediated protection from acoustic overexposure: relation to slow effects of olivocochlear stimulation. J Neurophysiol 73:506–514

22. Taranda J, Maison SF, Ballestero J, Katz E, Savino J, Vetter DE, Boulter J, Liberman MC, Fuchs PA, Elgoyhen AB (2009) A point mutation in the hair cell nicotinic cholinergic receptor prolongs cochlear inhibition and enhances noise protection. PLoS Biol 7:e18

23. Maison S, Usubuchi H, Liberman MC (2013) Efferent feedback minimizes cochlear neuropathy from moderate noise exposure. J Neurosci 33:5542–5552

24. Tong H, Kopp-Scheinpflug C, Pilati N, Robinson SW, Sinclair JL, Steinert JR, Barnes-Davies M, Allfree R, Grubb BD, Young SM, Forsythe ID (2013) Protection from noise-induced hearing loss by kv2.2 potassium currents in the central medial olivocochlear system. J Neurosci 33:9113–9121

25. Boero LE, Castagna VC, Di Guilmi MN, Goutman JD, Belén Elgoyhen A, Gómez-Casati ME (2018) Enhancement of the medial olivocochlear system prevents hidden hearing loss. J Neurosci. https://doi.org/10.1523/JNEUROSCI.0363-18.2018

26. Oatman LC (1976) Effects of visual attention on the intensity of auditory evoked potentials. Exp Neurol 51:41–53

27. Glenn J, Oatman L (1977) Effects of visual attention on the latency of auditory evoked potentials. Exp Neurol 40:34–40

28. Delano PH, Elgueda D, Hamame CM, Robles L (2007) Selective attention to visual stimuli reduces cochlear sensitivity in chinchillas. J Neurosci 27:4146–4153

29. Terreros G, Jorratt P, Aedo XC, Bele XA, Delano XPH, De Fisiología P, De Medicina F, De Otorrinolaringología D (2016) Selective attention to visual stimuli using auditory distractors is altered in alpha-9 nicotinic receptor subunit knock-out mice. J Neurosci 36:7198–7209

30. Chambers AR, Hancock KE, Maison S, Liberman MC, Polley DB (2012) Sound-evoked olivocochlear activation in unanesthetized mice. J Assoc Res Otolaryngol 13:209–217

31. Astl J, Popelář J, Kvašňák E, Syka J (1996) Comparison of response properties of neurons in the inferior colliculus of guinea pigs under different anesthetics. Int J Audiol 35:335–345

32. Cederholm JME, Froud KE, Wong ACY, Ko M, Ryan AF, Housley GD (2012) Differential actions of isoflurane and ketamine-based anaesthetics on cochlear function in the mouse. Hear Res 292:71–79

33. Gaese BH, Ostwald J (2001) Anesthesia changes frequency tuning of neurons in the rat primary auditory cortex. J Neurophysiol 86:1062–1066

34. Osanai H, Tateno T (2016) Neural response differences in the rat primary auditory cortex under anesthesia with ketamine versus the mixture of medetomidine, midazolam and butorphanol. Hear Res 339:69–79

35. Boyev KP, Liberman MC, Brown MC (2002) Effects of anesthesia on efferent-mediated adaptation of the DPOAE. J Assoc Res Otolaryngol 3:362–373

36. Yamamoto C, Chujo T (1978) Visualization of central neurons and recording of action potentials. Exp Brain Res 31:299–301

37. Takahashi T (1978) Intracellular recording from visually identified motoneurons in rat spinal cord slices. Proc R Soc Lond B Biol Sci 202:417–421

38. MacVicar BA (1984) Infrared video microscopy to visualize neurons in the in vitro brain slice preparation. J Neurosci Methods 12:133–139

39. Fujino K, Koyano K, Ohmori H (1997) Lateral and medial olivocochlear neurons have distinct electrophysiological properties in the rat brain slice. J Neurophysiol 77:2788–2804

40. Robertson D (1996) Physiology and morphology of cells in the ventral nucleus of trapezoid body and rostral periolivary regions of the rat superior olivary complex studied in slices. Audit Neurosci 2:15–31

41. Doudna JA, Charpentier E (2014) Genome editing. The new frontier of genome engineering with CRISPR-Cas9. Science 346:1258096

42. Robbins PD, Tahara H, Ghivizzani SC (1998) Viral vectors for gene therapy. Trends Biotechnol 16:35–40

43. Madisen L, Zwingman TA, Sunkin SM, Oh SW, Zariwala HA, Gu H, Ng LL, Palmiter RD, Hawrylycz MJ, Jones AR, Lein ES, Zeng H (2010) A robust and high-throughput Cre reporting and characterization system for the whole mouse brain. Nat Neurosci 13:133–140

44. Torres Cadenas L, Fischl MJ, Weisz CJC (2020) Synaptic inhibition of medial olivocochlear efferent neurons by neurons of the medial nucleus of the trapezoid body. J Neurosci 40:509–525

45. Clause A, Kim G, Sonntag M, Weisz CJC, Vetter DE, Rűbsamen R, Kandler K (2014) The precise temporal pattern of prehearing spontaneous activity is necessary for tonotopic map refinement. Neuron 82:822–835

46. Luo L, Callaway EM, Svoboda K (2018) Genetic dissection of neural circuits: a decade of progress. Neuron 98:256–281

47. Romero GE, Trussell LO (2021) Distinct forms of synaptic plasticity during ascending vs descending control of medial olivocochlear efferent neurons. Elife. https://doi.org/10.7554/eLife.66396

48. Sanes DH (1993) The development of synaptic function and integration in the central auditory system. J Neurosci 13:2627–2637

49. Kim G, Kandler K (2003) Elimination and strengthening of glycinergic/GABAergic connections during tonotopic map formation. Nat Neurosci 6:282–290

50. Weisz CJ, Rubio ME, Givens RS, Kandler K (2016) Excitation by axon terminal GABA spillover in a sound localization circuit. J Neurosci 36:911–925

51. Magnusson AK, Park TJ, Pecka M, Grothe B, Koch U (2008) Retrograde GABA signaling adjusts sound localization by balancing excitation and inhibition in the brainstem. Neuron 59:125–137

52. Sanes DH, Friauf E (2000) Development and influence of inhibition in the lateral superior olivary nucleus. Hear Res 147:46–58

53. Boyden ES, Zhang F, Bamberg E, Nagel G, Deisseroth K (2005) Millisecond-timescale, genetically targeted optical control of neural activity. Nat Neurosci 8:1263–1268

54. Zhao S, Ting JT, Atallah HE, Qiu L, Tan J, Gloss B, Augustine GJ, Deisseroth K, Luo M, Graybiel AM, Feng G (2011) Cell type–specific channelrhodopsin-2 transgenic mice for optogenetic dissection of neural circuitry function. Nat Methods. https://doi.org/10.1038/nmeth.1668

55. Moyer J, Brown T (2002) Patch-clamp techniques applied to brain slices. In: Walz W (ed) Neuromethods. Humana Press, Totowa, NJ, pp 135–194

56. Fischl MJ, Weisz CJC (2020) In vitro wedge slice preparation for mimicking in vivo neuronal circuit connectivity. J Vis Exp. https://doi.org/10.3791/61664

57. Armstrong CM, Gilly WF (1992) Access resistance and space clamp problems associated with whole-cell patch clamping. Methods Enzymol 207:100–122

58. Ferragamo MJ, Golding NL, Oertel D (1998) Synaptic inputs to stellate cells in the ventral cochlear nucleus. J Neurophysiol 79:51–63

59. Kandler K, Nguyen T, Noh J, Givens RS (2013) An optical fiber-based uncaging system. Cold Spring Harb Protoc 2013:118–121

Chapter 16

Auditory Brainstem Response (ABR) Measurements in Small Mammals

Ye-Hyun Kim, Katrina M. Schrode, and Amanda M. Lauer

Abstract

Auditory brainstem responses (ABRs) are evoked potentials that are measured from the scalp and reflect the activity of the auditory neurons in the auditory pathway in response to sound stimuli. ABR has been widely used in clinics and laboratories to assess neural pathologies and hearing status in subjects. Given its noninvasiveness, reliability, sensitivity, and relatively high-throughput nature, ABR test is an excellent tool for examining hearing status and integrity of the auditory pathway in laboratory mice. The goal of this chapter is to provide users with general ABR procedural methods and key considerations critical in obtaining reliable ABR measurements in small mammals. The protocol presented here focuses on mice as an example, but the general principles can be extended with minor modification to other species such as rats, hamsters, and bats.

Key words Auditory brainstem response (ABR), Evoked potentials, Animal hearing function, Auditory phenotype, Auditory system, Auditory pathway

1 Introduction/Overview

Auditory brainstem responses (ABRs) are evoked electrical potentials recorded from the scalp in response to onset of sound stimuli [1, 2]. ABRs are far field potentials that are recorded at a distance from their neural generator, the cochlea, and auditory brainstem nuclei. ABR test is a common hearing screening tool used both in clinics and in animal research laboratories. ABR reflects synchronous activity of auditory neurons in the ascending auditory pathway in response to transient sound stimuli. ABR captures neural activity occurring early in the ascending auditory pathway after stimulus onset, from the cochlea to lower auditory midbrain [3–5]. ABR measurements can thus provide information on the integrity of the ascending auditory pathway and allow inferences of the hearing status.

Andrew K. Groves (ed.), *Developmental, Physiological, and Functional Neurobiology of the Inner Ear*, Neuromethods, vol. 176, https://doi.org/10.1007/978-1-0716-2022-9_16,

ABR test is noninvasive and provides reliable and objective electrophysiological measurements of auditory brainstem activity in response to sound stimuli. The ABR test is highly sensitive in evaluating animals' hearing status, unlike Preyer's reflex which is a simple hearing screening test that measures pinnae or whole body startle reflex in response to a handclap or "click-box" stimuli [6–8]. Preyer's reflex is useful ABR quickly screening the presence or absence of profound hearing loss; however, it lacks sensitivity and specificity and its result can be affected by motor ability of the animals.

Furthermore, ABR test does not require animal training or attention to perform the test, in contrast to psychoacoustic experiments which require extensive animal training and are difficult to perform. Animal psychoacoustic experiments are considered the gold standard in the field and have contributed extensively to understanding of complex hearing functions, such as sound localization, frequency discrimination, and pitch perception processes in animals [9–12]. However, they may not be suitable for very young or hard to train animals, animal models with rapid progression of hearing loss or profound deafness, or testing large number of animals in short periods of time.

As ABR test is done under anesthesia in a relatively short time and does not require learning or complex tasks, it can be used to test hearing in animals with learning or motor deficits, or models of rapid progression of hearing loss or profound deafness. It is also suited for rapid studies that require high-throughput screening of auditory phenotype and hearing impairments [13–15]. In addition, ABR tests can be used to investigate auditory development [16], identify auditory nerve damage and cochlear synaptopathy [17, 18], or evaluate abnormality in auditory brainstem pathways [19, 20].

What is important to note, however, is that while ABR test can reveal information on some aspects of hearing function, it is not a perfect predictor of actual hearing ability. That is, ABR test results do not accurately reflect perceptual/behavioral hearing sensitivity. In human and mouse studies, it has been shown that ABR thresholds were 10–20 dB SPL higher than perceptual hearing thresholds measured using psychoacoustic tone detection tasks, depending on test frequency [21, 22]. This discrepancy is thought to arise due to differences in acoustic stimuli used and the general status of the animal during the testing. ABR test uses very brief (1–5 ms) acoustic stimuli with rapid onset, whereas psychoacoustic hearing tests use sound stimuli that are longer in duration and slower in onset. As sound detection improves with increasing duration of the sound, a process called temporal integration [23], the use of very brief ABR stimuli can impact the ABR threshold [21]. Another consideration is the status of the animal. Animals are anesthetized during ABR tests while animals in behavioral tests are awake and attending to sound stimuli. This difference in conscious responsiveness can contribute to potential divergence in hearing thresholds. Nevertheless, its noninvasiveness, specificity, reliability, and high-throughput nature makes ABR test an excellent tool for examining hearing function in laboratory mice.

2 Materials

Here we describe a detailed ABR test procedure in mice using Tucker-Davis Technologies (TDT) hardware integrated into a custom system. The general procedure, however, can be adapted to use for other systems, experimental objectives, and other small mammals.

Note: All protocols must be approved by the Institutional Animal Care and Use Committee (IACUC). All procedures and animal handing should be performed in compliance to Guide for the Care and Use of Laboratory Animals [24].

- 4–6 weeks old CBA/CaJ or C57BL/6 mice (For starting, we recommend using normal hearing mice to ensure the ABR equipment is working properly, and to familiarize with ABR waveforms and threshold detection. CBA/CaJ mice have stable hearing thresholds out to 1 year of age [25, 26] and are suitable for hearing research. On the other hand, C57BL/6 mice start to exhibit early onset high-frequency hearing loss at around 2–3 months of age [27]. If choosing C57BL/6, we recommend using young mice aged between 4 and 6 weeks-old, before the onset of high frequency hearing loss.)

- 100 mg/mL ketamine (ketamine hydrochloride injection, Dechra Veterinary Products)

- 20 mg/mL xylazine (AnaSed® INJECTION, Akorn)

- 0.9% sodium chloride for injection (10 mL in single dose plastic fliptop vial, Pfizer).

- ketamine (20 mg/mL) + xylazine (2 mg/mL) mix in 0.9% saline. (ketamine and xylazine are controlled drug substances, and must be stored in double-lock box when not in use. Check to make sure the stock is not expired. Use new syringe/needles each time you draw out the drugs and saline to make ketamine/xylazine ansthesia mix. For injection, give a dose of 0.1 mL of ketamine/xylazine mix per 20 g of mouse, yielding a dose of 100 mg/kg ketamine +10 mg/kg xylazine.)

- 1 mL Tuberculin (TB) syringes with short, small gauge needles (26 Gauge, 3/8″ syringe) (Becton Dickinson (BD).

- Gauze pad (~4 × 4″).

- Lubricating eye ointment (gel-type)/Cotton swabs.

- Electrical heating pad.
- Mouse rectal temperature probe/temperature monitor. (Any commercially available mouse rectal probe and temperature monitors can be used. Please make sure the rectal probe is small enough to fit the animal.)
- Vaseline.
- 3 Reusable genuine Grass® platinum subdermal needle electrodes (F-E2-24, Grass Technologies). (Platinum electrodes are recommended as they have better signal-to-noise ratio compared to stainless-steel subdermal needle electrodes.)
- Cage heating pad.
- 70% ethanol wipe.
- Scale.
- 0.9% sterile saline solution.

2.1 Equipment

- Computer.
- TDT Z-Bus system 3 (Tucker-Davis Technologies, TDT).
- RX6 Multi-function Processor (TDT).
- PA5 Programmable Attenuator (TDT).
- High frequency free-field dome tweeter (FD28D, Fostex).
- Speaker amplifier (Crown CH1, Crown Audio Inc).
- ABR preamplifier (ISO-80, Isolated Bio-Amplifier, World Precision Instruments).
- Filter (Krohn-Hite Model 3550, Krohn-Hite Corporation).
- Audio monitor (computer speaker) (Audio monitor is used to monitor animal's state such as breathing, heartbeat, and anesthesia state, during ABR testing.)
- Sound isolating chamber (Controlled Acoustic Environments, Industrial Acoustic Company (IAC)) lined with Sonex® foam (Pinta Acoustics).
- Custom MATLAB-based threshold analysis program.

3   Methods

3.1 General Setup

ABRs are recorded from the scalp and obtained by averaging the neural response over many repetitions of the sound stimuli. ABR stimuli normally consist of single brief clicks or tone-pips. Clicks are abrupt broadband stimuli with condensed spectral density, exciting large number of auditory neurons at a wide range of frequencies at once [28, 29]. Click-ABRs are useful in quickly assessing the overall integrity of the auditory function. Tone-burst stimuli are short

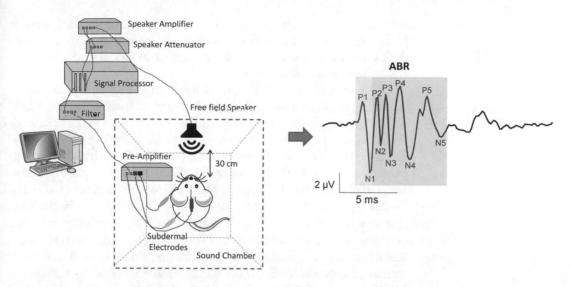

Fig. 1 Schematic for auditory brainstem response (ABR) setup. (*Left*) A general schematic for ABR test setup using a free-field speaker is shown. Note sound isolating chamber is illustrated as dotted line. ABR stimulus generation and data acquisition are controlled by a signal processor (e.g., TDT RX6 Multifunction Processor) and a custom software connected to a PC. ABR stimuli are amplified, attenuated, and delivered through a free field speaker. ABR is recorded from anesthetized animal's scalp using subdermal electrodes, placed at the vertex (red electrode), ventral edge of the left pinna (grey electrode), and on the hind leg (green electrode, ground). ABR signals are amplified, band-pass filtered, digitized, and averaged over 300 stimulus repetitions. (*Right*) An example of averaged ABR waveform recorded from a normal hearing mouse in response to 95-dB SPL click stimulus. Wave 1–5, each comprising of positive(P) and negative (N) peaks, are clearly identified. Wave 1 (shaded in blue) is generated by auditory nerve and spiral ganglion neurons and waves 2–5 (shaded in pink) are generated by multiple auditory brainstem nuclei in the ascending auditory pathway

(2–5 ms) frequency-specific sound stimuli, which allow stimulation of auditory neurons at a specific frequency location along the length of the cochlea. Therefore, tone-burst ABRs allow assessment of auditory function at a specific frequency.

In general, ABR stimuli are generated by a computer signal generator, amplified, and delivered through a speaker. There are two types of speaker system that can be used for ABR test: free-field or closed-field. The choice of speaker type depends on the goals of the experiment. Both systems require accurate calibration of the speaker output before ABR testing. The closed-field system, in which the speaker microphone is inserted in the animal's ear using a probe, allows for convenient testing of each ear. However, it requires complex calibration and bypasses the contribution of the pinna (external ear) and head in sound detection, which may affect hearing sensitivity [22]. In addition, strain or individual difference in ear canal shape can affect the speaker output and cause electrical distortions. A free-field speaker, on the other hand, is placed in front of the animal's head at a distance. A free-field speaker, which presents sounds at a distance at a zero-degree angle from the animal's head, is more suited for comparative hearing study [22]. When using a free-field speaker, it is especially critical to test ABRs in an acoustic chamber lined with foam to minimize acoustic reflections from the environment.

In our protocol, ABR stimuli and signal acquisition are controlled by a custom-made MATLAB program interfacing the TDT Z-Bus system 3 (Fig. 1). ABR stimuli consist of 1-ms clicks and 5-ms tone pips (0.5 ms rise/fall) of varying frequencies (8, 12, 16, 24, 32 kHz) and presented at a rate of 20/s with alternating polarity. ABR stimuli are amplified (Crown CH1, Crown Audio Inc.) and delivered through a free-field dome tweeter (FD28D, Fostex), placed at 0 degrees, 30 cm away from animal's head. The distance of free-field speaker can be chosen at experimenter's discretion, but it must be consistent during calibration and testing to ensure accurate threshold measurements. ABR stimuli are presented in descending 5–10 dB level increments (PA5 programmable attenuator, TDT) until no discernable ABR signal is observed. ABR signals, which are picked up by the subdermal platinum electrodes in the animal scalp, are amplified ×300,000 (ISO-80, Isolated Bio-Amplifier, World Precision Instruments), band pass filtered between 300–3000 Hz (Krohn-Hite Model 3550, Krohn-Hite Corporation), digitized at a sampling rate of 195 kHz (RX6 Multi-function Processor, TDT), and averaged across 300 presentations. Threshold analysis is done using a custom MATLAB-based threshold analysis program. Using this protocol, ABR testing should last about 40–60 min per animal, including the time for anesthesia injection.

3.2 ABR Protocol

Before anesthetizing a mouse for ABR testing, ensure that all equipment is in good working condition.

1. Turn on the computer and ABR equipment including the TDT system, filter, preamplifier, speaker amplifier, speaker, and ABR testing program. Check to make sure all connections are secure.

2. Turn on the heating pad (37 °C) inside the sound-isolating booth and the temperature monitor. Cover the heating pad with gauze to protect the animal's abdomen.

3. Unplug the preamplifier (ISO-80, Isolated Bio-Amplifier, World Precision Instruments) from the charging port, and connect to input.

4. Connect platinum subdermal needle electrodes to the preamplifier. Inspect the needles to make sure they are not bent or dull and turn on the preamplifier.

5. Check the speaker output level. Its output should be within the range of calibrated level.

Speaker output is periodically calibrated with a free-field 1/4″ microphone (Type 4939, Brüel & Kjær, Nærum, Denmark) placed at the position of the mice ear (30 cm) and the spectral output of the speaker is analyzed using a custom MATLAB-based software.

6. Open the ABR testing file. Confirm that your parameters are correct. Most common parameters include: Click or 5 ms tone, 20/s presentation rate, 20–30 ms recording duration, 300–500 averages. Check stimulus frequencies and levels.

7. Place electrodes in saline-soaked gauze. Check response to ABR click stimulus. The noise floor should be ~80–100 nanoVolts after 300–500 averages.

3.3 Animal Preparation/ Anesthesia

1. Weigh the animal on a scale and inspect it for any physical abnormalities or signs of illness. Do not test a sick animal.

2. Record animal ID, date of birth, sex, test date, physical observations, and any other relevant test information in the lab notebook. The experimenter should be blinded to animal and condition as much as possible.

3. Prepare the animal for anesthesia. Anesthesia mix contains ketamine (20 mg/mL) and xylazine (2 mg/mL) mix in 0.9% saline. Draw 0.1 cc for each 20 g of body weight using a 1-mL TB syringe (26G 3/8″). For sensitive strains or very small mice (<15 g), reduce the dosage to 1/2 to 2/3 as their anesthesia tolerability is lower.

4. Restrain the animal with its head pointing slightly downward and inject the anesthetic intra-peritoneally (i.p.). Put the mouse back in its cage until it is mostly immobile. Record the volume and time of the injection, as this information will be useful in case a booster is needed. Mice should lose consciousness in less than 10 min.

5. Check the state of anesthesia by pinching the animal's foot. The animal should not move in response to toe pinching. Apply eye ointment to the eyes with a cotton swab to prevent corneal abrasions during testing. Proceed quickly as the animal begins to lose body heat rapidly.

6. Place the animal on the heading pad inside of the chamber. The animal should be placed in a direction with its head pointing toward the speaker. Adjust the heating pad and animal so that the vertex of the animal's head is 30 cm from the speaker. (*Note*: Animal head distance from the speaker should be consistent to ensure accurate calibration.)

7. Wipe the temperature probe with 70% alcohol and apply Vaseline to the probe. Insert temperature probe into the rectum

until you feel resistance. If the animal's rectum is too small for the probe, place the probe underneath the animal's abdomen.

8. Insert the subdermal platinum electrodes just under the skin of the animal's scalp (Fig. 1). The electrode connected to the red lead should be inserted on the vertex of the skull. The electrode connected to the black lead should be inserted behind the ventral edge of the left pinna. The ground electrode should be inserted under the skin of the hind leg. To help with electrode placement, blunt forceps can be used to lift the skin during insertion. Be consistent with electrode placements across animals. (*Note*: To reduce electrical interference and prevent wires acting as antennae, braid the subdermal electrode wires loosely together, and keep the electrode cables far away from the speaker and speaker cables.)

9. Make sure the animal is not moving or whisking. Close the sound booth door and turn off the booth light.

3.4   ABR Testing

1. Turn on the audio monitor (computer speaker) to check for heartbeat. If there is no signal or noise in the heartbeat, readjust the electrode placement and make sure that the preamplifier connections in the booth are secure.

2. Open the ABR test file. Input animal ID and specify file folder name where the ABR data will be saved. Record file folder name in the lab notebook for reference.

3. Test click-ABR first. In normal hearing mice, a waveform with ~4–5 peaks will emerge between the 2–10 ms time points in response to a 1-ms click stimulus. The baseline noise should be less than 1 μV, and the ABR should be clearly discernible from the noise. If large noise is present, stop the test and readjust or replace the electrodes and repeat the test.

4. Stimulus level is presented in 10 dB attenuation steps. For each stimulus, 300 recordings are collected for each level, and the waveform saved is the continuous average of these recordings. The ABR waveform amplitude decreases and peak latency increases as the stimulus level decreases. Once the response gets closer to threshold, decrease the attenuation by 5 dB steps to get finer resolution for accurate threshold analysis.

5. Once click-ABR test is finished, save data and proceed with tone-burst ABR tests for 8, 16, 32 kHz at the minimum. Additional frequencies (4, 6, 12, 24 kHz) may be added depending on experimental goals. (*Note*: These tone frequency choices are specific for mice. Normal hearing mice are most sensitive at frequencies near 12–16 kHz [30]. If testing for other small rodents or mammals, make sure to choose appropriate test frequencies corresponding to species specific hearing range [31].)

6. Always start with click and proceed with desired stimulus frequencies in a semirandom order to reduce a confounding effect
of test frequency order.

7. Periodically monitor and record the animal's body temperature
during ABR testing.

8. If the animal begins to whisk or move, stop testing and give it a
supplemental anesthesia dose of 1/2 to 1/3 of the original
amount (without removing the electrodes, if possible). Seda-
tion levels and length of time vary across strains and age. Once
animal has attained deep anesthesia level, resume ABR testing.

9. Check to make sure all data are saved in the designated folder
location.

3.5 Recovery

1. After ABR testing is completed, remove the subdermal electrodes and temperature probe from the mouse and place the
animal in an empty clean cage. Put the cage on a heating pad
to maintain body temperature and promote recovery from
anesthesia.

2. Monitor the animal until it fully emerges from anesthesia.
Return the mouse to its home cage once it can freely move
around and drink from the waterspout.

3.6 After ABR Testing

1. Exit from testing program and turn off equipment.

2. Turn off the amplifier inside the sound booth and then discon-
nect the platinum electrodes. Inspect the electrodes for any
damage, clean the needles with alcohol wipes, and store them
in a secure place.

3. Wipe off the temperature probe with a Kimwipe and alcohol.
Remove the gauze from the heating pad and clean the sound
booth. Clean bench and scale with disinfectant.

3.7 ABR Threshold Analysis

ABR threshold is defined as the lowest sound level at which ABR
waveforms are no longer discernable from the background noise
(Fig. 2). Thresholds are typically determined by visual inspection of
stacked ABR waveforms, in which ABR waveforms are stacked from
highest to lowest sound intensity and aligned in the time axis for
each frequency. Thresholds can also be automatically determined
using statistic-based algorithms or methods, such as ABR growth
curve (ABR magnitude vs. stimulus level) regression analysis [32]
or cross-correlation of adjacent waveforms [33]. Automatic thresh-
olding methods are more robust in objective and systematic threshold determination compared to visual inspection, which are
inherently susceptible to interpretation subjectivity and experimenter bias. If opting for visual inspection method, implementa-
tion of objective criteria, thorough training, and cross validation are
essential for reliable threshold determination. In addition, the

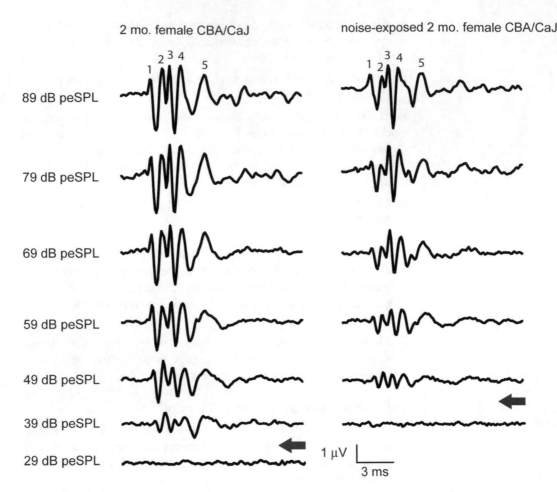

Fig. 2 ABR waveform traces from control and noise exposed CBA/CaJ mice. (*Left*) ABR waveforms recorded from a control CBA/CaJ (2 months old) in response to click stimuli are stacked in decreasing sound level in 10 dB steps. ABR wave 1–5 are clearly identified as indicated by number. As the sound level decreases, ABR waveform amplitudes decrease and peak latencies increase. Blue arrow indicates sound level near threshold. (*Right*) ABR waveforms recorded from a noise-exposed CBA/CaJ in response to click stimuli. The mouse was exposed to 100 dB SPL broadband noise for 2 h at 6 weeks of age and tested for ABR at 2 months of age. Compared to controls, the ABRs of noise exposed mice are smaller in amplitude, exhibit altered waveform characteristics, and have elevated threshold (red arrow)

experimenter should be blind to the animal information to minimize bias.

In our protocol, we use a custom MATLAB-based software based on ABR input/output function to interpolate the ABR thresholds [20, 32, 34]. Threshold is statistically determined as a stimulus level that produces peak-to peak ABR signals that are at least two standard deviations above the background noise level. ABR signal magnitudes are calculated by finding the largest peak-to-peak ABR voltage within a 10 ms time window, starting 2 ms after the stimulus onset. Background noise level is calculated from

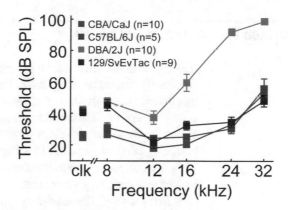

Fig. 3 ABR audiogram of different mouse strains tested at 2 months of age. CBA/CaJ, C57BL/6J, DBA/2J, and 129/SvEvTac mice are tested for click (clk) and tone-burst (8, 12, 16, 24, 32 kHz) ABRs at 2 months of age. Hearing sensitivity varies across different strains. ABR thresholds for each strain group are plotted as a function of ABR test frequency. Thresholds are indicated as mean ± standard error of means (SEM)

the last 5 ms of the 30 ms recording epoch, 20 ms after the stimulus offset. Threshold is determined for each stimulus frequency for each animal and averaged across groups. ABR thresholds can be plotted as a function of stimulus frequency (Fig. 3).

3.8 ABR Waveform Analysis

ABRs reflect synchronous activity of the auditory neurons along the ascending auditory pathway as the ABR sound stimuli propagate from the cochlea through the auditory brainstem. Normal ABR waveforms consist of four to five defined peaks and troughs (Wave 1–5), with each wave reflecting neural activity of different generators in the ascending auditory pathway (Fig. 1). ABR waveforms occur within the first 10 ms of stimulus onset and each wave is separated in the time domain.

In our protocol, ABR waveform analysis is done post-hoc, using a custom-made software to select peaks and calculate amplitude and latency of each wave. ABR amplitude is measured by the difference between the peak and throughs for each wave. Peak latency is measured as the time between stimulus onset and occurrence of a given peak. It can also be measured from the time the sound reaches the ear, by subtracting the travel time. In general, ABR waveform amplitude decreases and peak latency increases, with decreasing ABR stimulus intensity (Fig. 2).

3.9 Best Practices

Speaker calibration: It is highly important to calibrate the speaker regularly to ensure the speaker output level is correct and consistent across different tone frequencies and clicks. Ideally, the speaker output should be checked before each test session.

4   Information We Can Obtain from ABRs

Threshold: ABR thresholds provide an indirect estimate of hearing sensitivity. By analyzing ABR thresholds for clicks and tone frequencies, one can survey functional hearing sensitivity along the cochlear spiral. Threshold analysis is useful in assessing hearing status such as in development [16], noise-induced hearing loss [20, 35, 36], age-related hearing loss [27, 37], across mice strains [14, 15], across transgenic mouse models (e.g., [13, 38–40]), and in hearing recovery after treatments (e.g., [41–46]).

Integrity of the auditory pathway: ABR waveforms provide information on integrity of the auditory pathway and hearing sensitivity. ABR waveforms reflect synchronous activity of the auditory neurons along the ascending auditory pathway in response to sound stimuli.

Based on animal studies, Wave 1 is generated by auditory nerve and spiral ganglion neurons [3, 4, 47], and the subsequent waves have contributions from multiple auditory nuclei in the ascending pathway (Fig. 1). It has been shown that Wave 2 is dominated by cochlear nucleus (globular bushy cells), Wave 3 is mostly generated by medial nucleus of trapezoid body (MNTB) principal cells in the superior olivary complex, which receive input from the globular bushy cells and spherical cells in the cochlear nucleus. Wave 4 is mostly generated by medial superior olive (MSO) cells in the superior olivary complex, which receive input from contralateral spherical cells in the cochlear nucleus. Wave 5 is thought to be generated by cells in the lateral lemniscus and/or inferior colliculus [3, 4]. ABR wave 4 and 5 have binaural interactions. In mice, Wave 5 is not always identifiable [14].

Given that ABR waveforms reflect neural responses in the auditory pathway in response to onset of sound stimuli, waveform analysis, such as measurement of amplitude and latency, is useful in examining response characteristics and assessing abnormality in the auditory brainstem pathway (Fig. 2). Waveform analysis can also include measurement of peak duration, slope, area, and inter-peak latency.

For instance, by measuring the amplitude and latency of each wave as a function of stimulus intensity, one can examine differences in ABR waveform characteristics across strains [15], investigate changes in ABRs following noise exposure [18, 20, 35, 48], and acquire information on synchrony and auditory processing in transgenic mouse models [19, 40, 49]. Furthermore, by calculating relative peak amplitude ratios relative to peak 1, one can deduce cochlear neuropathy or central hyperactivity following noise exposure [20, 50]. ABR testing can be also adapted to evaluate complex sound processing, such as responses to various masking and gap in noise stimulus conditions in mice [51–54].

5   Considerations

5.1   Protocol Factors that Can Affect ABRs

Electrode placement: To obtain reliable ABR recordings, it is critical to keep the subdermal electrode placements consistent across animals. Different electrode configurations near the pinna can affect the shape of ABR waveforms [55]. Given that ABR is a far-field potential, changes in electrode location can change the relative contribution of the peripheral generator, the cochlea.

Animal body temperature: It is important to maintain the animal's body temperature during ABR recording, by placing the animal on a heating pad and monitoring its core body temperature. Anesthetized mice can quickly become hypothermic, and loss of body temperature can affect ABR results [56, 57] or cause death.

Level of anesthesia: It is essential to maintain a deep level of anesthesia for stable ABR recordings. Absent corneal reflex and whisker movements, presence of relaxed muscle tone, and deep and regular breathing indicate that the animal has entered surgical level of anesthesia suitable for ABR recording. If the mouse is lightly or insufficiently anesthetized, the excess myogenic responses, such as whisker movement or body twitching, can introduce electrical noise to ABR recording. These myogenic noises can mask the ABRs at low sound levels, resulting in higher ABR thresholds. If the animal begins to come out of the anesthesia during ABR testing, we recommend giving a ketamine/xylazine booster (1/3 to 1/2 of original dose) subcutaneously (s.c.) without moving the animal and electrodes and resume the recording once deep anesthesia state is achieved. Care should be taken with anesthesia, as species and strains vary in sensitivity. The dose of ketamine and particularly xylazine should be adjusted if mice are not achieving deep anesthesia levels or if you are observing high mortality rates. Isoflurane should not be used, as it alters ABR thresholds and waveforms [58, 59].

Electrical/acoustic noise: To minimize noise during ABR recording, it is best to test ABRs in a magnetically shielded sound-isolating chamber lined with foam to reduce acoustic reflections. Grounding the sound-isolating chamber and shielding electrical equipment in the room further helps to reduce any potential electromagnetic noise that can interfere with ABR recording. Light in the sound-isolating chamber can act as a source of electrical noise; thus, it is advised to have the light turned off during ABR recording. It is also a good practice to keep the testing room as quiet as possible, free of noise and vibration.

**5.2
Stimulus/Recording
Parameters that Can
Affect ABRs**

Stimulus Repetition rate: ABR stimulus repetition rate is usually set to <30/s. Increasing stimulus repetition rate shortens the overall ABR test duration; however, it will also increase noise as it gives less time for recovery in between stimulus presentation, leading to higher thresholds (5–10 dB SPL). It is also important to set the stimulus repetition rate so that it does not synchronize with 50–60 Hz mains from electric supply, which can seriously interfere with ABR recording.

Recording time window: The recording epoch should be long enough to include a prestimulus baseline period, a stimulus onset-ABR recording period, and a poststimulus period during which the response returns to baseline. In our ABR protocol, the recording epoch is 30 ms, in which the first 10 ms includes prestimulus period and ABR recording period, and the remaining 20 ms poststimulus period allows for ample time for recovery to baseline. A prestimulus baseline period of ~1 ms accounts for the time the stimulus travels from the free-field speaker to the animal's ear.

Number of response averages: The typical range of response averages is 256–1025. In our protocol, we average the response over 300 sweeps, which produces adequate signal quality and noise reduction and allows for testing of multiple mice in a session. Larger number of averages will increase ABR test duration. To ensure good signal quality and feasible ABR test schedule, we advise to take a conservative approach in setting the number of averages, while weighing other factors such as anesthesia tolerability, ABR test time, and hearing status of the mouse. Typically, mice with hearing loss require a greater number of averages as the ABR response amplitude is smaller compared to normal hearing mice. Number of response averages should be enough to account for the hearing status differences, yet it needs to be consistent across test animals.

Alternate polarity: In our protocol, ABR stimulus is presented in alternate polarity, consisting of condensation (compression) and rarefication (decompression) stimulus polarities, mimicking the way sound travels through the air and stimulates the tympanic membrane. Condensation stimulus polarity creates positive pressure and moves the tympanic membrane inward, while rarefication stimulus creates negative pressure and pulls the tympanic membrane outward. Stimulus polarity has effects on ABR waveform amplitude and latency [60–62]. Use of alternate stimulus polarity reduces stimulus artifact and contamination from cochlear microphonics [63].

**5.3   Subject
and Environmental
Factors that Can
Affect ABRs**

Hearing status/development: The onset of hearing in mice starts around postnatal day (P) 10–12. Mice start to respond to limited acoustic stimuli (<16 kHz) at P12 with overall thresholds exceeding 100 dB SPL [16, 27]. The hearing range and sensitivity improves over the next three weeks, reaching adult-like sensitivity around P18 or older [16]. The ABR waveforms also mature progressively during the first few weeks of postnatal development. ABR peak latencies shorten and amplitudes reach maturation by P35, reflecting auditory pathway development in the brain over this period. Therefore, ABR testing may not be suitable for neonatal mice as auditory development can confound the result.

Mice strain/background: Hearing sensitivity varies across different strains and genetic background. Commonly used inbred background strains such as C57 or 129-related strains (129/SvJ, 129J), which are widely used for targeted genetic mutation, exhibit elevated ABR threshold at 2–3 months of age [14, 27]. CD1 mice, an outbred strain also commonly used in research, show early-onset severe hearing loss by 3–4 weeks of age [64]. Use of these mice in hearing research is not recommended, as the strain or genetic background can inadvertently confound the interpretation of the ABR results. It is important to note that the ABR waveform generators also differ across species [5, 65], so care must be taken in comparing results with other species such as humans.

Age-related hearing loss: Hearing sensitivity declines as mice age, and the rate of progression and pattern differ by strain. For instance, C57BL/6 mice exhibit early onset high frequency hearing loss starting around 2–3 months of age and progressing to severe hearing loss in all frequency ranges by 12 months of age [25, 27]. On the other hand, CBA/CaJ mice maintain good hearing up to 2–3 years of age [14]. Given the difference in hearing sensitivity and progression profile across different strains, CBA/-CaJ is considered a favored mouse model in hearing research due to its stable and low ABR thresholds across age.

Noise level in the vivarium: Often overlooked, but highly crucial factors that can affect the ABR results are the environmental noises in the vivarium. Most of the noises present in the animal facilities are human generated [66, 67], but also include ventilated cage systems [68] or ultrasonic noise generated by mechanical or electrical equipment [69]. Prolonged exposure to noise can negatively impact hearing health of the mice [20, 35, 36, 70], and susceptibility to noise-induced hearing loss differs across strain and age [71–74]. To reduce experimental variability from unintended or indirect environmental noise exposure, we recommend housing the mice in a quiet, low-traffic room with controlled noise environment [66].

Acknowledgements

This work was funded by David M. Rubenstein Fund for Hearing Research and NIH grants R01 DC016641 and R01DC017620.

References

1. Jewett DL, Romano MN, Williston JS (1970) Human auditory evoked potentials: possible brain stem components detected on the scalp. Science 167(3924):1517–15178. https://doi.org/10.1126/science.167.3924.1517

2. Jewett DL, Williston JS (1971) Auditory-evoked far fields averaged from the scalp of humans. Brain 94(4):681–696. https://doi.org/10.1093/brain/94.4.681

3. Henry KR (1979) Auditory brainstem volume-conducted responses: origins in the laboratory mouse. J Am Aud Soc 4(5):173–178

4. Melcher JR, Guinan JJ Jr, Knudson IM, Kiang NY (1996) Generators of the brainstem auditory evoked potential in cat. II. Correlating lesion sites with waveform changes. Hear Res 93(1–2):28–51. https://doi.org/10.1016/0378-5955(95)00179-4

5. Melcher JR, Kiang NY (1996) Generators of the brainstem auditory evoked potential in cat. III: identified cell populations. Hear Res 93(1–2):52–71. https://doi.org/10.1016/0378-5955(95)00200-6

6. Cheatham MA, Pearce M, Richter CP, Onodera K, Shavit JA (2001) Use of the pinna reflex as a test of hearing in mutant mice. Audiol Neurootol 6(2):79–86. https://doi.org/10.1159/000046813

7. Jero J, Coling DE, Lalwani AK (2001) The use of Preyer's reflex in evaluation of hearing in mice. Acta Otolaryngol 121(5):585–589

8. Kiernan AE, Zalzman M, Fuchs H, Hrabe de Angelis M, Balling R, Steel KP, Avraham KB (1999) Tailchaser (Tlc): a new mouse mutation affecting hair bundle differentiation and hair cell survival. J Neurocytol 28(10–11):969–985. https://doi.org/10.1023/a:1007090626294

9. Heffner HE, Heffner RS (2001) Behavioral assessment of hearing in mice. In: Willott JF (ed) Handbook of mouse auditory research: from behavior to molecular biology. CRC Press, Boca Raton, FL, pp 19–29

10. Heffner HE, Koay G, Heffner RS (2006) Behavioral assessment of hearing in mice—conditioned suppression. In: Crawley J et al (eds) Current protocols in neuroscience. Suppl. 34. Wiley, New York, pp 8.21D.1–8.21D.15

11. Klump GM, Dooling RJ, Fay RR, Stebbins WC (eds) (1995) Methods in comparative psychoacoustics. Birkhauser Verlag, Basel

12. Prosen CA, Bath KG, Vetter DE, May BJ (2000) Behavioral assessments of auditory sensitivity in transgenic mice. J Neurosci Methods 97(1):59–67. https://doi.org/10.1016/s0165-0270(00)00169-2

13. Bowl MR, Simon MM, Ingham NJ, Greenaway S, Santos L, Cater H, Taylor S, Mason J, Kurbatova N, Pearson S, Bower LR, Clary DA, Meziane H, Reilly P, Minowa O, Kelsey L, International Mouse Phenotyping Consortium, Tocchini-Valentini GP, Gao X, Bradley A, Skarnes WC, Moore M, Beaudet AL, Justice MJ, Seavitt J, Dickinson ME, Wurst W, de Angelis MH, Herault Y, Wakana S, Nutter LMJ, Flenniken AM, McKerlie C, Murray SA, Svenson KL, Braun RE, West DB, Lloyd KCK, Adams DJ, White J, Karp N, Flicek P, Smedley D, Meehan TF, Parkinson HE, Teboul LM, Wells S, Steel KP, Mallon AM, Brown SDM (2017) A large scale hearing loss screen reveals an extensive unexplored genetic landscape for auditory dysfunction. Nat Commun 8(1):886. https://doi.org/10.1038/s41467-017-00595-4

14. Zheng QY, Johnson KR, Erway LC (1999) Assessment of hearing in 80 inbred strains of mice by ABR threshold analyses. Hear Res 130(1–2):94–107. https://doi.org/10.1016/s0378-5955(99)00003-9

15. Zhou X, Jen PH, Seburn KL, Frankel WN, Zheng QY (2006) Auditory brainstem responses in 10 inbred strains of mice. Brain Res 1091(1):16–26. https://doi.org/10.1016/j.brainres.2006.01.107

16. Song L, McGee J, Walsh EJ (2006) Frequency- and level-dependent changes in auditory brainstem responses (ABRS) in developing mice. J Acoust Soc Am 119(4):2242–2257. https://doi.org/10.1121/1.2180533

17. Bramhall NF, McMillan GP, Kujawa SG, Konrad-Martin D (2018) Use of non-invasive measures to predict cochlear synapse counts. Hear Res 370:1131–1119. https://doi.org/10.1016/j.heares.2018.10.006

18. Kujawa SG, Liberman MC (2009) Adding insult to injury: cochlear nerve degeneration after "temporary" noise-induced hearing loss. J Neurosci 29(45):14077–14085. https://doi.org/10.1523/JNEUROSCI.2845-09.2009

19. Ison JR, Allen PD, Oertel D (2017) Deleting the HCN1 subunit of hyperpolarization-activated ion channels in mice impairs acoustic startle reflexes, gap detection, and spatial localization. J Assoc Res Otolaryngol 18(3):4274–4240. https://doi.org/10.1007/s10162-016-0610-8

20. Schrode KM, Muniak MA, Kim YH, Lauer AM (2018) Central compensation in auditory brainstem after damaging noise exposure.

eNeuro 5(4):ENEURO.0250-18.2018. https://doi.org/10.1523/ENEURO. 0250-18.2018. Erratum in: eNeuro. 2019 Apr 5;6(2)

21. Gorga MP, Beauchaine KA, Reiland JK, Worthington DW, Javel E (1984) The effects of stimulus duration on ABR and behavioral thresholds. J Acoust Soc Am 76(2):616–619. https://doi.org/10.1121/1.391158

22. Heffner HE, Heffner RS (2003) Audition. In: Davis S (ed) Handbook of research methods in experimental psychology. Blackwell, pp 413–440

23. Watson CS, Gengel RW (1969) Signal duration and signal frequency in relation to auditory sensitivity. J Acoust Soc Am 46(4):989–997. https://doi.org/10.1121/1. 1911819

24. National Research Council (US) Committee for the Update of the Guide for the Care and Use of Laboratory Animals (2011) Guide for the care and use of laboratory animals, 8th edn. National Academies Press (US), Washington, DC

25. Kane KL, Longo-Guess CM, Gagnon LH, Ding D, Salvi RJ, Johnson KR (2012) Genetic background effects on age-related hearing loss associated with Cdh23 variants in mice. Hear Res 283(1–2):80–88. https://doi.org/10. 1016/j.heares.2011.11.007

26. Ohlemiller KK, Dahl AR, Gagnon PM (2010) Divergent aging characteristics in CBA/J and CBA/CaJ mouse cochleae. J Assoc Res Otolaryngol 11(4):605–623. https://doi.org/10. 1007/s10162-010-0228-1

27. Shnerson A, Pujol R (1981) Age-related changes in the C57BL/6J mouse cochlea. I. Physiological findings. Brain Res 254(1):65–75. https://doi.org/10.1016/ 0165-3806(81)90059-6

28. Durrant RD, Boston JR (2007) Stimuli for auditory evoked potential assessment. In: Burkard RF, Eggermont JJ, Don M (eds) Auditory evoked potentials, basic principle and clinical application. Lippincott Williams & Wilkins, Baltimore, pp 42–72

29. Beutelmann R, Laumen G, Tollin D, Klump GM (2015) Amplitude and phase equalization of stimuli for click evoked auditory brainstem responses. J Acoust Soc Am 137(1):EL717. https://doi.org/10.1121/1.4903921

30. Koay G, Heffner R, Heffner H (2002) Behavioral audiograms of homozygous med (J) mutant mice with sodium channel deficiency and unaffected controls. Hear Res 171(1–2):111–118. https://doi.org/10. 1016/s0378-5955(02)00492-6

31. Heffner HE, Heffner RS (2007) Hearing ranges of laboratory animals. J Am Assoc Lab Anim Sci 46(1):20–22

32. May BJ, Prosen CA, Weiss D, Vetter D (2002) Behavioral investigation of some possible effects of the central olivocochlear pathways in transgenic mice. Hear Res 171(1–2):142–157. https://doi.org/10.1016/s0378-5955(02) 00495-1

33. Suthakar K, Liberman MC (2019) A simple algorithm for objective threshold determination of auditory brainstem responses. Hear Res 381:107782. https://doi.org/10.1016/j. heares.2019.107782

34. McGuire B, Fiorillo B, Ryugo DK, Lauer AM (2015) Auditory nerve synapses persist in ventral cochlear nucleus long after loss of acoustic input in mice with early-onset progressive hearing loss. Brain Res 1605:22–30. https:// doi.org/10.1016/j.brainres.2015.02.012

35. Lauer AM, May B (2011) The medial olivocochlear system attenuates the developmental impact of early noise exposure. J Assoc Res Otolaryngol 12(3):329–343. https://doi. org/10.1007/s10162-011-0262-7

36. May BJ, Lauer AM, Roos MJ (2011) Impairments of the medial olivocochlear system increase the risk of noise-induced auditory neuropathy in laboratory mice. Otol Neurotol 32(9):1568–1578. https://doi.org/10.1097/ MAO.0b013e31823389a1

37. Sha SH, Kanicki A, Dootz G, Talaska AE, Halsey K, Dolan D, Altschuler R, Schacht J (2008) Age-related auditory pathology in the CBA/J mouse. Hear Res 243(1–2):87–94. https://doi.org/10.1016/j.heares.2008. 06.001

38. Cunningham CL, Qiu X, Wu Z, Zhao B, Peng G, Kim YH, Lauer A, Müller U (2020) TMIE defines pore and gating properties of the mechanotransduction channel of mammalian cochlear hair cells. Neuron 107(1):1261–43. e8. https://doi.org/10.1016/j.neuron.2020. 03.033

39. Schwander M, Sczaniecka A, Grillet N, Bailey JS, Avenarius M, Najmabadi H, Steffy BM, Federe GC, Lagler EA, Banan R, Hice R, Grabowski-Boase L, Keithley EM, Ryan AF, Housley GD, Wiltshire T, Smith RJ, Tarantino LM, Müller U (2007) A forward genetics screen in mice identifies recessive deafness traits and reveals that pejvakin is essential for outer hair cell function. J Neurosci 27(9):2163–2175. https://doi.org/10.1523/ JNEUROSCI.4975-06.2007

40. Yang YM, Aitoubah J, Lauer AM, Nuriya M, Takamiya K, Jia Z, May BJ, Huganir RL, Wang LY (2011) GluA4 is indispensable for driving fast neurotransmission across a high-fidelity

central synapse. J Physiol 589(17):4209–4227. https://doi.org/10.1113/jphysiol.2011.208066

41. Akil O, Seal RP, Burke K, Wang C, Alemi A, During M, Edwards RH, Lustig LR (2012) Restoration of hearing in the VGLUT3 knock-out mouse using virally mediated gene therapy. Neuron 75(2):283–293. https://doi.org/10.1016/j.neuron.2012.05.019

42. Askew C, Rochat C, Pan B, Asai Y, Ahmed H, Child E, Schneider BL, Aebischer P, Holt JR (2015) Tmc gene therapy restores auditory function in deaf mice. Sci Transl Med 7(295):295ra108. https://doi.org/10.1126/scitranslmed.aab1996

43. Chowdhury S, Owens KN, Herr RJ, Jiang Q, Chen X, Johnson G, Groppi VE, Raible DW, Rubel EW, Simon JA (2018) Phenotypic optimization of urea-thiophene carboxamides to yield potent, well tolerated, and orally active protective agents against aminoglycoside-induced hearing loss. J Med Chem 61(1):849–847. https://doi.org/10.1021/acs.jmedchem.7b00932

44. Gao X, Tao Y, Lamas V, Huang M, Yeh WH, Pan B, Hu YJ, Hu JH, Thompson DB, Shu Y, Li Y, Wang H, Yang S, Xu Q, Polley DB, Liberman MC, Kong WJ, Holt JR, Chen ZY, Liu DR (2018) Treatment of autosomal dominant hearing loss by in vivo delivery of genome editing agents. Nature 553(7687):217–221. https://doi.org/10.1038/nature25164

45. Isgrig K, Shteamer JW, Belyantseva IA, Drummond MC, Fitzgerald TS, Vijayakumar S, Jones SM, Griffith AJ, Friedman TB, Cunningham LL, Chien WW (2017) Gene therapy restores balance and auditory functions in a mouse model of usher syndrome. Mol Ther 25(3):7807–7891. https://doi.org/10.1016/j.ymthe.2017.01.007

46. Lentz JJ, Jodelka FM, Hinrich AJ, McCaffrey KE, Farris HE, Spalitta MJ, Bazan NG, Duelli DM, Rigo F, Hastings ML (2013) Rescue of hearing and vestibular function by antisense oligonucleotides in a mouse model of human deafness. Nat Med 19(3):345–350. https://doi.org/10.1038/nm.3106

47. Melcher JR, Knudson IM, Fullerton BC, Guinan JJ Jr, Norris BE, Kiang NY (1996) Generators of the brainstem auditory evoked potential in cat. I. An experimental approach to their identification. Hear Res 93(1–2):1–27. https://doi.org/10.1016/0378-5955(95)00178-6

48. Lin HW, Furman AC, Kujawa SG, Liberman MC (2011) Primary neural degeneration in the Guinea pig cochlea after reversible noise-induced threshold shift. J Assoc Res Otolaryngol 12(5):605–616. https://doi.org/10.1007/s10162-011-0277-0

49. Kim YH, Holt JR (2013) Functional contributions of HCN channels in the primary auditory neurons of the mouse inner ear. J Gen Physiol 142(3):207–223. https://doi.org/10.1085/jgp.201311019

50. Hickox AE, Liberman MC (2014) Is noise-induced cochlear neuropathy key to the generation of hyperacusis or tinnitus? J Neurophysiol 111(3):552–564. https://doi.org/10.1152/jn.00184.2013

51. Lina IA, Lauer AM (2013) Rapid measurement of auditory filter shape in mice using the auditory brainstem response and notched noise. Hear Res 298:73–79. https://doi.org/10.1016/j.heares.2013.01.002

52. Lowe AS, Walton JP (2015) Alterations in peripheral and central components of the auditory brainstem response: a neural assay of tinnitus. PLoS One 10(2):e0117228. https://doi.org/10.1371/journal.pone.0117228

53. Mehraei G, Hickox AE, Bharadwaj HM, Goldberg H, Verhulst S, Liberman MC, Shinn-Cunningham BG (2016) Auditory brainstem response latency in noise as a marker of cochlear synaptopathy. J Neurosci 36(13):3755–3764. https://doi.org/10.1523/JNEUROSCI.4460-15.2016

54. Song L, McGee J, Walsh EJ (2008) Development of cochlear amplification, frequency tuning, and two-tone suppression in the mouse. J Neurophysiol 99(1):344–355. https://doi.org/10.1152/jn.00983.2007

55. Shaheen LA, Valero MD, Liberman MC (2015) Towards a diagnosis of cochlear neuropathy with envelope following responses. J Assoc Res Otolaryngol 16(6):727–745. https://doi.org/10.1007/s10162-015-0539-3

56. Gold S, Cahani M, Sohmer H, Horowitz M, Shahar A (1985) Effects of body temperature elevation on auditory nerve-brain-stem evoked responses and EEGs in rats. Electroencephalogr Clin Neurophysiol 60(2):146–153. https://doi.org/10.1016/0013-4694(85)90021-5

57. Williston JS, Jewett DL (1982) The Q10 of auditory brain stem responses in rats under hypothermia. Audiology 21(6):457–465. https://doi.org/10.3109/00206098209072758

58. Cederholm JM, Froud KE, Wong AC, Ko M, Ryan AF, Housley GD (2012) Differential actions of isoflurane and ketamine-based anaesthetics on cochlear function in the mouse.

Hear Res 292(1–2):71–79. https://doi.org/10.1016/j.heares.2012.08.010

59. Ruebhausen MR, Brozoski TJ, Bauer CA (2012) A comparison of the effects of isoflurane and ketamine anesthesia on auditory brainstem response (ABR) thresholds in rats. Hear Res 287(1–2):25–29. https://doi.org/10.1016/j.heares.2012.04.005

60. Fowler CG (1992) Effects of stimulus phase on the normal auditory brainstem response. J Speech Hear Res 35(1):167–174. https://doi.org/10.1044/jshr.3501.167

61. Kumar K, Bhat JS, D'Costa PE, Srivastava M, Kalaiah MK (2014) Effect of stimulus polarity on speech evoked auditory brainstem response. Audiol Res 3(1):e8. https://doi.org/10.4081/audiores.2013.e8

62. Peake WT, Kiang NY (1962) Cochlear responses to condensation and rarefaction clicks. Biophys J 2(1):23–34. https://doi.org/10.1016/s0006-3495(62)86838-6

63. Maurer K, Schäfer E, Leitner H (1980) The effect of varying stimulus polarity (rarefaction Vs. condensation) on early auditory evoked potentials (EAEPs). Electroencephalogr Clin Neurophysiol 50(3–4):332–334. https://doi.org/10.1016/0013-4694(80)90162-5

64. Shone G, Raphael Y, Miller JM (1961) Hereditary deafness occurring in cd/1 mice. Hear Res 57(1):153–156. https://doi.org/10.1016/0378-5955(91)90084-m

65. Fullerton BC, Levine RA, Hosford-Dunn HL, Kiang NY (1987) Comparison of cat and human brain-stem auditory evoked potentials. Electroencephalogr Clin Neurophysiol 66(6):547–570. https://doi.org/10.1016/0013-4694(87)90102-7

66. Lauer AM, May BJ, Hao ZJ, Watson J (2009) Analysis of environmental sound levels in modern rodent housing rooms. Lab Anim 38(5):154–160. https://doi.org/10.1038/laban0509-154

67. Milligan SR, Sales GD, Khirnykh K (1993) Sound levels in rooms housing laboratory animals: an uncontrolled daily variable. Physiol Behav 53(6):1067–1076. https://doi.org/10.1016/0031-9384(93)90361-i

68. Perkins SE, Lipman NS (1996) Evaluation of microenvironmental conditions and noise generation in three individually ventilated rodent caging systems and static isolator cages. Contemp Top Lab Anim Sci 35(2):61–65

69. Sales GD, Wilson KJ, Spencer KE, Milligan SR (1988) Environmental ultrasound in laboratories and animal houses: a possible cause for concern in the welfare and use of laboratory animals. Lab Anim 22(4):369–375. https://doi.org/10.1258/002367788780746188

70. Turner JG, Parrish JL, Hughes LF, Toth LA, Caspary DM (2005) Hearing in laboratory animals: strain differences and nonauditory effects of noise. Comp Med 55(1):12–23

71. Davis RR, Newlander JK, Ling X, Cortopassi GA, Krieg EF, Erway LC (2001) Genetic basis for susceptibility to noise-induced hearing loss in mice. Hear Res 155(1–2):82–90. https://doi.org/10.1016/s0378-5955(01)00250-7

72. Kendall A, Schacht J (2014) Disparities in auditory physiology and pathology between C57BL/6J and C57BL/6N substrains. Hear Res 318:18–22. https://doi.org/10.1016/j.heares.2014.10.005

73. Ohlemiller KK, Wright JS, Heidbreder AF (2000) Vulnerability to noise-induced hearing loss in 'middle-aged' and young adult mice: a dose-response approach in CBA, C57BL, and BALB inbred strains. Hear Res 149(1–2):239–247. https://doi.org/10.1016/s0378-5955(00)00191-x

74. Yoshida N, Hequembourg SJ, Atencio CA, Rosowski JJ, Liberman MC (2000) Acoustic injury in mice: 129/SvEv is exceptionally resistant to noise-induced hearing loss. Hear Res 141(1–2):97–106. https://doi.org/10.1016/s0378-5955(99)00210-5

Chapter 17

Behavioral Models Loudness, Hyperacusis, and Sound Avoidance

Richard Salvi, Connor Mauche, Hannah Thorner, Guang-Di Chen, and Senthilvelan Manohar

Abstract

Loudness is the subjective perception of an acoustic stimulus that is most closely correlated with sound intensity. The dynamic range of loudness perception spans more than 120 dB, but how the auditory system is able to encode this information remains largely a mystery. Hearing loss and other medical conditions can disrupt intensity coding, often leading to loudness recruitment or in some cases hyperacusis, a sometimes debilitating condition in which moderate intensity sounds are perceived as intolerably loud, which evoke fear or instill sound avoidance. Insights into the neural mechanisms that give rise to normal or aberrant loudness perception can be aided by the development of animal models that can "tell the auditory physiologist" just how loud, annoying, or aversive a sound is. These behavioral measures of loudness can then be correlated with various electrophysiological or brain imaging metrics to the acoustic properties of the sound. In this chapter, we describe an operant behavioral technique for measuring reaction time-intensity functions that has proved to be highly effective in assessing the normal growth of loudness and two abnormal forms of loudness growth, recruitment, and hyperacusis. We also describe another behavioral technique to test for sound avoidance that takes advantage of a rodent's innate aversion to brightly illuminated open spaces. Novel methods for inducing loudness recruitment, hyperacusis, and sound aversion are also described.

Key words Hyperacusis, Loudness intolerance, Sound avoidance, Hearing loss, Fear, Amygdala, Salicylate, Noise-induced hearing loss

Abbreviations

ABR	Auditory brainstem response
ASAP	Active sound avoidance paradigm
HL	Hearing level
RT	Reaction time
RT-I	Reaction time-intensity
SPL	Sound pressure level
ULL	Uncomfortable loudness level

Andrew K. Groves (ed.), *Developmental, Physiological, and Functional Neurobiology of the Inner Ear*, Neuromethods, vol. 176,
https://doi.org/10.1007/978-1-0716-2022-9_17,

1   Introduction

Loudness is one of the most fundamental aspects of auditory perception, but exactly how this sensory dimension is coded in the central nervous system is still poorly understood and remains a central question in auditory neuroscience, psychophysics, and clinical efforts to develop better auditory prosthetic devices. Advances in neuroscience have often relied on the development of animal models used to glean new scientific insights, develop disease models, and test and evaluate theories of sensory coding. The purpose of this chapter is to describe some animal behavioral techniques that auditory neuroscientists can use to assess the normal growth of loudness, loudness intolerance disorders, and sound avoidance behaviors. Before describing these methods, we will first describe some aspects of normal loudness perception followed by three abnormal forms of loudness perception referred to as loudness recruitment, hyperacusis and avoidance (fear) hyperacusis.

1.1 Loudness Dynamic Range

Loudness is the subjective perception of an acoustic stimulus most closely related to sound pressure level. Low-intensity sounds are subjectively described as quiet whereas intense sounds are referred to as loud. While sound intensity is the dominant acoustic variable, the sensation of loudness is affected by other acoustic parameters as well as psychological variables. The sone scale of loudness was developed for humans using magnitude estimation procedures, a procedure in which a subject typically assigns a number or value that represents the loudness of the acoustic stimulus [1, 2]. The dynamic range of loudness in humans is approximately 130 dB at the mid-frequencies; the dynamic range extends from sounds that are barely audible around, approximately 0 dB SPL, to stimuli near the aural threshold of pain around 130 dB SPL [3]. The dynamic range for loudness varies with frequency and is much smaller near the lower (~20 Hz) and upper frequency limits (~20,000 Hz) of audibility. Psychologists, engineers, and neuroscientists have devoted considerable effort trying to understand how the auditory system is able to encode loudness over such a large range of sound intensities [4, 5]. Much of this effort has focused on identifying the neural correlates of loudness in animal models with normal hearing and various hearing impairments [6–9]. The changes in loudness perception that emerge in subjects with hearing loss and some neurological disorders may provide important clues as to where and how loudness is encoded in the central nervous system.

1.2 Loudness Recruitment

In most subjects with cochlear hearing loss, the dynamic range of loudness is compressed because the threshold for hearing is increased while loudness perception remains largely unchanged at high sound intensities. At intensities just above threshold, loudness increases at a faster than normal rate, but the rate of growth gradually decreases so that at high intensities, the loudness in the impaired ear matches the loudness in the normal ear. This phenomenon, generally referred to as loudness recruitment, is the most common loudness disorder that develop in listeners with sensorineural hearing loss [10–13]. The rapid growth of loudness that occurs with recruitment can be considered beneficial to the hearing-impaired listener. Instead of perceiving speech or music as muffled or muted, the intensity-dependent neural amplification provided by loudness recruitment boosts the weak signals into the normal loudness range. Models of loudness coding need to consider how the central auditory system deals with the dynamic range compression associated with loudness recruitment.

1.3 Loudness Hyperacusis

Some individuals suffer from hyperacusis, a sensory hypersensitivity disorder in which everyday sounds of moderate intensity are perceived as uncomfortable or intolerably loud [14]. The prevalence of loudness hyperacusis in adults is estimated to be approximately 9% [15], but this may be an underestimate because many individuals are unaware that they have a loudness tolerance problem [16]. Normal hearing listeners report that sounds around 100 dB hearing level (HL) are uncomfortably loud. However, uncomfortable loudness levels (ULL) are much lower in individuals with hyperacusis [17–19]. In many individuals suffering from hyperacusis, the dynamic range for loudness is often compressed at both low and high intensities because hearing thresholds are elevated and ULL are reduced. However, in a small subgroup of individuals with hyperacusis, patients present with clinically normal thresholds. In cases such as this, the dynamic range is only compressed at high intensities because only the ULLs are reduced [16, 18].

1.4 Sound Avoidance Hyperacusis

Some individuals develop fear hyperacusis and try to avoid being exposed to sounds that they feel will be too loud [17, 20, 21]. Auditory stimuli can evoke sound avoidance behaviors because they instinctively elicit negative emotions or fear. For example, most individuals with Williams syndrome and autism spectrum disorder have an intrinsic aversion to moderate intensity sounds [22–24]. Other individuals may develop sound avoidance behaviors because the acoustic stimuli acquire negative emotions (e.g., loud sounds experienced during a dangerous military operation). In other cases, an individual may develop sound avoidance behaviors if everyday sounds are linked to negative emotions [25].

2   Measuring Loudness: Recruitment and Hyperacusis

2.1 Reaction Time-Intensity Functions

Assessing loudness growth in nonverbal animals using psychophysical loudness scaling procedures is a difficult, if not impossible, task [26, 27]. Fortunately, human studies have shown that reaction time (RT), the amount of time it takes for a subject to respond to an acoustic stimulus, decreases with increasing intensity. Investigators have successfully used reaction time versus intensity (RT-I) functions to assess loudness growth in humans [28–30] as well as nonhuman primates [31–34].

2.2 Subjects

In this section, we describe how RT-I intensity functions can be used to assess the growth of loudness in normal hearing rats (>10 weeks of age). RT-I functions can be measured with tone bursts to obtain frequency-specific information or with broadband noise bursts to obtain a global assessment of loudness growth over a large frequency range [9, 35–37].

2.3 Equipment

Most of the operant conditioning equipment used to measure RT-I functions is commercially available. Each operant chamber (Med Associates, ENV-008, Length: 29.2 cm; width 24.1 cm; height: 21 cm) has a grid floor, acrylic side walls, one of which opens to serve as an entry door, and front and back walls with channels into which aluminum panels or test equipment can be inserted (Fig. 1). The center channel of the front wall contains a nose-poke hole for detecting the animal's response. A house light, which can be switched on or off, is located in a channel near the nose-poke hole. A pellet dispenser (Med Associates; ENV-203-45) mounted on the right front of the operant chamber delivers a food pellet

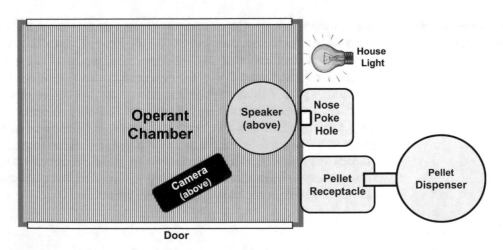

Fig. 1 Schematic of operant conditioning apparatus. Rat initiates a trial by inserting and holding its nose in the nose-poke hole. When a noise burst or tone burst is presented to the speaker on a Go-Sound trial, the rat withdraws its nose from the nose-poke hole that signals a response

(Bio-Serv, 45 mg pellets) via a plastic tube connected to a food pellet receptacle (Med Associates, ENV-200R2M0) mounted to the right of the nose-poke. A loudspeaker (Fostex FT 28D) is mounted on the roof of the enclosure just in front of the nose-poke hole.

Each operant chamber is housed in a custom-made sound booth (Booth outer dimensions: length = 76 cm; width = 71 cm; height = 76 cm) made of dense double-walled sound-attenuating board (MDF, 2.5 cm thickness) with a layer of polystyrene foam insulation (1.27 cm) sandwiched between the inner and outer walls of the booth. A double-walled door on the front of the sound booth provides access to the interior of the sound booth. The interior walls of the sound booth are lined with sound absorbing foam (Sonex Acoustics, thickness 5 cm). Cables connecting the speaker, pellet feeder, house-light, and nose-poke response sensor are routed through a small opening in the wall of the sound booth.

The equipment used to generate the acoustic stimuli, record the behavioral response, turn on/off the house light and deliver food pellets to the operant testing chamber were purchased from Tucker-Davis Technologies (TDT RX6 Multifunction Processor, TDT PA5 Programmable Attenuator); the TDT hardware were connected to a personal computer. The rat's performance on the RT-I task is monitored in real time on a computer monitor which displays the number of trials and the numbers of Hits, Misses, False Alarms, and Correct Rejections [36, 38]. On trials when an acoustic stimulus is presented, the computer measures the time between the onset of the stimulus and the onset of the behavioral response. This information together with the stimulus parameters is stored on a personal computer. The RT-I data and other information were stored on the computer and analyzed offline. Solid state switches (Omega Engineering; DR-IO-OCD-R0-060; DR-IO-IDC-R0-028N; DR-IO-IDC-R0-028P) are used to interface the TTL input/output ports on the TDT RX6 with input/output ports on Med Associates hardware (pellet feeder, house light, sensor on noise-poke hole). Custom Matlab software is used to generate and control the timing and delivery of the acoustic stimuli, response intervals, food pellet dispenser, and house lights, and to detect the response from the nose-poke sensor. The behavior of the rat in the operant testing chamber is monitored with a security camera (Zmodo, KT7204AD) mounted above the operant chamber and output is displayed on a monitor. Sound pressure levels in the operant chamber are calibrated using a sound level meter (Larson Davis System 824) equipped with a microphone (one-half inch free field microphone, model 2520, Larson-Davis) placed at the approximate position of the animal's ear with its head in the nose-poke hole. The calibration data is stored in the computer and used to set the stimulus levels during RT-I testing.

2.4 Procedures   The incremental steps used to shape the behavior of the rat on the operant Go-/No-Go procedure used to measure RT-I functions are outlined below. The rats are individually housed in the lab animal facility and maintained on a light–dark cycle (7 am on:7 pm off). The rats are given free access to water, but food intake is restricted to maintain body weight between 85 and 100% of free-feeding weight. Each daily test session (~1 h) occurs between 8 am and 6 pm. Rats are generally tested 6–7 days per week. At the end of each daily test session, the rat is returned to its home cage in the animal facility and a small amount of rat chow is put in its food tray. (1) During the Acclimatization phase (3–5 days), the rat is brought to the lab to interact with and acclimate to the experimenter and the testing environment. Body weight is recorded in a daily log file along with other pertinent behavioral or health information. The experimenter handles the rat for 10–15 min/day. (2) During Go-Sound training, the rat is put in the operant chamber for approximately 60 min with the house light on. If the rat inserts its nose into the hole and holds it there for 0.5 s, a noise burst (60 dB SPL, broadband noise bursts, 300 ms duration, 5 ms/rise fall time) is presented. If the rat withdraws its nose from the hole within a defined response interval (2–3 s) following the sound stimulus, a 45 mg food pellet is delivered and the response is scored as a Hit. If the rat does not remove its nose from the hole by the end of the 2-s response interval, the trial is scored as a Miss. During this phase of training, the goal is to train the rat to keep its nose in the hole for least 0.5 s, and then immediately withdraw it when the sound is presented to obtain the food reward. Once the rat consistently pokes and withdraws to the sound to receive the food reward, the experimenter gradually increases the amount of time the rat must hold its nose in the hole from 0.5 to 4 s. (3) The third step involves training the rat to distinguish Go-Sound trials from No-Go-Silence trials. Go-Sound trials are randomly interleaved with No-Go-Silence trials at a ratio of 9 to 1 respectively. Daily testing continues on the 9:1 ratio until the rat achieves a criterion with a Hit rate of >85% on Go-Sound trials and a False Alarm rate of <15% on No-Go-Silent trial. Once this criterion is met, the ratio of Go-Sound trials to No-Go-Silence trials is changed to 8:2. Once the Hit rate is >85% and False Alarm rate is <15%, the ratio of Go-Sound and No-Go-Silence trials is set to 7:3. Training continues with the 7:3-ratio until the Hit rate is >85% and the False Alarm rate is <15% and at least 200 trials are run per daily session. (4) During the fourth stage, the intensity of the noise burst is randomly presented at one of seven intensities from 30 to 90 dB SPL (30, 40, 50, 60, 70, 80, or 90 dB SPL). Once performance reaches a Hit rate of >85% on Go-Sound trails and <15% False Alarm rate on No-Go-Silence trials, training is considered complete and formal data collection with the RT-I paradigm begins. Approximately 30% of the trials are No-Go-Silence catch trials. During the RT-I data acquisition phase, the target stimuli were presented in a quasi-random order according to the psychophysical method of constant. RT-I functions were constructed from trials on which a rat correctly responded on a stimulus trial (i.e., a Hit).

A useful starting point is to measure RT-I loudness growth functions using noise bursts (300 ms duration, 5 ms rise/fall time) presented at seven intensities (30, 40, 50, 60, 70, 80, or 90 dB SPL) with a 7:3-ratio of Go-Sound trials to No-Go-Silence trials. During each day of testing, data are collected from a total of 200–400 trials (Go-Stimulus plus No-Go-Silence trials) collected over approximately 1 h of testing. Typically, data from the first or last 200 trails is used for RT-I analyses. RTs (stimulus onset to removal of the nose from the poke- hole) are measured from at least 20 Go-Stimulus trials at each intensity. RTs are only used on days when the False-Alarm rate is <30%. The RT data on which the stimulus is detected (i.e., Hit) are used to construct RT-I functions for noise-bursts. The RT-I function schematized in Fig. 2a shows a gradual reduction in RT as the intensity of the noise burst increases from 30 to 90 dB SPL.

2.5 Effects of Stimulus Duration, Bandwidth, and Frequency on Loudness Growth

The loudness of a sound not only increases with intensity but is also influenced by other factors such as stimulus frequency, bandwidth, and duration. It is well known from psychophysical studies that loudness increases with stimulus duration out to approximately 300–400 ms, a phenomenon referred to a temporal integration of loudness [27, 39]. The RT-I procedures described above can be used to test for temporal integration of loudness by simply measuring RT-I functions at several noise burst durations. When the duration of the broadband noise burst is increased from 20 ms to 300 ms, there is a systematic reduction in reaction time as schematized in Fig. 2b. Temporal integration of loudness measured in rats with the RT-I technique is consistent with data obtained from humans [35].

It is well known from human studies that when the overall intensity of sound is held constant, but the bandwidth of the stimulus increases, the perceived loudness increases [1, 40]. This increase in loudness with increasing stimulus bandwidth is most noticeable at intensities well above threshold. Loudness increases as the energy of the stimulus extends beyond the critical band. The RT-I procedure can be used to demonstrate the spectral summation of loudness by measuring RT-I functions for a single tone, a narrow band noise and then a broadband noise as schematized in Fig. 2c. The total energy for the tone, narrow band noise, and broadband noise was the same for each intensity shown on the x-axis. The RT-I function for pure tones lies above that for the narrow band noise while the shortest RTs are obtained for broadband noise. The spectral integration for loudness measured in rats using the RT-I technique is similar to that observed in humans [35].

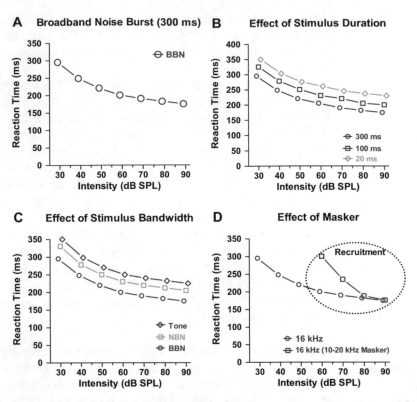

Fig. 2 (**a**) Schematic of reaction time-intensity (RT-I) function for long duration (300 ms) broadband noise bursts . Note orderly decrease in reaction time with increasing intensity. (**b**) Schematic illustrating the decrease in reaction time with increasing intensity for short (20 ms), medium (100 ms), and long duration (300 ms) broadband noise bursts. Note overall decrease in reaction time as the duration of the noise burst increases. Little further decrease occurs for durations greater than 300 ms. (**c**) Schematic illustrating the reaction time-intensity functions for tone bursts, narrow band noise bursts, and broadband noise bursts. Note the overall decrease in reaction time as stimulus bandwidth increases from tone burst through narrowband noise to finally broadband noise. (**d**) Schematic illustrating the reaction time-intensity function for a 16 kHz tone measured in quiet compared to the reaction times measured in the presence of 10–20 kHz octave band noise. Because the 10–20 kHz background noise caused an increase in threshold for the 16 kHz tone, the reaction times could only be measured from 60 to 90 dB SPL . In the presence of the background noise, reaction times around 60 dB SPL are much longer than when measured in quiet; however, reaction times rapidly decrease with intensity and return to normal reaction times at high stimulus levels

If a tone burst is presented to a human listener in the presence of moderately intense background noise, the subject will be unable to detect the tone burst until the intensity of the tone burst exceeds threshold. Once the tone burst becomes detectable, the loudness of the tone initially increases rapidly with intensity and RTs become shorter. However, the growth of loudness gradually slows with increasing intensity so that the loudness of the tone in the background noise is similar to the loudness it would be in quiet [41]. The rapid growth of loudness that occurs in background noise resembles a clinical condition known as loudness recruitment, a phenomenon often seen in patients with cochlear sensorineural hearing loss [42, 43]. RT-I functions can be measured in background noise to simulate loudness recruitment seen in patients with cochlear hearing loss.

When tone burst RT-I functions are measured in rats in the presence of moderate intensity background noise (Fig. 2d), RTs rapidly decrease once the signal is above threshold until RTs catch up to the normal RT-I function at high stimulus levels. The shape of the RT-I function in the presence of the masker is recruitment-like (i.e., rapid loudness growth just above threshold followed by normal growth rate at high intensities).

Taken together, these results indicate that RT-I functions obey the same rules of loudness growth observed in traditional human psychoacoustic studies. Therefore, RT-I functions provide researchers with a powerful tool to test for evidence of recruitment and hyperacusis in animal models.

3   Animal Models of Recruitment and Hyperacusis

3.1   Noise-Induced RT-I Functions with Recruitment-Like Features

Hearing loss resulting from aging, exposure to intense noise and ototoxic drugs has long been known to induce loudness recruitment. At frequencies within the region of hearing loss, sounds that are just above the threshold of hearing are barely audible and not very loud. However, small increases in sound intensity cause a large increase in loudness. Consequently, loudness initially grows at a faster than normal rate just above threshold, but at higher intensities loudness growth slows to a normal rate so that listeners with hearing loss perceive intense sounds at the same loudness as normal listeners. In cases of loudness recruitment, the dynamic range for loudness is compressed into a narrower range due to the hearing loss.

Researchers have used tone bursts to measure RT-I in animals with cochlear hearing losses induced by intense noise exposure [32, 44] and ototoxic drugs [45]. A common finding in these studies was that RT-I functions were recruitment-like at the high frequencies where hearing thresholds were elevated. RTs decreased rapidly at intensities just above thresholds, but the rate of decline slowed so that RTs at high intensities were normal.

3.2   Genetic and Noise-Induced RT-I Functions with Hyperacusis-like Features

We are aware of only one study in birds which found evidence of RT-I functions with hyperacusis-like features [46]. Belgian Waterslager canaries (BWS) suffer from a genetic high frequency hearing loss. When RT-I functions in BWS canaries were compared with those from normal hearing canaries at different frequencies, clear differences emerged. At the high frequencies where BWS canaries suffer from ~40 dB hearing loss, RTs in the BWS canaries were longer than normal at intensities just above threshold. However, as intensity increased RTs rapidly decreased and eventually became much shorter than normal at high sound levels, clear evidence of RT-I with hyperacusis-like features. A somewhat different pattern emerged at low frequencies where the Waterslager canaries had a very mild hearing loss. At these low frequencies, RTs were nearly normal at low intensities, but as intensity increased, RTs became much shorter than normal at moderate and high intensities. These results suggested that hyperacusis might have a spectral profile, with hyperacusis being more pronounced at frequencies below the region of maximum hearing loss. The authors reported that the BWS canaries hopped erratically around their cage when an intense sound was presented and speculated that the BWS canaries might be stressed by the acoustic stimuli.

One recent study found that extremely long-duration exposure to intense high-frequency noise induced a high-frequency hearing loss (Fig. 3a) and hyperacusis-like RT-I functions at low frequencies where hearing was normal and at frequencies along the low-frequency border of the high-frequency hearing loss (Fig. 3b). However, recruitment-like RT-I functions were present at the high frequencies in the region of maximum hearing loss (Fig. 3c) [9]. These results suggest that low-frequency hyperacusis may occur in individuals with high-frequency noise-induced hearing, particular after intense, long-duration noise exposures that may be extremely stressful.

3.3 Transient Ototoxic Drug-Induced Hyperacusis

Rheumatoid arthritis used to be routinely treated with high doses of aspirin. Patients were instructed to keep increasing the dose of aspirin until their ears started to ring and then lower the dose until the ringing tinnitus disappeared [47]. High doses of salicylate, the active ingredient in aspirin, have been used experimentally for many years to induce transient hearing loss and tinnitus in animal models [48–50]. High doses of salicylate not only increase the threshold of hearing but also greatly reduce the neural output of the cochlea. Paradoxically, electrophysiological studies indicate that salicylate induces neural hyperactivity in the central auditory pathway. The degree of hyperactivity increasing from the cochlear nucleus to the auditory cortex [51, 52]. The sound-evoked hyperactivity raised the possibility that salicylate might induce hyperacusis. Indeed, when rats were treated with high doses of salicylate ($\geq$150 mg/kg) known to induce tinnitus [50], RT-I functions became hyperacusis-like [36]. The most dramatic changes were observed with broadband noise bursts (Fig. 4a). Posttreatment RTs were longer at low intensities because of the ~20 dB hearing loss induced by the drug. However, RTs rapidly declined with intensity so that posttreatment RTs were shorter than normal from 60 to 90 dB SPL, behavioral evidence of hyperacusis. When rats were treated with a low dose of salicylate ($\leq$100 mg/kg), posttreatment RTs were unchanged. The effects of salicylate on tone bursts RT-I

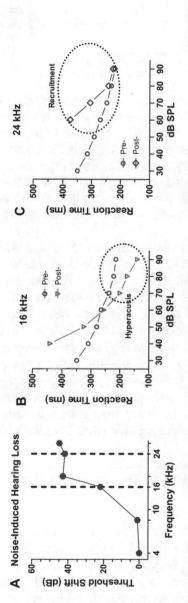

Fig. 3 Schematic of reaction time-intensity (RT-I) function measured in rats with permanent noise-induced high-frequency hearing loss. (**a**) Permanent noise-induced threshold shift following prolonged high-frequency noise exposure. (**b**) Schematic of pre- and postexposure RT-I at 16 kHz. RT-I is hyperacusis-like with a steeper than normal slope and RTs that are below normal from 70 to 90 dB SPL. (**c**) Schematics of pre- and postexposure RT-I at 24 kHz; postexposure RT-I is recruitment-like with a steep slope from 60 to 80 dB SPL and normal RTs at 80 and 90 dB SPL

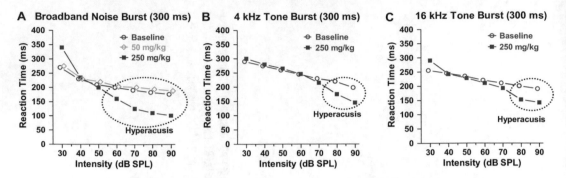

Fig. 4 Schematic of reaction time-intensity (RT-I) functions before (baseline) and several hours after treatment with sodium salicylate. (**a**) Schematic of RT-I functions obtained with broadband noise bursts (300 ms) at baseline and several hours after a low-dose (50 mg/kg) or high dose (250 mg/kg) of sodium salicylate. RTs become much shorter than baseline at moderate to high sound levels (60–90 dB SPL) only for the 250 mg/kg dose of salicylate, behaviors indicative of hyperacusis. The 50 mg/kg dose of salicylate does not cause hearing loss and does not change the RT-I function. Schematic of RT-I functions measured with tone bursts presented at (**b**) 4 kHz or (**c**) 16 kHz; results shown at baseline and several hours postsalicylate. RTs at 80 and 90 dB SPL are shorter than normal several hours postsalicylate, indicative of hyperacusis

functions were less pronounced as schematized in Fig. 4b, c. At 4 kHz and 16 kHz, the RTs decreased below baseline values only at 80 and 90 dB SPL, whereas RTs were normal at low intensities. When salicylate treatment was discontinued, RT-I functions returned to baseline. A major advantage of using salicylate is that hyperacusis can be turned on and off by administering high doses of the drug. Thus, investigators can compare electrophysiological, biochemical or brain imaging data before and after administering a high dose of salicylate [52–54]. Stress is believed to play an important role in hyperacusis [55]. Consistent with this hypothesis, high-dose of salicylate that induce hyperacusis cause a dramatic increase in corticosterone stress hormone [51].

4   Active Sound Avoidance Paradigm (ASAP) to Assess Fear or Avoidance Hyperacusis

Humans with fear hyperacusis try to avoid being exposed to loud sounds or to escape from situations where sounds might become too loud [17, 56]. What type of behavior could be used to test for sound aversion in an animal model? Some species have innate sensory aversions to visual, auditory or olfactory stimuli that could be used to test for stimulus aversion [57, 58]. For example, rodents exhibit an aversion to brightly illuminated open spaces and tend to move into a dark enclosed space [59]. We took advantage of this light/dark preference and introduced noise into the dark box to determine what noise intensity would drive the rat from the dark enclosure into a bright, open space.

4.1 Subjects   The ASAP procedure has been used to assess sound avoidance behavior in rats with normal hearing compared to rats with high-frequency hearing loss [60]. The ASAP procedure has also been used to detect strong sound-avoidance behavior in Fragile X knockout rats, an animal model of autism spectrum disorder with putative sensory hypersensitivity disorders [61].

4.2 Equipment   The custom designed ASAP apparatus differs from commercial light–dark boxes because of the need to reduce the transmission of sound from the Dark box to the brightly illuminated open chamber. The Dark box is constructed of dense, sound-attenuating board (MDF, 2.54 cm thickness) (Fig. 5a). A hinge on the upper edge of the front door provides access to the inside of the Dark box. The interior walls and ceiling of the Dark box are lined with sound absorbing foam (Sonex Acoustics, thickness 5 cm). An enclosure (Med Associates, ENV-007, 29.5 × 23.5 × 27.3 cm) with a grid floor is located near the front door of the Dark Box. A loudspeaker (Fostex FT 28D) is mounted on the roof of the enclosure; placing the speaker on the roof of the small enclosure reduces the variability of the noise intensity inside the box. A square opening (6.4 × 6.4 cm) on the acrylic wall of the grid floor enclosure is connected to an entry/exit port (6.4 × 6.4 cm) on the door of the Dark box. The entry/exit port leads to a runway with a plastic floor and wire mesh wall along the front of the Dark box. The runway connects to the entry port on the Light box, which is constructed of clear acrylic. Two bright lights located above the apparatus illuminated the runway and Light Box (350 lux). A camera (Digital USB 2.0 CMOS Camera, Vari-Focal lens, 2.9–8.2 mm, Stoelting Co.) connected to Any-maze software (Stoelting) is used to monitor the movement of the rat between the Dark box and Light box and to automatically record the time the rats spends in the Dark box versus the runway and Light-box.

The loudspeaker on the roof of the grid floor enclosure is connected to a power amplifier (Crown amplifier XLS202). The noise stimuli are digitally synthesized (Adobe Audition Software) and delivered to the sound card in a personal computer which is connected to the power amplifier and loudspeaker. The digitally synthesized noises used for assessing the sound avoidance in rats are broadband noise (2–20 kHz), low-frequency noise (2–8 kHz), and high frequency (16–20 kHz). AnyMaze software is used to control the presentation order and duration of the noise presented through the loudspeaker. Sound levels are measured at the height of the rat's head at several locations in the Light box, runway, and the grid floor enclosure in the Dark box. Measurements are made with a sound level meter (Larson Davis, model 824) and half-inch condenser microphone (Larson Davis, model 2540). Because of the sound attenuating properties of the Dark box, the sound levels in

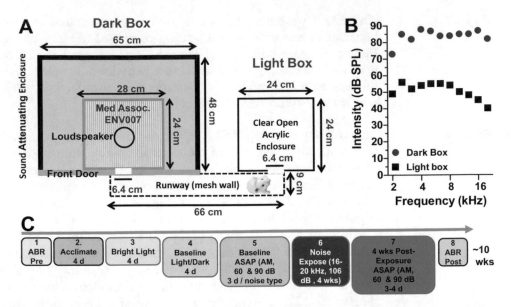

Fig. 5 (**a**) Overhead view of schematic of ASAP apparatus with dimensions (Note: height of sound attenuating enclosure: 55 cm). Height of ENV007 enclosure with grid floor: 10 cm. Camera (not shown) mounted outside and above the apparatus and connected to AnyMaze software. Software is used to monitor the amount of time the rat spends in the runway and Light box versus time in the Dark box. (**b**) Sound levels in the Light box are 25 to 45 dB below the sound levels in the Dark box. The largest intensity differences occur above 8 kHz. (**c**) Schematic timeline: (1) preexposure auditory brainstem response (ABR), (2) acclimate to apparatus in dim light, (3) acclimate to bright light, (4) measure baseline light/dark avoidance, (5) obtain baseline ASAP, (6) Noise exposure: 16–20 kHz noise, 106 dB SPL, 4 weeks, (7) postexposure ASAP test, and (8) postexposure ABR test

the Light box are 30–40 dB below those in the grid floor enclosure (Fig. 5b). Attenuation in the runaway is still considerable at 25–35 dB.

4.3 Procedures

The eight stages of the ASAP procedure used to test for sound avoidance in rats with normal hearing and high-frequency hearing loss are schematized in the timeline shown Fig. 5c. The rats are maintained on a regular light–dark cycle (12 h:12 h). Testing occurs between 8 am and 5 pm. (1) The first step is to screen for preexisting hearing loss by measuring tone burst-evoked auditory brainstem response (ABR) thresholds [62]. Rats with significantly elevated thresholds are eliminated from further testing. (2) The second step is to acclimate the rat to the ASAP equipment. The equipment is thoroughly cleaned before an animal is placed in the apparatus. The experimenter handles the rat for approximately five minutes to reduce anxiety and to familiarize the animal with the experimenter. Afterward, the rat is placed into the Light box for 30 minutes to allow the rat to explore the testing environment. (3) During the third stage, the bright lights (~350 lux) are turned on and the rat is placed in the Light box for 30 min. (4) During stage 4, the rat gradually increases the percentage of time it spends in the Dark box from approximately 60% on the first day to more than 90% on the fourth day. (5) Baseline ASAP testing is carried out over 3 days for each type of noise being used in the study. Baseline testing consists of three 10 min trials/day using one of three randomly selected conditions, that is, ambient (Am) room noise (~35 dB SPL), 60 or 90 dB SPL noise. For each 10-min trial, the percent time spent in the Dark box is determined for the Am, 60 and 90 dB SPL conditions. The mean data for each stimulus is computed over the 3 test days for each animal. (6) To determine if high-frequency noise-induced hearing loss affects sound avoidance behavior, the rat is exposed to intense noise. In this example, the rat is exposed to 16–20 kHz noise for 4 weeks at an intensity of 106 dB SPL. This exposure is expected to cause a significant hearing loss above 12 kHz while preserving normal hearing below 12 kHz. This long-duration noise exposure is performed in a dedicated noise-exposure room located in laboratory animal facility [63, 64]. (7) During stage 7, 1 month after the noise exposure, the animal is retested on the same ASAP paradigm used in stage 5. (8) After completing ASAP testing, the animal moves to stage 8 where ABR thresholds are measured again as in stage 1. The differences between ABR thresholds measured in stage 8 versus stage 1 are used to determine the magnitude of the noise-induced hearing loss.

4.4   Analysis of Results

The schematics in Fig. 6 illustrate how the results from the ASAP paradigm are organized and interpreted. Figure 6a illustrates the average baseline data for a group of normal hearing rats evaluated with a 2–20 kHz noise and a 2–8 kHz noise. For the Am condition, normal hearing rats spend approximately 95% of the time in the Dark box. As the intensity of the noise increases to 60 dB SPL and then 90 dB SPL, the rat spends less time in the dark. Percent time spent in the dark box decreases in an intensity-dependent manner for both the 2–20 kHz noise and the 2–8 kHz. Under ambient background noise, the rats have a strong innate preference for the Dark box. However, as the intensity of the noise in the Dark box increases, the rats spend progressively less time in the Dark box and escape from the noise by moving to the Light box.

Prolonged exposure to intense noise can exacerbate sound avoidance [60] as schematized in Fig. 6b, c, which shows ASAP measures obtained before and after a prolonged, continuous, and stressful high-frequency noise exposure (16–20 kHz, 106 dB SPL, 4-weeks). The noise exposure induced a high frequency hearing loss (~40 dB hearing loss above 12 kHz) whereas low frequency hearing (<12 kHz) remained normal. The noise-exposed rats spend more than 90% of the time in the Dark box, essentially the same amount of time as when their hearing was normal. These results indicate that the prolonged noise exposure and the resulting

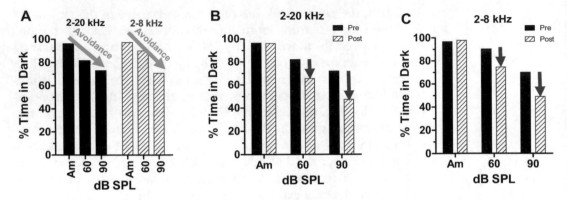

Fig. 6 (**a**) Schematic of mean percent time spent in Dark box as a function of noise intensity (Am: ambient sound, 60 or 90 dB SPL) for 2–20 kHz and 2–8 kHz test stimuli. Green arrows indicate intensity-dependent escape from the noise in the Dark box, evidence of active sound avoidance. (**b**, **c**) Schematic showing ASAP results preexposure and several months postexposure after rats develop a high-frequency noise-induced hearing loss. In Am condition, no change occurs in percent time in Dark box between pre-and postexposure conditions. The noise-induced, high-frequency hearing loss leads to decreased percent time in Dark box (blue arrows) during 60 and 90 dB SPL. test conditions, behaviors indicating that the high-frequency hearing loss exacerbated sound-avoidance

high-frequency hearing loss did not alter the dark preference of the rats. However, when the 2–20 kHz and 2–8 kHz noise bands were presented at 60 or 90 dB SPL, the rats spend less time in the Dark box after they developed a high-frequency hearing loss. These results suggest that the noise-exposed rats perceived the 60 and 90 dB SPL noises as more aversive, annoying, or fear evoking than when their hearing was normal. This sound avoidance behavior could develop because the intense noise used to induce the high frequency hearing loss is aversive and induces negative emotions. This could occur when neural activity evoked by the intense noise exposure is relayed from the auditory pathway to emotional centers in the limbic system such as the amygdala, hippocampus, and hypothalamus. Thus, ASAP could provide a method for assessing the emotional aspects of an acoustic stimulus [22, 52, 65, 66].

5   Summary

One of the lofty goals of auditory neuroscience is to link neurophysiological activity in the brain to sensory processes such as loudness. The loudness of a sound and the positive and negative emotions that it evokes is a multidimensional process affected by many features of the acoustic stimulus, environmental context, innate factors, hearing impairment, and prior experience. Auditory RT-I functions provide neuroscientists with a powerful tool to test models of loudness encoding in relation to intensity because RT-I functions change in different ways when a subject develops recruitment versus hyperacusis. For example, if a model predicts that loudness is tightly correlated with discharge rate of a neuron or group of neurons, then the maximum discharge rate at high intensities should be greater than normal in animals that have hyperacusis-like RT-functions. In animal with transient, salicylate-induced hyperacusis, higher than normal discharge rates and evoked response amplitudes have been observed at higher levels of the auditory pathway [53, 67]. However, in animals with permanent noise-induced hyperacusis, the maximum firing rates in large populations of neurons were below normal, that is, neural gain along the central auditory pathway was insufficient to account for behavioral evidence of hyperacusis [9]. This discrepancy suggests that in cases of permanent hearing loss where considerable neuro-degeneration has occurred, other neurophysiological metrics besides discharge rates need to be considered in order to account for hyperacusis. RT-I functions vary in a systematic manner as stimulus duration increases out to approximately 300 ms consistent with the temporal integration of loudness. This time-dependent change in reaction time and loudness would seem to require some form of temporal integration of neural activity with a time constant that is consistent with the RT-I data. RT-I functions also vary in a systematic fashion as stimulus bandwidth increases consistent with the spectral integration of loudness. Identifying a neurophysiological metric that varies in manner consistent with the RT data could provide new insight into the neural coding of loudness.

In addition to evoking the sensation of loudness, acoustic stimuli such as an alarm or explosion can also evoke fear and anxiety, causing a subject to avoid being exposed to the sounds. Some of the emotional responses to a sound may be innate, learned, or triggered by damage to the auditory system. The ASAP procedure provides researchers with a behavioral tool to quantify sound avoidance and aversion and then seek out neural metrics in emotional regions of the brain such as the amygdala that may be associated with the behaviorally quantified sound aversion.

References

1. Zwicker A, Flottorp G, Stevens S (1957) Critical bandwidth in loudness summation. J Acoust Soc Am 29:548–557

2. Robinson DW, Dadson RS (1956) A re-determination of the equal loudness relations for pure tones. Br J Appl Phys 7:166–181

3. Franks JR, Stephenson MMD, Merry CJ (1996) Preventing occupational hearing loss: a practical guide. DHHS(NIOSH) publication, vol 96-110. U.S. Dept. of Health and Human Services, Public Health Service, Centers for Disease Control and Prevention, National Institute for Occupational Safety and Health, Division of Biomedical and Behavioral Science, Physical Agents Effects Branch, Bethesda, MD

4. Simpson AJ, Reiss JD (2013) The dynamic range paradox: a central auditory model of intensity change detection. PLoS One 8(2): e57497. https://doi.org/10.1371/journal. pone.0057497

5. Moore BC, Glasberg BR (2004) A revised model of loudness perception applied to cochlear hearing loss. Hear Res 188 (1–2):70–88. https://doi.org/10.1016/ S0378-5955(03)00347-2

6. Zeng FG, Kong YY, Michalewski HJ, Starr A (2005) Perceptual consequences of disrupted auditory nerve activity. J Neurophysiol 93(6):3050–3063. https://doi.org/10.1152/jn.00985.2004

7. Dean I, Harper NS, McAlpine D (2005) Neural population coding of sound level adapts to stimulus statistics. Nat Neurosci 8(12):1684–1689. https://doi.org/10.1038/nn1541

8. Zwislocki JJ (1995) Cochlear precursors of neural pitch and loudness codes. Ann Otol Rhinol Laryngol Suppl 166:12–15

9. Radziwon K, Auerbach BD, Ding D, Liu X, Chen GD, Salvi R (2019) Noise-induced loudness recruitment and hyperacusis: insufficient central gain in auditory cortex and amygdala. Neuroscience 422:212–227. https://doi.org/10.1016/j.neuroscience.2019.09.010

10. Fowler EP (1965) Some attributes of "loudness recruitment" and "loudness decruitment". Trans Am Otol Soc 53:78–84

11. Smith PA, Ferguson MA (1994) Comparison of measures of frequency resolution and recruitment in patients undergoing neuro-otological investigation. Br J Audiol 28(3):155–167

12. Fritze W (1980) The graph of loudness recruitment (ABLB-test). Arch Otorhinolaryngol 226(1–2):11–13

13. Knight KK, Margolis RH (1984) Magnitude estimation of loudness. II: loudness perception in presbycusic listeners. J Speech Hear Res 27(1):28–32

14. Baguley DM (2003) Hyperacusis. J R Soc Med 96(12):582–585. https://doi.org/10.1258/jrsm.96.12.582

15. Andersson G, Lindvall N, Hursti T, Carlbring P (2002) Hypersensitivity to sound (hyperacusis): a prevalence study conducted via the Internet and post. Int J Audiol 41(8):545–554. https://doi.org/10.3109/14992020209056075

16. Gu JW, Halpin CF, Nam EC, Levine RA, Melcher JR (2010) Tinnitus, diminished sound-level tolerance, and elevated auditory activity in humans with clinically normal hearing sensitivity. J Neurophysiol 104(6):3361–3370. https://doi.org/10.1152/jn.00226.2010

17. Tyler RS, Pienkowski M, Roncancio ER, Jun HJ, Brozoski T, Dauman N, Dauman N, Andersson G, Keiner AJ, Cacace AT, Martin N, Moore BC (2014) A review of hyperacusis and future directions: part I. Definitions and manifestations. Am J Audiol 23(4):402–419. https://doi.org/10.1044/2014_AJA-14-0010

18. Sheldrake J, Diehl PU, Schaette R (2015) Audiometric characteristics of hyperacusis patients. Front Neurol 6:105. https://doi.org/10.3389/fneur.2015.00105

19. Sherlock LP, Formby C (2005) Estimates of loudness, loudness discomfort, and the auditory dynamic range: normative estimates, comparison of procedures, and test-retest reliability. J Am Acad Audiol 16(2):85–100. https://doi.org/10.3766/jaaa.16.2.4

20. Blaesing L, Kroener-Herwig B (2012) Self-reported and behavioral sound avoidance in tinnitus and hyperacusis subjects, and association with anxiety ratings. Int J Audiol 51(8):611–617. https://doi.org/10.3109/14992027.2012.664290

21. Sander K, Frome Y, Scheich H (2007) FMRI activations of amygdala, cingulate cortex, and auditory cortex by infant laughing and crying. Hum Brain Mapp 28(10):1007–1022. https://doi.org/10.1002/hbm.20333

22. Levitin DJ, Menon V, Schmitt JE, Eliez S, White CD, Glover GH, Kadis J, Korenberg JR, Bellugi U, Reiss AL (2003) Neural correlates of auditory perception in Williams syndrome: an fMRI study. NeuroImage 18(1):74–82. https://doi.org/10.1006/nimg.2002.1297

23. Levitin DJ, Cole K, Lincoln A, Bellugi U (2005) Aversion, awareness, and attraction: investigating claims of hyperacusis in the Williams syndrome phenotype. J Child Psychol Psychiatry 46(5):514–523. https://doi.org/10.1111/j.1469-7610.2004.00376.x

24. Moller AR (2007) Neurophysiologic abnormaliteis in autism. Nova Science Publishers

25. LeDoux JE, Cicchetti P, Xagoraris A, Romanski LM (1990) The lateral amygdaloid nucleus: sensory interface of the amygdala in fear conditioning. J Neurosci 10(4):1062–1069

26. Brand T, Hohmann V (2002) An adaptive procedure for categorical loudness scaling. J Acoust Soc Am 112(4):1597–1604. https://doi.org/10.1121/1.1502902

27. Anweiler AK, Verhey JL (2006) Spectral loudness summation for short and long signals as a function of level. J Acoust Soc Am 119(5 Pt 1):2919–2928. https://doi.org/10.1121/1.2184224

28. Marshall L, Brandt JF (1980) The relationship between loudness and reaction time in normal hearing listeners. Acta Otolaryngol 90(3–4):244–249. https://doi.org/10.3109/00016488009131721

29. Seitz PF, Rakerd B (1997) Auditory stimulus intensity and reaction time in listeners with

longstanding sensorineural hearing loss. Ear Hear 18(6):502–512

30. Schlittenlacher J, Ellermeier W, Arseneau J (2014) Binaural loudness gain measured by simple reaction time. Atten Percept Psychophys 76(5):1465–1472. https://doi.org/10.3758/s13414-014-0651-1

31. Moody DB (1970) Reaction time as an index of sensory function in animals. In: Stebbins WC (ed) Animal pscyhophysics: the design and conduct of sensory experiments. Appleton-Century-Crofts, New York, pp 277–301

32. Moody DB (1973) Behavioral studies of noise-induced hearing loss in primates: loudness recruitment. Adv Otorhinolaryngol 20: 82–101

33. Pfingst BE, Hienze R, Kimm J, Miller J (1975) Reaction-time procedure for measuring hearing. I. Suprathreshold functions. J Acoust Soc Am 57:421–430

34. Stebbins WC (1966) Auditory reaction time and the derivation of equal loudness contours for the monkey. J Exp Anal Behav 9(2):135–142. https://doi.org/10.1901/jeab.1966.9-135

35. Radziwon K, Salvi R (2020) Using auditory reaction time to measure loudness growth in rats. Hear Res 395:108026. https://doi.org/10.1016/j.heares.2020.108026

36. Radziwon K, Holfoth D, Lindner J, Kaier-Green Z, Bowler R, Urban M, Salvi R (2017) Salicylate-induced hyperacusis in rats: dose- and frequency-dependent effects. Hear Res 350:133–138. https://doi.org/10.1016/j.heares.2017.04.004

37. Hayes SH, Radziwon KE, Stolzberg DJ, Salvi RJ (2014) Behavioral models of tinnitus and hyperacusis in animals. Front Neurol 5:179. https://doi.org/10.3389/fneur.2014.00179

38. Radziwon KE, Stolzberg DJ, Urban ME, Bowler RA, Salvi RJ (2015) Salicylate-induced hearing loss and gap detection deficits in rats. Front Neurol 6:31. https://doi.org/10.3389/fneur.2015.00031

39. Buus S, Florentine M, Poulsen T (1997) Temporal integration of loudness, loudness discrimination, and the form of the loudness function. J Acoust Soc Am 101(2):669–680

40. Cacace AT, Margolis RH (1985) On the loudness of complex stimuli and its relationship to cochlear excitation. J Acoust Soc Am 78(5):1568–1573

41. Zwislocki JJ, Jordan HN (1986) On the relations of intensity jnd's to loudness and neural noise. J Acoust Soc Am 79(3):772–780

42. Derleth RP, Dau T, Kollmeier B (2001) Modeling temporal and compressive properties of the normal and impaired auditory system. Hear Res 159(1–2):132–149. https://doi.org/10.1016/s0378-5955(01)00322-7

43. Moore BC (1997) A compact disc containing simulations of hearing impairment. Br J Audiol 31(5):353–357. https://doi.org/10.3109/03005364000000029

44. Moody DB, Winger G, Woods JH, Stebbins WC (1980) Effect of ethanol and of noise on reaction time in the monkey: variation with stimulus level. Psychopharmacology 69(1):45–51. https://doi.org/10.1007/BF00426520

45. Stebbins WC, Coombs S (1975) Behavioral assessment of ototoxicity in nonhuman primates. In: Weiss B, Laties VG (eds) Behavioral toxicology. Springer, Boston, pp 401–427. https://doi.org/10.1007/978-1-4684-2859-9_15

46. Lauer AM, Dooling RJ (2007) Evidence of hyperacusis in canaries with permanent hereditary high-frequency hearing loss. Semin Hear 28(4):319–326

47. Mongan E, Kelly P, Nies K, Porter WW, Paulus HE (1973) Tinnitus as an indication of therapeutic serum salicylate levels. JAMA 226(2):142–145

48. Myers EN, Bernstein JM (1965) Salicylate ototoxicity; a clinical and experimental study. Arch Otolaryngol 82(5):483–493. https://doi.org/10.1001/archotol.1965.00760010485006

49. Jastreboff PJ, Brennan JF, Sasaki CT (1988) An animal model for tinnitus. Laryngoscope 98(3):280–286. https://doi.org/10.1288/00005537-198803000-00008

50. Lobarinas E, Sun W, Cushing R, Salvi R (2004) A novel behavioral paradigm for assessing tinnitus using schedule-induced polydipsia avoidance conditioning (SIP-AC). Hear Res 190 (1–2):109–114. https://doi.org/10.1016/S0378-5955(04)00019-X

51. Jiang C, Luo B, Manohar S, Chen GD, Salvi R (2017) Plastic changes along auditory pathway during salicylate-induced ototoxicity: hyperactivity and CF shifts. Hear Res 347:28–40. https://doi.org/10.1016/j.heares.2016.10.021

52. Chen YC, Li X, Liu L, Wang J, Lu CQ, Yang M, Jiao Y, Zang FC, Radziwon K, Chen GD, Sun W, Krishnan Muthaiah VP, Salvi R, Teng GJ (2015) Tinnitus and hyperacusis involve hyperactivity and enhanced connectivity in auditory-limbic-arousal-cerebellar network. elife 4:e06576. https://doi.org/10.7554/eLife.06576

53. Auerbach BD, Radziwon K, Salvi R (2019) Testing the central gain model: loudness

growth correlates with central auditory gain enhancement in a rodent model of hyperacusis. Neuroscience 407:93–107. https://doi.org/10.1016/j.neuroscience.2018.09.036

54. Wong E, Radziwon K, Chen GD, Liu X, Manno FA, Manno SH, Auerbach B, Wu EX, Salvi R, Lau C (2020) Functional magnetic resonance imaging of enhanced central auditory gain and electrophysiological correlates in a behavioral model of hyperacusis. Hear Res 389:107908. https://doi.org/10.1016/j.heares.2020.107908

55. Hasson D, Theorell T, Bergquist J, Canlon B (2013) Acute stress induces hyperacusis in women with high levels of emotional exhaustion. PLoS One 8(1):e52945. https://doi.org/10.1371/journal.pone.0052945

56. Pienkowski M, Tyler RS, Roncancio ER, Jun HJ, Brozoski T, Dauman N, Coelho CB, Andersson G, Keiner AJ, Cacace AT, Martin N, Moore BC (2014) A review of hyperacusis and future directions: part II. Measurement, mechanisms, and treatment. Am J Audiol 23(4):420–436. https://doi.org/10.1044/2014_AJA-13-0037

57. Mazurek B, Haupt H, Joachim R, Klapp BF, Stover T, Szczepek AJ (2010) Stress induces transient auditory hypersensitivity in rats. Hear Res 259(1–2):55–63. https://doi.org/10.1016/j.heares.2009.10.006

58. Ferrero DM, Lemon JK, Fluegge D, Pashkovski SL, Korzan WJ, Datta SR, Spehr M, Fendt M, Liberles SD (2011) Detection and avoidance of a carnivore odor by prey. Proc Natl Acad Sci U S A 108(27):11235–11240. https://doi.org/10.1073/pnas.1103317108

59. Crawley J, Goodwin FK (1980) Preliminary report of a simple animal behavior model for the anxiolytic effects of benzodiazepines. Pharmacol Biochem Behav 13(2):167–170. https://doi.org/10.1016/0091-3057(80)90067-2

60. Manohar S, Spoth J, Radziwon K, Auerbach BD, Salvi R (2017) Noise-induced hearing loss induces loudness intolerance in a rat Active Sound Avoidance Paradigm (ASAP). Hear Res 353:197–203. https://doi.org/10.1016/j.heares.2017.07.001

61. Radziwon K, Auerbach BD, Kolisetti R, Beadle M, Salvi R (2018) Auditory hypersensitivity and temporal integration deficits in Fragile X rats. Paper presented at the Assoc. Res. Otolaryngology, San Diego, 9–14 Feb 2018

62. Chen GD, Decker B, Krishnan Muthaiah VP, Sheppard A, Salvi R (2014) Prolonged noise exposure-induced auditory threshold shifts in rats. Hear Res 317:1–8. https://doi.org/10.1016/j.heares.2014.08.004

63. Liu X, Li L, Chen GD, Salvi R (2020) How low must you go? Effects of low-level noise on cochlear neural response. Hear Res 392:107980. https://doi.org/10.1016/j.heares.2020.107980

64. Zhao DL, Sheppard A, Ralli M, Liu X, Salvi R (2018) Prolonged low-level noise exposure reduces rat distortion product otoacoustic emissions above a critical level. Hear Res 370:209–216. https://doi.org/10.1016/j.heares.2018.08.002

65. Knipper M, Van Dijk P, Nunes I, Ruttiger L, Zimmermann U (2013) Advances in the neurobiology of hearing disorders: recent developments regarding the basis of tinnitus and hyperacusis. Prog Neurobiol 111:17–33. https://doi.org/10.1016/j.pneurobio.2013.08.002

66. Mazurek B, Stover T, Haupt H, Klapp BF, Adli M, Gross J, Szczepek AJ (2010) The significance of stress: its role in the auditory system and the pathogenesis of tinnitus. HNO 58(2):162–172. https://doi.org/10.1007/s00106-009-2001-5

67. Chen YC, Chen GD, Auerbach BD, Manohar S, Radziwon K, Salvi R (2017) Tinnitus and hyperacusis: contributions of paraflocculus, reticular formation and stress. Hear Res 349:208–222. https://doi.org/10.1016/j.heares.2017.03.005

Correction to: A Manual Technique for Isolation and Single-Cell RNA Sequencing Analysis of Cochlear Hair Cells and Supporting Cells

Cody West, Paul T. Ranum, Ryotaro Omichi, Yoichiro Iwasa, Miles J. Klimara, Daniel Walls, Jin-Young Koh, and Richard J. H. Smith

Correction to:
Chapter 7 in: Andrew K. Groves (ed.), *Developmental, Physiological, and Functional Neurobiology of the Inner Ear*, Neuromethods, vol. 176, https://doi.org/10.1007/978-1-0716-2022-9_7

The original version of this chapter was inadvertently published with wrong spelling of author name "Miles J. Kilmara". Now, the correct spelling "Miles J. Klimara" has been updated in this chapter and Table of Contents.

The updated online version of this chapter can be found at
https://doi.org/10.1007/978-1-0716-2022-9_7

Andrew K. Groves (ed.), *Developmental, Physiological, and Functional Neurobiology of the Inner Ear*, Neuromethods, vol. 176, https://doi.org/10.1007/978-1-0716-2022-9_18,
© The Author(s), under exclusive license to Springer Science+Business Media, LLC, part of Springer Nature 2022

INDEX

Andrew K. Groves (ed.), *Developmental, Physiological, and Functional Neurobiology of the Inner Ear*, Neuromethods, vol. 176,
https://doi.org/10.1007/978-1-0716-2022-9,

Printed in the United States
by Baker & Taylor Publisher Services